CAMBRIDGE TRACTS IN MATHEMATICS

General Editors

199 Combinatorics of Minuscule Representations

CAMBRIDGE TRACTS IN MATHEMATICS

A complete list of books in the series can be found at www.cambridge.org/mathematics. Recent titles include the following:

166. The Lévy Laplacian. By M. N. Feller
167. Poincaré Duality Algebras, Macaulay's Dual Systems, and Steenrod Operations. By D. Meyer and L. Smith
168. The Cube-A Window to Convex and Discrete Geometry. By C. Zong
169. Quantum Stochastic Processes and Noncommutative Geometry. By K. B. Sinha and D. Goswami
170. Polynomials and Vanishing Cycles. By M. Tibăr
171. Orbifolds and Stringy Topology. By A. Adem, J. Leida, and Y. Ruan
172. Rigid Cohomology. By B. Le Stum
173. Enumeration of Finite Groups. By S. R. Blackburn, P. M. Neumann, and G. Venkataraman
174. Forcing Idealized. By J. Zapletal
175. The Large Sieve and its Applications. By E. Kowalski
176. The Monster Group and Majorana Involutions. By A. A. Ivanov
177. A Higher-Dimensional Sieve Method. By H. G. Diamond, H. Halberstam, and W. F. Galway
178. Analysis in Positive Characteristic. By A. N. Kochubei
179. Dynamics of Linear Operators. By F. Bayart and É. Matheron
180. Synthetic Geometry of Manifolds. By A. Kock
181. Totally Positive Matrices. By A. Pinkus
182. Nonlinear Markov Processes and Kinetic Equations. By V. N. Kolokoltsov
183. Period Domains over Finite and p-adic Fields. By J.-F. Dat, S. Orlik, and M. Rapoport
184. Algebraic Theories. By J. Adámek, J. Rosický, and E. M. Vitale
185. Rigidity in Higher Rank Abelian Group Actions I: Introduction and Cocycle Problem. By A. Katok and V. Niţică
186. Dimensions, Embeddings, and Attractors. By J. C. Robinson
187. Convexity: An Analytic Viewpoint. By B. Simon
188. Modern Approaches to the Invariant Subspace Problem. By I. Chalendar and J. R. Partington
189. Nonlinear Perron–Frobenius Theory. By B. Lemmens and R. Nussbaum
190. Jordan Structures in Geometry and Analysis. By C.-H. Chu
191. Malliavin Calculus for Lévy Processes and Infinite-Dimensional Brownian Motion. By H. Osswald
192. Normal Approximations with Malliavin Calculus. By I. Nourdin and G. Peccati
193. Distribution Modulo One and Diophantine Approximation. By Y. Bugeaud
194. Mathematics of Two-Dimensional Turbulence. By S. Kuksin and A. Shirikyan
195. A Universal Construction for R-free Groups. By I. Chiswell and T. Müller
196. The Theory of Hardy's Z-Function. By A. Ivić
197. Induced Representations of Locally Compact Groups. By E. Kaniuth and K. F. Taylor
198. Topics in Critical Point Theory. By K. Perera and M. Schechter
199. Combinatorics of Minuscule Representations. By R. M. Green
200. Singularities of the Minimal Model Program. By J. Kollár
201. Coherence in Three-Dimensional Category Theory. By N. Gurski

Combinatorics of Minuscule Representations

R. M. GREEN
University of Colorado Boulder

CAMBRIDGE
UNIVERSITY PRESS

CAMBRIDGE
UNIVERSITY PRESS

University Printing House, Cambridge CB2 8BS, United Kingdom

Cambridge University Press is part of the University of Cambridge.

It furthers the University's mission by disseminating knowledge in the pursuit of education, learning and research at the highest international levels of excellence.

www.cambridge.org
Information on this title: www.cambridge.org/9781107026247

First published 2013

A catalogue record for this publication is available from the British Library

ISBN 978-1-107-02624-7 Hardback

Contents

Introduction

Highest weight modules play a key role in the representation theory of several classes of algebraic objects occurring in Lie theory, including Lie algebras, Lie groups, algebraic groups, Chevalley groups and quantized enveloping algebras. In many of the most important situations, the weights may be regarded as points in Euclidean space, $\mathbb{R}^n$, and there is a finite group (called a Weyl group) that acts on the set of weights by linear transformations. The minuscule representations are those for which the Weyl group acts transitively on the weights, and the highest weight of such a representation is called a minuscule weight. The term "minuscule weight" is a translation of Bourbaki's term *poids minuscule* [8, VIII, section 7.3]; the spelling "miniscule" is also found in the literature, although less commonly, and Russian-speaking authors often call minuscule weights *microweights*. The list of minuscule representations is as follows: all fundamental representations in type A_n, the natural representations in types C_n and D_n, the spin representations in types B_n and D_n, the two 27-dimensional representations in type E_6 and the 56-dimensional representation in type E_7.

Minuscule weights and minuscule representations are important because they occur in a wide variety of contexts in mathematics and physics, especially in representation theory and algebraic geometry. Minuscule representations are the starting point of Standard Monomial Theory developed by Lakshmibai, Seshadri and others [42], and they play a key role in the geometry of Schubert varieties.

One of the advantages of minuscule representations is that they are often much easier to understand than typical representations of the same objects. For example, Seshadri [70] proves that standard monomials give a basis for the homogeneous coordinate ring of G/P in the minuscule case, which is significantly more tractable than the general case. Littelmann's path model for representations of Lie algebras [45] is also much simpler when minuscule representations are involved. Minuscule representations have many other algebraic applications: they are useful in the theory of Chevalley groups, where they are used to construct minuscule weight geometries [5, section 6], and they also appear in the context of Bruhat decompositions [86]. Certain questions in the representation theory of algebraic groups can be reduced to questions about minuscule representations [80]. In the theory of Macdonald polynomials [85, section 6], minuscule weights are used to define the Macdonald operators.

Minuscule representations also have applications to areas more remote from representation theory, including special functions theory (see [25] and references therein), random walks [2, 44], and conformal field theory: the paper [22] considers a correspondence between minimal fluxes and minuscule weights of Lie algebras in types A, D and E. Another application to physics appears in the paper [71], which constructs a set of lattices by decorating the root lattices of various Lie algebras with their minuscule representations. The paper studies a family of Hamiltonians of fermions hopping on these lattices in the presence of a background gauge field; in this context, the Hamiltonians are themselves elements of the Lie algebras acting in their minuscule representations.

As its title suggests, the focus of this book is on combinatorial properties of minuscule representations. This primarily refers to combinatorial properties of the weights of the representations, especially the action of the Weyl group by orthogonal transformations on the set of weights. Most of the literature on minuscule representations says very little about this action, other than that it is transitive. However, the details of the action turn out to be fascinating. For example, we shall see in Chapter 8 that the failure of this action in general to be doubly transitive gives the Weyl group an interesting structure as a permutation group, which is intimately related to (a) branching rules of minuscule representations, (b) the combinatorics of the polytope formed by the convex hull of the weights and (c) various well-known families of graphs. An important special case is when the weights are closed under negation and the Weyl group acts as a rank 4 permutation group. In this case, the action of the Weyl group on pairs of opposite weights is doubly transitive, but the failure in general of the action to be triply transitive leads to the interesting combinatorial features of the 28 bitangents to a plane quartic curve and the 27 lines on a cubic surface; these features include examples of structures such as 2-graphs and generalized quadrangles, as well as the rich combinatorics of Steiner complexes and Schläfli double sixes.

A main object of interest in this book is a certain type of infinite labelled poset known as a full heap. Full heaps are defined from generalized Cartan matrices using only combinatorics, and they can be used to construct affine analogues of minuscule representations. These give rise to faithful permutation representations of affine Weyl groups (some of which are familiar from other contexts) as well as minuscule-like representations of various affine Kac–Moody algebras. These representations are closely related to the combinatorics of the associated root system; among other things, this allows a construction of Chevalley bases for Lie algebras using the combinatorics of heaps, as discussed in Chapter 7. The heap-theoretic approach also makes certain features of minuscule representations of Lie algebras easy to understand, such as the construction of certain invariant symplectic and orthogonal forms. We also remark on the invariant cubic forms in type E_6 and invariant quartic forms in type E_7.

The final chapter contains a survey of some important combinatorial properties related to minuscule representations, including minuscule elements, Gaussian posets, and the jeu de taquin approach to Schubert calculus.

This book is not, and is not intended to be, a general introduction to Lie algebras or Weyl groups of finite and affine type. There are already plenty of good books on these topics, and we will refer to these when necessary instead of developing the relevant theory from scratch. For Lie algebras, we recommend the books by Erdmann and Wildon [23] for a beginner, Carter [11] for an intermediate reader, and Kac [37] for an expert. For Weyl groups and Coxeter groups, we recommend the books by Humphreys [36] for a beginner, Björner and Brenti [4] for a reader interested in combinatorics, and Geck and Pfeiffer [26] for a reader interested in computation or representation theory. In contrast, the approach of this book is to develop both Lie algebras and Weyl groups using a single, example driven, combinatorial approach that makes explicit calculations easy. In particular, it should be possible, without a computer, to understand all (or almost all) the examples and exercises in this book.

A reader familiar with linear algebra, groups, rings and point set topology should be able in principle to read this book from cover to cover, treating it as a romp through a circle of algebraic and combinatorial ideas with minuscule representations at the centre. However, the material is arranged to make the book useful as a reference. The reader who wishes to browse is advised to start by looking at the many examples and exercises, particularly those in the last four chapters. Readers already familiar with some of the objects of study may like to try to solve the exercises using their own methods. Each chapter of this book concludes with a section of Notes and references, which includes historical notes, references to the literature and directions for further reading.

The chapters are structured as follows. Chapter 1 introduces Lie algebras and Weyl groups of types A_n, B_n, C_n and D_n and shows how (with the exception of the Lie algebra of type B_n) these can be constructed, in their natural representations, using the combinatorics of heaps. Chapter 2 develops the theory of heaps over Dynkin diagrams. These heaps can be thought of (and drawn) as partially ordered sets whose elements are labelled by vertices of the Dynkin diagram, subject to certain rules. This combinatorial theory underpins the approach of the rest of the book. Chapter 3 explains how to associate algebraic objects with heaps, including faithful permutation representations of finite and affine Weyl groups. Chapter 4 develops and recalls the basic ideas of Lie theory as it applies to Lie algebras and Weyl groups. Sections 4.2 and 4.3 summarize key results and definitions that are needed in the sequel. These sections are not self-contained, and are primarily based on the books of Carter [11], Humphreys [36] and Kac [37].

Chapter 5 defines minuscule representations in terms of the combinatorics of heaps. Chapter 6 proves the classification of full heaps over affine Dynkin diagrams. These heaps give rise to representations of affine Kac–Moody algebras that can be thought of as infinite dimensional analogues of minuscule representations. Chapter 7 constructs Chevalley bases and structure constants for simple Lie algebras over $\mathbb{C}$ in terms of heaps. Chapter 8 examines some combinatorial properties of the Weyl group in its action on the weights of a minuscule representation, and explicitly describes the orbits of ordered pairs under this action. We also discuss some combinatorial properties of the weight polytope, that is, the convex hull of the weights of a minuscule representation.

The jewel in the crown of minuscule representations is the 56-dimensional representation in type E_7, and Chapters 9 and 10 are devoted to a study of it and its close relatives. Chapter 9 discusses the combinatorics of the 28 bitangents to a plane quartic curve, which can be identified with opposite pairs of weights in the 56-dimensional representation, $L(E_7, \omega_6)$. Embedded in this is the combinatorics of the famous configuration of 27 lines on a cubic surface, which we discuss in the first two sections of Chapter 10. It also turns out that $L(E_7, \omega_6)$ is the largest member of a tower of three minuscule representations, the others being a spin representation in type D_6 and the middle fundamental representation in type A_5. This gives rise to an interesting tower of three 2-graphs (in the sense of G. Higman) and an interesting tower of small but nontrivial generalized quadrangles. Chapter 10 ends with a discussion of some remarkable invariant forms related to the minuscule representations in types E_6 and E_7.

We conclude in Chapter 11 by surveying some important contributions to the combinatorics of minuscule representations by Proctor, Stembridge, Thomas and Yong, among others. Appendix A recalls the basic terminology associated with partially ordered sets, graphs and categories, and Appendix B includes Lie theoretic data, including a list of Dynkin diagrams together with our numbering conventions for the vertices.

The diagrams in this book were produced using TikZ by Till Tantau. The plane partitions were produced using a package by Jang Soo Kim, and the skew Young tableaux were produced using the `youngtab` package by Volker Börchers and Stefan Gieseke.

I thank my research assistant Keli Parker for making many corrections to an earlier version of this book. I am also grateful to Hugh Denoncourt and the two anonymous referees for their constructive criticism, and to Robert Marsh for his editorial work on the papers [28, 29, 30], on which large parts of this book are based. I would like to thank everyone who has invited me to give talks on the material in this book at their seminars and conferences, especially Tim Penttila and the late Bob Liebler for giving me multiple opportunities to speak at the Rocky Mountain Algebraic Combinatorics seminar. I also thank the University of Colorado Boulder for granting me a sabbatical in Spring 2010, during which I wrote most of this book. Finally, I thank my wife Tara, and children Annabel, Emma and Harrison, for allowing me the time to complete it.

It should be assumed, unless explicitly stated to the contrary, that the results presented in this book are not original. Any errors that remain are of course my own.

This material is based upon work supported by the National Science Foundation under Grant Number 0905768. Any opinions, findings, and conclusions or recommendations expressed in this material are those of the author and do not necessarily reflect the views of the National Science Foundation.

1

Classical Lie algebras and Weyl groups

One of the main goals of the first half of this book is to construct Lie algebras and Weyl groups from certain labelled partially ordered sets known as heaps. The formal definition of heaps is in terms of categories, which will be defined in Chapter 2. The purpose of Chapter 1 is to show how particular examples of heaps can be used to give combinatorial constructions of algebraic objects.

In Section 1.1, we summarize the basic properties of Lie algebras. In Section 1.2, we define the Lie algebras of types A_n, B_n, C_n and D_n in terms of matrices; that is, the Lie algebras $\mathfrak{sl}(n+1, V)$, $\mathfrak{so}(2n+1, V)$, $\mathfrak{sp}(2n, V)$ and $\mathfrak{so}(2n, V)$, respectively. Except in type B_n, the Lie algebra representations given by these matrices will turn out to be "minuscule". Because of this, they may easily be constructed in terms of heaps (shown in Figures 1.1, 1.2 and 1.3) as we explain in Section 1.3. Finally, in Section 1.4, we define the classical Weyl groups of types A_n, B_n and D_n as abstract groups. We also explain how to construct the groups using the same three infinite families of heaps shown in Section 1.3.

1.1 Lie algebras

A *Lie algebra* is a vector space $\mathfrak{g}$ over a field k equipped with a bilinear map $[\,,\,] : \mathfrak{g} \times \mathfrak{g} \to \mathfrak{g}$ (the *Lie bracket*) satisfying the conditions

$$[x, x] = 0,$$
$$[[x, y], z] + [[y, z], x] + [[z, x], y] = 0,$$

for all $x, y, z \in \mathfrak{g}$. (These conditions are known respectively as *antisymmetry* and the *Jacobi identity*.)

Example 1.1.1 Let A be any associative algebra over k. Then A can be made into a Lie algebra by replacing the associative multiplication $\circ$ by the Lie bracket

$$[x, y] := x \circ y - y \circ x.$$

Exercise 1.1.2 Verify that the Lie bracket defined in Example 1.1.1 does indeed satisfy the antisymmetry property and the Jacobi identity.

Example 1.1.3 One of the most important examples of a Lie algebra is $\mathfrak{gl}_n(k)$. This is the Lie algebra obtained from the associative matrix algebra $M_n(k)$ of all $n \times n$

matrices with entries in k by applying the construction of Example 1.1.1; in other words, the elements of $\mathfrak{gl}_n(k)$ are all $n \times n$ matrices over k and the Lie bracket $[A, B]$ is defined by $[A, B] = AB - BA$.

If $\mathfrak{g}_1$ and $\mathfrak{g}_2$ are Lie algebras over a field k, then a *homomorphism* of Lie algebras from $\mathfrak{g}_1$ to $\mathfrak{g}_2$ is a k-linear map $\phi : \mathfrak{g}_1 \to \mathfrak{g}_2$ such that $\phi([x, y]) = [\phi(x), \phi(y)]$ for all $x, y \in \mathfrak{g}_1$. An *isomorphism* of Lie algebras is a bijective homomorphism.

If V is any vector space over k then the Lie algebra $\mathfrak{gl}(V)$ is the k-vector space of all k-linear maps $T : V \to V$, equipped with the Lie bracket satisfying

$$[T, U] := T \circ U - U \circ T,$$

where $\circ$ is composition of maps.

A *representation* of a Lie algebra $\mathfrak{g}$ over k is a homomorphism $\rho : \mathfrak{g} \to \mathfrak{gl}(V)$ for some k-vector space V. In this case, we call V a (left) *module* for the Lie algebra $\mathfrak{g}$ (or a $\mathfrak{g}$-module, for short) and we say that V *affords* ρ. If $x \in \mathfrak{g}$ and $v \in V$, we write $x.v$ to mean $\rho(x)(v)$. The *dimension* of a module (or of the corresponding representation) is the dimension of V. If ρ is the zero map, then the representation ρ and the module V are said to be *trivial*.

A *submodule* of a $\mathfrak{g}$-module V is a k-subspace W of V such that $x.w \in W$ for all $x \in \mathfrak{g}$ and $w \in W$. If V has no submodules other than itself and the zero submodule, then V is said to be *irreducible*. If W is a submodule of V, the quotient vector space V/W acquires a well-defined $\mathfrak{g}$-module structure via the condition $x.(v + W) = (x.v) + W$; this is known as a *quotient module*.

If V_1 and V_2 are $\mathfrak{g}$-modules, then a k-linear map $f : V_1 \to V_2$ is called a *homomorphism of* $\mathfrak{g}$*-modules* if $f(x.v) = x.f(v)$ for all $x \in \mathfrak{g}$ and $v \in V_1$. An *isomorphism of* $\mathfrak{g}$*-modules* is an invertible homomorphism of $\mathfrak{g}$-modules.

A subspace $\mathfrak{h}$ of $\mathfrak{g}$ is called a *subalgebra* of $\mathfrak{g}$ if $[\mathfrak{h}, \mathfrak{h}] \subseteq \mathfrak{h}$. If, furthermore, we have $[\mathfrak{g}, \mathfrak{h}] \subseteq \mathfrak{h}$ (or, equivalently, $[\mathfrak{h}, \mathfrak{g}] \subseteq \mathfrak{h}$) then $\mathfrak{h}$ is said to be an *ideal* of $\mathfrak{g}$. We write $\mathfrak{h} \leq \mathfrak{g}$ (respectively, $\mathfrak{h} \trianglelefteq \mathfrak{g}$) to mean that $\mathfrak{h}$ is a subalgebra (respectively, an ideal) of $\mathfrak{g}$. If S is a subset of $\mathfrak{g}$, then the smallest subalgebra of $\mathfrak{g}$ containing S is called the *subalgebra generated by* S, and the elements of S are called *generators* of the subalgebra. If $\mathfrak{h}$ is an ideal of $\mathfrak{g}$, then $\mathfrak{h}$ becomes a $\mathfrak{g}$-module via the action $g.h = [g, h]$. The quotient module $\mathfrak{g}/\mathfrak{h}$ then inherits a well-defined Lie algebra structure via the condition $[x + \mathfrak{h}, y + \mathfrak{h}] = [x, y] + \mathfrak{h}$; we call such a Lie algebra a *quotient Lie algebra* of $\mathfrak{g}$.

If $\mathfrak{g}$ has no ideals other than itself and the zero ideal, then $\mathfrak{g}$ is said to be *simple*. A simple Lie algebra of dimension 1 is called a *trivial simple Lie algebra*. The *derived algebra*, $\mathfrak{g}'$, of a Lie algebra $\mathfrak{g}$ is the subalgebra generated by all elements $\{[x_1, x_2] : x_1, x_2 \in \mathfrak{g}\}$. It can be shown that $\mathfrak{g}'$ is an ideal of $\mathfrak{g}$.

Example 1.1.4 Let V be an n-dimensional k-vector space. Since we can identify $M_n(k)$ with $\mathrm{End}(V)$, we can also identify $\mathfrak{gl}_n(k)$ with $\mathfrak{gl}(V)$; note that both identifications rely on the choice of a basis for V. This identification, which is an isomorphism of Lie algebras $\rho : \mathfrak{gl}_n(k) \to \mathfrak{gl}(V)$, endows V with the structure of a $\mathfrak{gl}_n(k)$-module. More precisely, we have

$$\rho([A, B]) = \rho(A) \circ \rho(B) - \rho(B) \circ \rho(A),$$

where the map $\circ$ on the right hand side refers to composition in the associative algebra $\mathrm{End}(V)$. The $\mathfrak{gl}_n(k)$-module structure on V satisfies $A.v = \rho(A)(v)$, and can be identified with left multiplication of the vector v by the matrix A, once a basis of V has been chosen. The dimension of the module V (or the representation ρ) is n.

Recall that the *trace* of a matrix is the sum of its diagonal entries. A standard property of matrices (see Exercise 1.1.5 below) is that $\mathrm{tr}(AB) = \mathrm{tr}(BA)$ when A and B are square matrices of the same size. It follows that conjugate matrices have the same trace, and thus that we may speak of the trace of an endomorphism of a vector space without reference to a basis. Using the formula for $\rho([A, B])$ in the previous paragraph, we see that

$$\mathrm{tr}\rho([A, B]) = \mathrm{tr}(AB) - \mathrm{tr}(BA) = 0.$$

It follows from this that if we define $\mathfrak{sl}_n(k)$ to be the subspace of $\mathfrak{gl}_n(k)$ consisting of all matrices A with $\mathrm{tr}(A) = 0$, then $\mathfrak{sl}_n(k) \trianglelefteq \mathfrak{gl}_n(k)$. (It is not too hard to show that $\mathfrak{sl}_n(k)$ is equal to the derived algebra of $\mathfrak{gl}_n(k)$, by considering a suitable basis of $\mathfrak{sl}_n(k)$.) Since $\mathfrak{sl}_n(k)$ has dimension $n^2 - 1$ and $\mathfrak{gl}_n(k)$ has dimension n^2, the quotient module $\mathfrak{gl}_n(k)/\mathfrak{sl}_n(k)$ is one-dimensional, and has the trivial module structure. In the case where $k = \mathbb{C}$, it can be shown that $\mathfrak{sl}_n(\mathbb{C})$ is a simple Lie algebra. The identification of $\mathfrak{gl}_n(k)$ with $\mathfrak{gl}(V)$ of Example 1.1.4 identifies the subspace $\mathfrak{sl}_n$ with a subspace of $\mathfrak{gl}(V)$, which we will call $\mathfrak{sl}(n, V)$.

Exercise 1.1.5 Prove that if A and B are two $n \times n$ matrices then we have $\mathrm{tr}(AB) = \mathrm{tr}(BA)$. Deduce that if P is invertible, then $\mathrm{tr}(P^{-1}AP) = \mathrm{tr}(A)$.

We can make the Lie algebra $\mathfrak{g}$ into a module over itself, in which $x.y := [x, y]$. This is known as the *adjoint module*, and the corresponding representation is called the *adjoint representation*. If $x \in \mathfrak{g}$, we define the linear map $\mathrm{ad}_x : \mathfrak{g} \to \mathfrak{g}$ by $\mathrm{ad}_x(y) := [x, y]$. (We may also write $(\mathrm{ad}\, x)(y)$ for $\mathrm{ad}_x(y)$ to avoid excessive subscripts.)

If $\mathfrak{g}$ is a Lie algebra over k, then a *derivation* of $\mathfrak{g}$ is a k-linear map $D : \mathfrak{g} \to \mathfrak{g}$ satisfying Leibniz's law, namely

$$D([x, y]) = [D(x), y] + [x, D(y)].$$

The Jacobi identity guarantees that for each $x \in \mathfrak{g}$, the map $D_x : \mathfrak{g} \to \mathfrak{g}$ given by

$$D_x(y) = \mathrm{ad}_x(y) = [x, y]$$

is a derivation.

If B is a k-basis for the Lie algebra $\mathfrak{g}$, then for any $b_i, b_j \in B$, we may write

$$[b_i, b_j] = \sum_{i \in B} \lambda_{ij}^k b_k,$$

where we have $\lambda_{ij}^k \in k$. The scalars λ_{ij}^k are known as the *structure constants* of $\mathfrak{g}$ with respect to B. (The concept of structure constants applies to any k-algebra with a distinguished basis.)

Exercise 1.1.6 Suppose that $\mathfrak{g}$ is a simple Lie algebra. Show that we have $\mathfrak{g}' = 0$ if $\mathfrak{g}$ is trivial, and $\mathfrak{g}' = \mathfrak{g}$ otherwise.

Exercise 1.1.7 Verify that the adjoint module for $\mathfrak{g}$ is indeed a module.

Exercise 1.1.8 Show that if $\mathfrak{g}$ is a Lie algebra and $x, y \in \mathfrak{g}$, then

$$\mathrm{ad}_{[x,y]} = (\mathrm{ad}_x \circ \mathrm{ad}_y) - (\mathrm{ad}_y \circ \mathrm{ad}_x).$$

Exercise 1.1.9 Show that if $V_1, V_2, \ldots, V_n$ are $\mathfrak{g}$-modules over the field k, then the tensor product $V_1 \otimes_k V_2 \otimes_k \cdots \otimes_k V_n$ becomes a $\mathfrak{g}$-module under the action

$$x.(v_1 \otimes v_2 \otimes \cdots \otimes v_n) = \sum_{i=1}^{n} (v_1 \otimes \cdots \otimes v_{i-1} \otimes x.v_i \otimes v_{i+1} \otimes \cdots \otimes v_n).$$

1.2 The classical Lie algebras

Some other important examples of Lie algebras can be defined as subalgebras of $\mathfrak{gl}(V)$ that respect certain invariant bilinear forms. In this section, we will explain exactly what this means.

Definition 1.2.1 Let $\mathfrak{g}$ be a Lie algebra over a field k and let V be a $\mathfrak{g}$-module. A bilinear map $B : V \otimes_k V \to k$ is said to be $\mathfrak{g}$-*invariant* if for all $x \in \mathfrak{g}$ and $v \in V$, we have

$$B(x.v, v) + B(v, x.v) = 0.$$

The *radical*, $\mathrm{rad}(B)$ of B is defined to be the subset of V given by

$$\{r \in V : B(r, v) = 0 \text{ for all } v \in V\}.$$

If the $\mathrm{rad}(B)$ is zero, we call B *nondegenerate*. If k does not have characteristic 2, a nondegenerate bilinear form B is said to be *orthogonal* (respectively, *symplectic*) if for all $v_1, v_2 \in V$ we have $B(v_1, v_2) = \epsilon B(v_2, v_1)$, where $\epsilon = 1$ (respectively, $\epsilon = -1$).

Exercise 1.2.2 Show that if the elements x and y satisfy the invariance conditions of Definition 1.2.1, then so does $[x, y]$.

Exercise 1.2.3 Let B be a bilinear form on an $\mathfrak{g}$-module V. Show that $\mathrm{rad}(B)$ is a $\mathfrak{g}$-submodule of V. Deduce that if V is a simple $\mathfrak{g}$-module and B is not the zero map, then B is nondegenerate.

If J is an $n \times n$ matrix over the field k, we may use J to define a bilinear form by

$$(x, y) := x^{\mathrm{T}} J y,$$

where T denotes transpose.

If V is an n-dimensional $\mathfrak{g}$-module with associated representation ρ, the condition for the aforementioned bilinear form to be $\mathfrak{g}$-invariant is to have

$$(g.x, y) + (x, g.y) = 0$$

for all g, x and y. This is equivalent to the condition

$$x^{\mathrm{T}} \rho(g)^{\mathrm{T}} J y + x^{\mathrm{T}} J \rho(g) y = 0$$

for all x and y; in other words, $\rho(g)^{\mathrm{T}} J + J\rho(g) = 0$ for all matrices $\rho(g)$.

Proposition 1.2.4 *Let V be an n-dimensional vector space over k and let J be an $n \times n$ matrix over k. Then the elements $g \in \mathfrak{gl}(V)$ that satisfy*

$$(g.x, y) + (x, g.y) = 0$$

for all $x, y \in V$ form a Lie subalgebra of $\mathfrak{gl}(V)$.

Proof Let g and h be elements of $\mathfrak{gl}(V)$ that satisfy the hypotheses. We need to show that $[g, h]$ has the same property; in other words, that

$$(g.(h.x), y) - (h.(g.x), y) + (x, g.(h.y)) - (x, h.(g.y)) = 0.$$

It follows from the hypotheses that

$$(h.x, y) + (x, h.y) = 0,$$

which in turn implies that

$$(g.(h.x), y) + (h.x, g.y) + (g.x, h.y) + (x, g.(h.y)) = 0.$$

Reversing the roles of g and h gives

$$(h.(g.x), y) + (g.x, h.y) + (h.x, g.y) + (x, h.(g.y)) = 0.$$

Subtracting these last two equations from each other completes the proof. □

Definition 1.2.5 Let V be an n-dimensional vector space over k and let J be an $n \times n$ matrix over k. We define the Lie subalgebra $\mathfrak{gl}_J(V)$ of $\mathfrak{gl}(V)$ to be the set of elements $g \in \mathfrak{gl}(V)$ that satisfy

$$(g.x, y) + (x, g.y) = 0$$

for all $x, y \in V$.

Example 1.2.6 Let I_n be the $n \times n$ identity matrix, and let J_C be the $2n \times 2n$ block matrix of the form

$$\begin{pmatrix} 0 & I_n \\ -I_n & 0 \end{pmatrix}.$$

Let V be a $2n$-dimensional vector space over k, and let g be a typical element of $\mathfrak{gl}(V)$. As a block matrix, we may write

$$g = \begin{pmatrix} A & B \\ C & D \end{pmatrix}.$$

As mentioned previously, a necessary and sufficient condition for $g \in \mathfrak{gl}_{J_C}(V)$ is for

$$g^{\mathrm{T}} J_C + J_C g = 0,$$

that is,

$$\begin{pmatrix} -C^{\mathrm{T}} & A^{\mathrm{T}} \\ -D^{\mathrm{T}} & B^{\mathrm{T}} \end{pmatrix} + \begin{pmatrix} C & D \\ -A & -B \end{pmatrix} = \begin{pmatrix} 0 & 0 \\ 0 & 0 \end{pmatrix}.$$

This means that B and C are symmetric $n \times n$ matrices, and that $A = -D^{\mathrm{T}}$. Since the space of symmetric $n \times n$ matrices has dimension $n(n+1)/2$, it follows that the dimension of $\mathfrak{gl}_{J_C}(V)$ is given by

$$2.\frac{n(n+1)}{2} + n^2 = n(2n+1).$$

If we equip V with the obvious ordered basis

$$(e_1, e_2, \dots, e_n, f_1, f_2, \dots, f_n),$$

then the bilinear form B_C associated with J_C has the following properties:

$$\begin{aligned}(e_i, e_j) &= 0 \text{ for all } i, j,\\ (f_i, f_j) &= 0 \text{ for all } i, j,\\ (e_i, f_j) &= \delta_{i,j},\\ (f_i, e_j) &= -\delta_{i,j},\end{aligned}$$

where δ is the Kronecker delta, meaning that $\delta_{i,j} = 1$ if $i = j$, and 0 otherwise. The fact that J_C is invertible means that this bilinear form is nondegenerate. It follows by bilinearity that the form B_C is alternating; in other words, we have $(x, x) = 0$ for all $x \in V$. Expanding $(x + y, x + y)$ then shows that the form is *skew-symmetric*; in other words, we have $(x, y) = -(y, x)$ for all $x, y \in V$. It follows that B_C is symplectic.

Definition 1.2.7 The Lie algebra $\mathfrak{gl}_{J_C}(V)$ of Example 1.2.6 is known as the *symplectic Lie algebra*, and is denoted by $\mathfrak{sp}(2n, V)$.

Example 1.2.8 Let I_n be the $n \times n$ identity matrix, and let J_D be the $2n \times 2n$ block matrix of the form

$$\begin{pmatrix} 0 & I_n \\ I_n & 0 \end{pmatrix}.$$

Let V be a $2n$-dimensional vector space over k, and let g be a typical element of $\mathfrak{gl}(V)$. As in Example 1.2.6, we may write

$$g = \begin{pmatrix} A & B \\ C & D \end{pmatrix}.$$

The necessary and sufficient condition for $g \in \mathfrak{gl}_{J_D}(V)$, namely

$$g^T J_D + J_D g = 0,$$

now becomes

$$\begin{pmatrix} C^T & A^T \\ D^T & B^T \end{pmatrix} + \begin{pmatrix} C & D \\ A & B \end{pmatrix} = \begin{pmatrix} 0 & 0 \\ 0 & 0 \end{pmatrix}.$$

This means that B and C are skew-symmetric $n \times n$ matrices, and that $A = -D^T$. Since the space of skew-symmetric $n \times n$ matrices has dimension $n(n-1)/2$, it follows that the dimension of $\mathfrak{gl}_{J_D}(V)$ is given by

$$2.\frac{n(n-1)}{2} + n^2 = n(2n-1).$$

If we equip V with the obvious ordered basis

$$(e_1, e_2, \dots, e_n, f_1, f_2, \dots, f_n),$$

then the bilinear form B_{D} associated with J_{D} has the following properties:

$$\begin{aligned}(e_i, e_j) &= 0 \text{ for all } i, j,\\ (f_i, f_j) &= 0 \text{ for all } i, j,\\ (e_i, f_j) &= \delta_{i,j},\\ (f_i, e_j) &= \delta_{i,j},\end{aligned}$$

where δ is the Kronecker delta. The fact that J_{D} is invertible means that the form B_{D} is nondegenerate. It follows by bilinearity that B_{D} is symmetric; in other words, we have $(x, y) = (y, x)$ for all $x \in V$. It follows that B_{D} is orthogonal.

Definition 1.2.9 The Lie algebra $\mathfrak{gl}_{J_{\text{D}}}(V)$ of Example 1.2.8 is known as the *special orthogonal Lie algebra*, and is denoted by $\mathfrak{so}(2n, V)$.

There is also an orthogonal Lie algebra $\mathfrak{so}(2n+1, V)$ associated with an odd-dimensional vector space V. For our purposes, this example is less important than the examples above, but we outline the details below.

Exercise 1.2.10 Let I_n be the $n \times n$ identity matrix, and let J_{B} be the $2n+1 \times 2n+1$ block matrix of the form

$$\begin{pmatrix} 1 & 0 & 0 \\ 0 & 0 & I_n \\ 0 & I_n & 0 \end{pmatrix}.$$

Let V be a $2n+1$-dimensional vector space over k.

(i) Show that a typical element of $\mathfrak{gl}_{J_{\text{B}}}(V)$ may be written in the form

$$\begin{pmatrix} 0 & B & C \\ -C^{\text{T}} & D & E \\ -B^{\text{T}} & F & -D^{\text{T}} \end{pmatrix},$$

where B and C are arbitrary $1 \times n$ matrices, E and F are arbitrary skew-symmetric $n \times n$ matrices, and D is an arbitrary $n \times n$ matrix.

(ii) Show that the dimension of $\mathfrak{gl}_{J_{\text{B}}}(V)$ is $n(2n+1)$.

(iii) Show that the associated bilinear form B_{B} is nondegenerate and symmetric, and therefore orthogonal.

Definition 1.2.11 The Lie algebra $\mathfrak{gl}_{J_{\text{B}}}(V)$ of Exercise 1.2.10 is known as the *special orthogonal Lie algebra*, and is denoted by $\mathfrak{so}(2n+1, V)$.

Definition 1.2.12 The Lie algebras $\mathfrak{sl}(n+1, V)$, $\mathfrak{so}(2n+1, V)$, $\mathfrak{sp}(2n, V)$ and $\mathfrak{so}(2n, V)$ are known as the *classical Lie algebras* of types A_n, B_n, C_n and D_n, respectively. Their dimensions are $n(n+2)$, $n(2n+1)$, $n(2n+1)$ and $n(2n-1)$ respectively. If V is a k-vector space, we will also use the parallel notations $\mathfrak{sl}_{n+1}(k)$, $\mathfrak{so}_{2n+1}(k)$, $\mathfrak{sp}_{2n}(k)$ and $\mathfrak{so}_{2n}(k)$.

1.3 Classical Lie algebras and partially ordered sets

In this book, we will describe constructions of various algebraic objects (including the classical Lie algebras) in terms of certain labelled partially ordered sets called *heaps*.

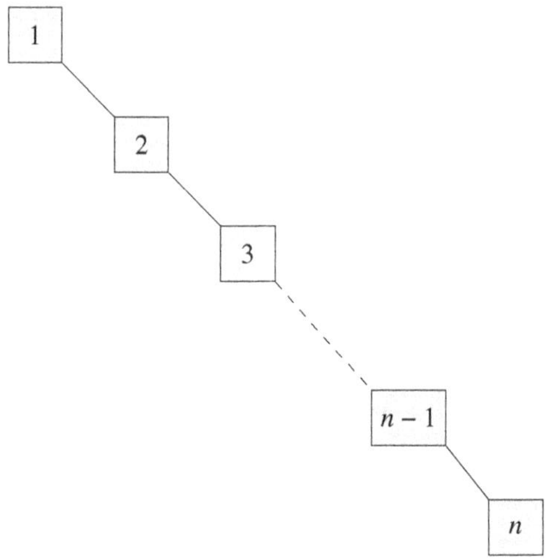

Figure 1.1 Partially ordered set corresponding to $\mathfrak{sl}(n+1, V)$

We will postpone the rigorous definition of heaps to Chapter 2; for now, we will think of a heap in terms of labelled Hasse diagrams such as the one in Figure 1.1.

Let E be the partially ordered set shown in Figure 1.1. In this case, E is a chain, with elements $x_1 > x_2 > x_3 > \cdots > x_n$, where the label of the element x_i is i. An ideal of E is a subset $I \subseteq E$ with the property that whenever $y \in I$ and $x \leq y$, we have $x \in I$. The set of ideals of E contains the subsets of the form

$$v_i := \{x_i, x_{i+1}, x_{i+2}, \ldots, x_n\}$$

for $1 \leq i \leq n$, together with the empty subset, $v_{n+1} := \emptyset$.

We now define V to be the k-vector space with basis $\{v_1, v_2, \ldots, v_{n+1}\}$, and for $1 \leq i \leq n$, we define linear operators E_i, F_i and H_i by their effect on basis elements, as follows.

Definition 1.3.1 We define $E_i(v_j) = v_k$ if the set-theoretic complement $v_k \backslash v_j$ is a single element of E with label i, or $E_i(v_j) = 0$, otherwise.

We define $F_i(v_j) = v_k$ if the set-theoretic complement $v_j \backslash v_k$ is a single element of E with label i, or $F_i(v_j) = 0$, otherwise.

We define $H_i(v_j) = v_j$ if v_j has a maximal element with label i; in other words, if $F_i(v_j)$ is nonzero. We define $H_i(v_j) = -v_j$ if $E \backslash v_j$ has a minimal element with label i; in other words, if $E_i(v_j)$ is nonzero. If neither of these conditions holds, we define $H_i(v_j) = 0$.

Remark 1.3.2 It is not immediately obvious that the above linear operators are well-defined, but this follows from various properties of the labelling. In particular, the operators H_i are well-defined because no two consecutive elements of the poset E have identical labels, and the operators E_i and F_i are well-defined because there is no pair of incomparable elements of E whose labels are the same. Eventually, these properties will follow axiomatically from Definition 2.2.2.

Example 1.3.3 We can regard the operators E_i, F_i and H_i as matrices acting with respect to the obvious ordered basis $(v_1, v_2, \ldots, v_{n+1})$ of V. We will write $E_{i,j}$ for the matrix unit whose effect is

$$E_{i,j}(v_k) = \delta_{j,k} v_i,$$

where δ is the Kronecker delta. Under these identifications, we have $E_i = E_{i,i+1}$, $F_i = E_{i+1,i}$ and $H_i = E_{i,i} - E_{i+1,i+1}$. Observe that all these matrices have zero trace, and can therefore be regarded as elements of $\mathfrak{sl}(n+1, V)$. The operators E_i, F_i, H_i for $1 \leq i \leq n$ therefore generate a Lie algebra that is a subalgebra of $\mathfrak{sl}(n+1, V)$. Once we have introduced the concepts of root system and Cartan decomposition in Section 4.2, it will follow that the subalgebra generated by the E_i, F_i and H_i is in fact the whole subalgebra $\mathfrak{sl}(n+1, V)$, in other words, a Lie algebra of type A_n. (See Exercise 5.3.21.)

Two features of this construction should be noted. The first is that the Lie algebra of type A_n can be constructed using only the labelled poset and the definitions of the operators E_i, F_i and H_i, and that these operators can be defined similarly for other labelled posets, provided that the labelling satisfies certain properties, such as those in Remark 1.3.2. The second feature is that this construction of the Lie algebra of type A_n not only constructs the Lie algebra as an abstract algebra, but also constructs it in its representation as a Lie algebra of matrices with zero trace.

It will turn out that the representations that can be constructed in this way are the minuscule representations of the title. The orthogonal Lie algebra $\mathfrak{so}(2n+1, V)$ in its natural representation on a $2n+1$-dimensional vector space is not minuscule, but we will see later that this Lie algebra has another representation, called the spin representation, which does have the property of being minuscule.

The natural representations of the other classical Lie algebras, $\mathfrak{sp}(2n, V)$ and $\mathfrak{so}(2n, V)$, do turn out to be minuscule. We can adapt the construction of $\mathfrak{sl}(n+1, V)$ above to deal with these cases, as we now explain.

Example 1.3.4 Consider the Lie algebra $\mathfrak{sp}(2n, V)$. For reasons that will eventually become clear in Section 3.3, the relevant labelled partially ordered set E in this situation is the one shown in Figure 1.2.

The poset E is a chain of length $2n-1$, which means that its set of ideals is a chain of length $2n$. We denote the ideals by

$$v_1 \supset v_2 \supset \cdots \supset v_{n-1} \supset v_n \supset v_{\overline{n}} \supset v_{\overline{n-1}} \supset \cdots \supset v_{\overline{2}} \supset v_{\overline{1}} = \emptyset.$$

We define V to be the k-vector space with (ordered) basis

$$\{v_1, v_{\overline{2}}, v_3, v_{\overline{4}}, \ldots, v_{\overline{1}}, v_2, v_{\overline{3}}, v_4, \ldots\},$$

and define linear operators E_i, F_i and H_i on V (for $1 \leq i \leq n$) using the same rules as Definition 1.3.1.

We can calculate the action of the generators E_i, F_i, H_i in terms of matrix units. If $i < n$, we have

$$\begin{aligned} E_i &= E_{i,i+1} + E_{\overline{i+1},\overline{i}}, \\ F_i &= E_{i+1,i} + E_{\overline{i},\overline{i+1}}, \quad \text{and} \\ H_i &= E_{i,i} - E_{i+1,i+1} - E_{\overline{i},\overline{i}} + E_{\overline{i+1},\overline{i+1}}, \end{aligned}$$

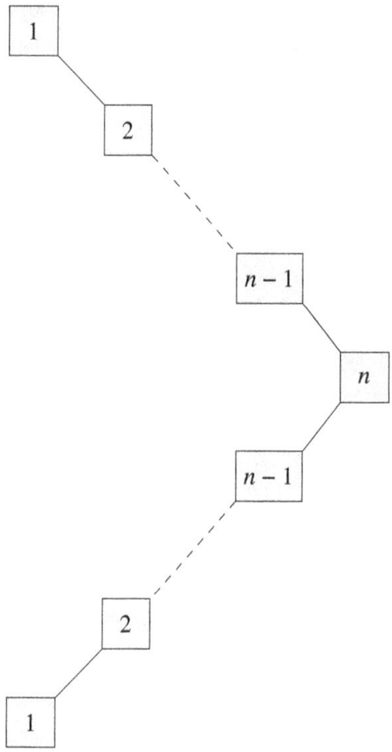

Figure 1.2 Partially ordered set corresponding to $\mathfrak{sp}(2n, V)$

and if $i = n$, we have

$$\begin{aligned} E_n &= E_{n,\overline{n}}, \\ F_n &= E_{\overline{n},n}, \quad \text{and} \\ H_n &= E_{n,n} - E_{\overline{n},\overline{n}}. \end{aligned}$$

Notice that, as in Example 1.2.6, all six of these $2n \times 2n$ matrices are of the form

$$\begin{pmatrix} A & B \\ C & D \end{pmatrix},$$

where the $n \times n$ matrices A, B, C and D have the properties that $A = -D^{\mathrm{T}}$, and B and C are symmetric. As in Example 1.3.3, it is not obvious that the given generators generate $\mathfrak{sp}(2n, V)$, but this will follow from later theory; see Exercise 5.3.25.

Example 1.3.5 Now consider the Lie algebra $\mathfrak{so}(2n, V)$. The relevant labelled partially ordered set in this situation is shown in Figure 1.3.

In this case, neither the poset E nor its set of ideals form a chain. However, we can adapt the description of the ideals in Example 1.3.4. There are two incomparable ideals, which we denote by v_n and $v_{\overline{n}}$; these are distinguished by the properties that the maximal element of v_n is labelled by n, and the maximal element of $v_{\overline{n}}$ is labelled

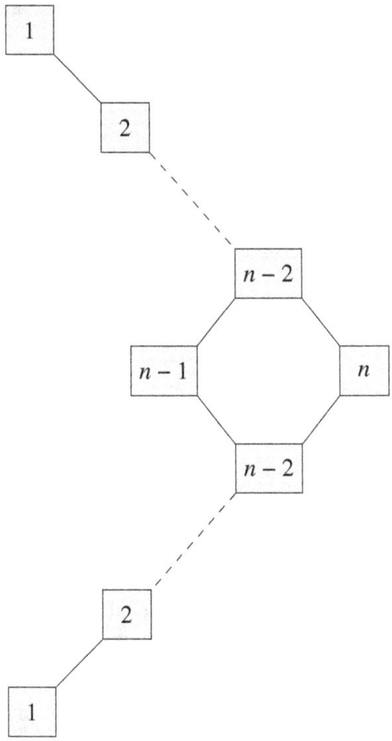

Figure 1.3 Partially ordered set corresponding to $\mathfrak{so}(2n, V)$

by $n - 1$. The other ideals are labelled so that we have

$$v_1 \supset v_2 \supset \cdots \supset v_{n-1} \supset v_n, v_{\overline{n}} \supset v_{\overline{n-1}} \supset \cdots \supset v_{\overline{2}} \supset v_{\overline{1}} = \emptyset.$$

We define V to be the k-vector space with (ordered) basis

$$\{v_1, v_2, \ldots, v_n, v_{\overline{1}}, v_{\overline{2}}, \ldots, v_{\overline{n}}\},$$

and define linear operators E_i, F_i and H_i on V (for $1 \leq i \leq n$) using the same rules as Definition 1.3.1.

The action of the generators E_i, F_i and H_i in terms of matrix units is given as follows. If $i < n$, we have

$$\begin{aligned} E_i &= E_{i,i+1} + E_{\overline{i+1},\overline{i}}, \\ F_i &= E_{i+1,i} + E_{\overline{i},\overline{i+1}}, \quad \text{and} \\ H_i &= E_{i,i} - E_{i+1,i+1} - E_{\overline{i},\overline{i}} + E_{\overline{i+1},\overline{i+1}}, \end{aligned}$$

and if $i = n$, we have

$$\begin{aligned} E_n &= E_{n-1,\overline{n}} + E_{n,\overline{n-1}}, \\ F_n &= E_{\overline{n},n-1} + E_{\overline{n-1},n}, \quad \text{and} \\ H_n &= E_{n-1,n-1} + E_{n,n} - E_{\overline{n-1},\overline{n-1}} - E_{\overline{n},\overline{n}}. \end{aligned}$$

Now for each odd i with $1 \leq i \leq n$, let us replace the basis element $v_{\bar{i}}$ by $-v_{\bar{i}}$. If $i < n$, we then obtain

$$\begin{aligned} E_i &= E_{i,i+1} - E_{\overline{i+1},\bar{i}}, \\ F_i &= E_{i+1,i} - E_{\bar{i},\overline{i+1}} \text{ and} \\ H_i &= E_{i,i} - E_{i+1,i+1} - E_{\bar{i},\bar{i}} + E_{\overline{i+1},\overline{i+1}}, \end{aligned}$$

and if $i = n$, we have

$$\begin{aligned} E_n &= \pm(E_{n-1,\bar{n}} - E_{n,\overline{n-1}}), \\ F_n &= \pm(E_{\bar{n},n-1} - E_{\overline{n-1},n}), \text{ and} \\ H_n &= E_{n-1,n-1} + E_{n,n} - E_{\overline{n-1},\overline{n-1}} - E_{\bar{n},\bar{n}}. \end{aligned}$$

Notice that, as in Example 1.2.8, all six of these $2n \times 2n$ matrices are of the form

$$\begin{pmatrix} A & B \\ C & D \end{pmatrix},$$

where the $n \times n$ matrices A, B, C and D have the properties that $A = -D^{\mathrm{T}}$, and B and C are skew-symmetric. As in Example 1.3.3, it is not obvious that the given generators generate $\mathfrak{so}(2n, V)$, but this will follow from later theory; see Exercise 5.3.28.

1.4 Classical Weyl groups and partially ordered sets

The classical Weyl groups are certain finite groups that play a central role in Lie theory. A classical Weyl group is said to be of type A_n (with $n \geq 1$), or of type B_n (with $n \geq 2$), or of type D_n (with $n \geq 4$). As we shall see later, a Weyl group is defined to be a group with a particular kind of presentation, but for now, we will ignore this and consider the groups as abstract groups.

Definition 1.4.1 As an abstract group, the Weyl group $W(A_n)$ of type A_n is isomorphic to the symmetric group $\mathcal{S}_{n+1}$.

The Weyl group $W(B_n)$ of type B_n is isomorphic to the wreath product $\mathbb{Z}_2 \wr \mathcal{S}_n$; that is, the semidirect product

$$\underbrace{(\mathbb{Z}_2 \times \mathbb{Z}_2 \times \cdots \times \mathbb{Z}_2)}_{n} \rtimes \mathcal{S}_n,$$

where the action of $\mathcal{S}_n$ is by permutation of the $\mathbb{Z}_2$-coordinates. The group $W(B_n)$ can be thought of as a group of signed permutations of n objects such as coins, which can be permuted and/or flipped over.

The Weyl group $W(D_n)$ of type D_n is isomorphic to the index 2 subgroup of $W(B_n)$ consisting of those signed permutations that effect an even number of sign changes.

Although these definitions may seem arbitrary at this point, the motivation for them will eventually become clear. There is also a Weyl group of type C_n, but it is the same as the Weyl group of type B_n; it is traditional to refer to this group as $W(B_n)$.

Remarkably, these Weyl groups may be defined in terms of the same partially ordered sets that were used to construct Lie algebras of types A_n, C_n and D_n in

Section 1.3. The key to this definition is to adapt Definition 1.3.1, which defined the action of various operators on the span of a vector space with basis indexed by the ideals of a labelled partially ordered set, E. The adapted definition is as follows.

Definition 1.4.2 If the set-theoretic complement $v_k \backslash v_j$ is a single element of E with label i, we define $S_i(v_j) = v_k$.

If the set-theoretic complement $v_j \backslash v_k$ is a single element of E with label i, we define $S_i(v_j) = v_k$.

If neither of the above two conditions applies to v_j, we define $S_i(v_j) = v_j$.

Remark 1.4.3 It will turn out from later definitions that the partially ordered sets E appearing in the situation of Definition 1.4.2 never have two elements covering each other with the same label, and this guarantees that the first two conditions of Definition 1.4.2 do not overlap.

It is immediate from Definition 1.4.2 that the action of S_i on the elements v_j is a permutation action. We now examine this more closely in each of the three examples given in Section 1.3.

Example 1.4.4 Consider the labelled partially ordered set E of Figure 1.1. Recall from Example 1.3.3 that the ideals of E are denoted by

$$(v_1, v_2, \dots, v_{n+1}),$$

where the maximal element of v_i is labelled by i. Identifying v_i with the symbol i for short, and applying Definition 1.4.2, we now find that the permutation action of the operator S_i (for $1 \leq i \leq n$) is the simple transposition $(i, i+1)$. (Recall that a transposition (i, j) is called *simple* if we have $j = i \pm 1$.)

The group generated by the simple transpositions

$$\{(1, 2), (2, 3), \dots, (n, n+1)\}$$

is well known to be isomorphic to the symmetric group $\mathcal{S}_{n+1}$. In fact, the Weyl group $W(A_n)$ is defined to be the symmetric group $\mathcal{S}_{n+1}$ with this particular set of generators, and with a particular set of relations that we will come to later.

Example 1.4.5 Consider the labelled partially ordered set E of Figure 1.2. Recall from Example 1.3.4 that the ideals of E are given by

$$v_1 \supset v_2 \supset \cdots \supset v_{n-1} \supset v_n \supset v_{\overline{n}} \supset v_{\overline{n-1}} \supset \cdots \supset v_{\overline{2}} \supset v_{\overline{1}} = \emptyset.$$

We denote v_i by i for short, and $v_{\overline{i}}$ by $\overline{i}$.

Applying Definition 1.4.2, we now find that the permutation action of the operator S_i (for $1 \leq i < n$) is a product of two simple transpositions,

$$(i, i+1)(\overline{i+1}, \overline{i}).$$

The action of the operator S_n is given by the simple transposition $(n, \overline{n})$. We leave it as an exercise to show that the operators S_i (for $1 \leq i \leq n$) generate a group isomorphic to $W(B_n) \cong \mathbb{Z}_2 \wr \mathcal{S}_n$. As in Example 1.4.4, these generators for $W(B_n)$ are part of the definition of the group.

Exercise 1.4.6 Verify that the operators S_i (for $1 \le i \le n$) in Example 1.4.5 generate a group isomorphic to $W(B_n)$. (Hint: think of the operators as permutations of signed objects. If $i < n$, show that S_i corresponds to transposing the signed objects at positions i and $i+1$. If $i = n$, show that S_n corresponds to changing the sign of the n-th object.)

Example 1.4.7 Consider the labelled partially ordered set E of Figure 1.3. Recall from Example 1.3.5 that the ideals of E are given by

$$v_1 \supset v_2 \supset \cdots \supset v_{n-1} \supset v_n, v_{\overline{n}} \supset v_{\overline{n-1}} \supset \cdots \supset v_{\overline{2}} \supset v_{\overline{1}} = \emptyset,$$

where the incomparable ideals v_n and $v_{\overline{n}}$ have maximal elements labelled by n and $n-1$ respectively. As in Example 1.4.5, we denote v_i by i for short, and $v_{\overline{i}}$ by $\overline{i}$.

Applying Definition 1.4.2, we now find that the permutation action of the operator S_i (for $1 \le i < n$) is a product of two simple transpositions,

$$(i, i+1)(\overline{i+1}, \overline{i}).$$

The action of the operator S_n is also given by a product of two simple transpositions, namely

$$(n-1, \overline{n})(n, \overline{n-1}).$$

We leave it as an exercise to show that the operators S_i (for $1 \le i \le n$) generate a group isomorphic to $W(D_n)$. As in Example 1.4.4, these generators for $W(D_n)$ are part of the definition of the group.

Exercise 1.4.8 Verify that the operators S_i (for $1 \le i \le n$) in Example 1.4.7 generate a group isomorphic to $W(D_n)$. (Hint: think of the operators as permutations of signed objects. If $i < n$, show that S_i corresponds to transposing the signed objects at positions i and $i+1$. If $i = n$, show that S_n corresponds to transposing the signed objects at positions $n-1$ and n, combined with changing the sign of each of them.)

1.5 Notes and references

1. Our treatment of the matrix Lie algebras $\mathfrak{sl}(n+1, V)$, $\mathfrak{so}(2n+1, V)$, $\mathfrak{sp}(2n, V)$ and $\mathfrak{so}(2n, V)$ is based on that of Erdmann and Wildon [23]; see [23, sections 12.2, 12.3, 12.5, 12.4], respectively.
2. Over the complex numbers, the Lie algebras of types A_n, B_n, C_n, D_n are examples of simple Lie algebras. This will be discussed more fully in Theorem 5.1.1.
3. We will revisit the Lie algebra constructions in Examples 1.3.3, 1.3.4 and 1.3.5 in Example 4.1.8 and Exercises 4.1.9 and 4.1.10, respectively.
4. The generators E_i, F_i and H_i for the Lie algebras given here form part of the standard Serre presentation for the algebra; see Definition 4.1.3 and the Notes and References of Chapter 4 for details.
5. We will revisit the Weyl group constructions in Examples 1.4.4, 1.4.5 and 1.4.7 in Example 3.3.9 and Exercises 3.3.10 and 3.3.14, respectively.
6. The generators S_i for the Weyl group form part of the usual presentation for the group. For the relations, see Definition 3.1.9 and the Notes and References of Chapter 4.
7. The symplectic and orthogonal forms discussed here can be constructed in a systematic way using only the structure of the heap. This will be explained in detail in Section 5.6.

2

Heaps over graphs

The main combinatorial framework we will use in this book to study minuscule representations is that of heaps. A heap is a certain function from a partially ordered set to the set of vertices of a graph. We will often think of the heap as the Hasse diagram of the underlying poset, where the elements of the poset are labelled by vertices of the graph. For most purposes in this book, thinking of heaps pictorially as labelled partially ordered sets (as in Figure 2.7 below) will be adequate. However, for later applications, we will need concepts such as those of subheaps and quotient heaps, and it is unwieldy to describe these in terms of pictures. For such applications, it turns out to be concise and natural to define heaps in terms of categories, and in Section 2.1, we introduce the category of heaps and explain the connection between finite heaps and commutation monoids. A reader not comfortable with categories may skip most of Chapter 2 at a first reading, except Section 2.2, which introduces the key notion of "full heap".

For the applications in this book, the most important kind of heap is those for which the associated graph is a Dynkin diagram, particularly one of finite or affine type. Dynkin diagrams are certain graphs that may have multiple and directed edges. The information in a Dynkin diagram may also be captured using a matrix with integer entries, called the generalized Cartan matrix of a Dynkin diagram. The main aim of Section 2.2 is to define the notion of a full heap over a Dynkin diagram; this is a type of heap over the underlying graph of the Dynkin diagram that is compatible in some sense with the multiple edges and decorations of the diagram. Remarkably, it will turn out that the minuscule representations of the title and many other interesting combinatorial objects may be constructed in terms of full heaps. A main purpose of this book is to describe these constructions in detail.

Any partially ordered set gives rise to a distributive lattice, namely the set of all ideals of the partially ordered set under the operations of intersection ("meet") and union ("join"). If the partially ordered set has labelled vertices, as is the case for a heap, then the corresponding distributive lattice has labelled covering relations. The structure of the labelled distributive lattice corresponding to a full heap turns out to be the key to the applicability of the concept. In Section 2.3, we prove the main theorem of Chapter 2, the Local Structure Theorem for full heaps (Theorem 2.3.15), which describes the local structure of this distributive lattice. It turns out that the distributive lattice itself can be very complicated compared to the full heap itself, so it is more convenient to describe constructions in terms of the heap and its ideals.

The category construction is also very helpful in defining the notion of quotient heaps in Section 2.4, which turns out to be important for certain later constructions.

2.1 Basic definitions

In Section 2.1, we define heaps in terms of categories. The reader unfamiliar with categories should either read Sections A.1 and A.2 of the Appendix first, or skip to Section 2.2 and think of heaps in terms of the pictures given. Throughout this book, we will typically be interested in isomorphism classes of heaps rather than heaps themselves, and heap morphisms provide a convenient framework to define the notion of isomorphism. Morphisms are also useful for giving concise definitions of subheaps later in this section, and quotient heaps in Section 2.4.

We define a heap over a graph Γ without loops to be a partially ordered set $(E, \leq)$, together with a function $\varepsilon : E \to \Gamma$ (the labelling function) from E to the vertices of Γ, satisfying the following two properties:

(H1) for every vertex x and every edge $\{a, b\}$ of Γ, the subsets $\varepsilon^{-1}(x)$ and $\varepsilon^{-1}(\{a, b\})$ of E are chains with respect to the partial order $\leq$;

(H2) the partial order on E is the minimal partial order extending the given partial order on the above chains $\varepsilon^{-1}(x)$ and $\varepsilon^{-1}(\{a, b\})$.

We call the chains $\varepsilon^{-1}(x)$ and $\varepsilon^{-1}\{a, b\}$ appearing in (H1) above *vertex chains* and *edge chains* respectively.

Remark 2.1.1 The word "minimal" in Axiom (H2) means "minimal as a subset of $E \times E$", as in Section A.1. Although this axiom may at first seem mysterious, the characterization of the partial order that it gives may be expressed in a more elementary (but less concise) way, as in Lemma 2.1.5 (i) below.

There is a category **Heap** whose objects are heaps, in which a morphism f from a heap $(E_1, \leq_1)$ over a graph Γ_1 to a heap $(E_2, \leq_2)$ over a graph Γ_2 consists of a pair (f_E, f_Γ) in which

(i) f_E is a morphism of partially ordered sets (i.e., $x \leq y \Rightarrow f_E(x) \leq f_E(y)$);

(ii) f_Γ is a morphism of graphs (i.e., if a and b are adjacent, then the vertices $f_\Gamma(a)$ and $f_\Gamma(b)$ are adjacent or equal) and

(iii) the following diagram commutes:

$$\begin{array}{ccc} E_1 & \xrightarrow[\varepsilon_1]{} & \Gamma_1 \\ {\scriptstyle f_E}\big\downarrow & & \big\downarrow{\scriptstyle f_\Gamma} \\ E_2 & \xrightarrow[\varepsilon_2]{} & \Gamma_2 \end{array}$$

in other words, we have $f_\Gamma \circ \varepsilon_1 = \varepsilon_2 \circ f_E$.

We may refer to a heap over a graph Γ either by the poset E or by the labelling function $\varepsilon : E \to \Gamma$ and will often depict a locally finite heap by the Hasse diagram of E with vertices x labelled by $\varepsilon(x)$. This gives the motivation for the following.

The morphism f is said to be *injective* on vertices (respectively, *surjective*, *bijective*, *the identity* on vertices) if f_E is injective (respectively, surjective, bijective, the identity).

The morphism f is said to be *injective* on labels (respectively, *surjective*, *bijective*, *the identity* on labels) if f_Γ is injective (respectively, surjective, bijective, the identity).

Let Γ be a graph. The category **Heap** has a subcategory, **Heap**(Γ), whose objects are heaps over Γ. A morphism f of **Heap**(Γ) is a morphism $f \in \mathrm{Hom}_{\mathbf{Heap}}(A, B)$, where A and B are objects of **Heap**(Γ), and where f is the identity on labels. If $\varepsilon : E \to \Gamma$ is a heap, the group $\mathrm{Aut}_{\mathbf{Heap}(\Gamma)}(E)$ is the group of all isomorphisms $f : E \to E$ in the category **Heap**(Γ) under composition of maps. If $\mathrm{Aut}_{\mathbf{Heap}(\Gamma)}(E)$ is infinite cyclic, we call the heap E *periodic*.

Let E and F be heaps over Γ. We say that F is a *subheap* of E if there exists a morphism

$$f \in \mathrm{Hom}_{\mathbf{Heap}(\Gamma)}(F, E)$$

that is injective on vertices. We will often identify F with the subset of E given by the image of f on vertices. We say that the subheap F is *convex* if it corresponds to a convex subset of E, and that F is a *chain* if it is a chain in E considered as a partially ordered set.

A heap E is said to be *trivial* if its underlying poset has the property that $x \leq y$ implies $x = y$. A heap E is said to be *locally finite* if its underlying poset is locally finite (as defined in Appendix A.1). The *size*, $|E|$, of a heap E is the cardinality of the underlying poset. The *dual heap*, E^*, of E is the heap obtained from $\varepsilon : E \to \Gamma$ by reversing the partial order on E. There is a natural identification between the vertices of E and those of E^*, which we denote by the function $* = *_E : E \to E^*$. The *support* of a heap $\varepsilon : E \to \Gamma$ is the subgraph of Γ whose vertices are $\varepsilon(E)$.

Example 2.1.2 Figure 2.1 shows a heap E of size 5 over a graph Γ with three vertices. In this case, the labelling function $\varepsilon : E \to \Gamma$ satisfies $\varepsilon(a) = \varepsilon(d) = 1$, $\varepsilon(c) = 2$ and $\varepsilon(b) = \varepsilon(e) = 3$. The support of E is the whole of Γ.

The vertex chains of E are $\varepsilon^{-1}(1) = \{a, d\}$, $\varepsilon^{-1}(2) = \{c\}$ and $\varepsilon^{-1}(3) = \{b, e\}$.

The edge chains of E are $\varepsilon^{-1}(\{1, 2\}) = \{a, c, d\}$ and $\varepsilon^{-1}(\{2, 3\}) = \{b, c, e\}$.

The dual heap, E^*, has the same underlying set and labelling function, but the relations $d < c < a$ and $e < c < b$ in E become $a <^* c <^* d$ and $b <^* c <^* e$. In this particular case, E and E^* are isomorphic by the function $f : E \to E$ for which $f(a) = d$, $f(b) = e$, $f(c) = c$, $f(d) = a$ and $f(e) = b$.

Figure 2.2 shows a heap F over the same graph Γ. In this case, the labelling function $\varepsilon : F \to \Gamma$ satisfies $\varepsilon(x) = 1$ and $\varepsilon(y) = 3$. The support of F is the disconnected subgraph of Γ whose vertices are 1 and 3.

The vertex chains of F are $\varepsilon^{-1}(1) = \{x\}$, $\varepsilon^{-1}(2) = \emptyset$ and $\varepsilon^{-1}(3) = \{y\}$.

The edge chains of F are $\varepsilon^{-1}(\{1, 2\}) = \{x\}$ and $\varepsilon^{-1}(\{2, 3\}) = \{y\}$.

Since no two distinct elements of F are comparable, F is an example of a trivial heap (although E is not).

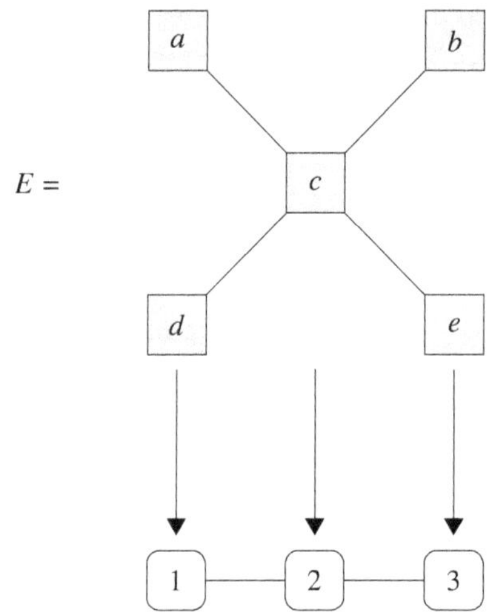

Figure 2.1 A heap E over a graph Γ

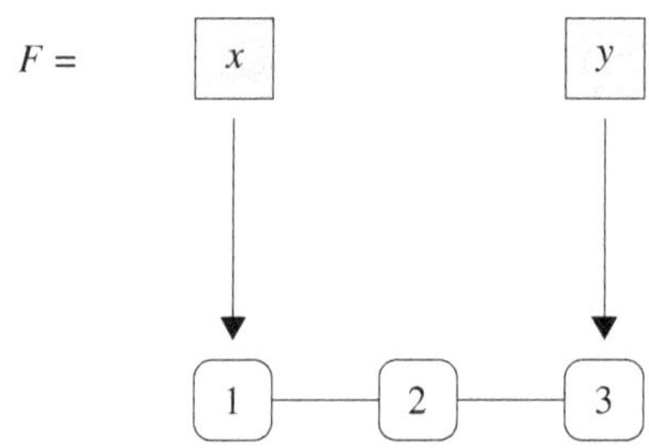

Figure 2.2 A trivial heap F over Γ

We define a function f_E from F to E sending x to a and y to e. This can be extended to a morphism f in **Heap**(Γ) from F to E that is the identity on labels. It is routine to check that f is a morphism of **Heap**(Γ), and since f is injective on vertices, this shows that F is a subheap of E.

Note that the partial order on F does *not* agree with the restriction of the partial order on E to the subset $f_E(F) = \{a, e\}$. (See Exercise 2.1.7 below.) Instead, the partial order on F is the minimal partial order that extends the vertex and edge chains of $\{a, e\}$ given by the partial order E. In the case of F, the vertex and edge chains are all singletons, which implies that the partial order on F is trivial, even though the elements a and e are comparable as elements of E.

We can make F into a subheap of E in a total of four ways: we may send x to a or d, and we may send y to b or e.

We are often only interested in the isomorphism type of a given heap. For this reason, it is convenient to introduce a shorthand notation for heaps up to isomorphism, in which each vertex x in the Hasse diagram of the heap E is labelled by $\varepsilon(x)$

$E =$

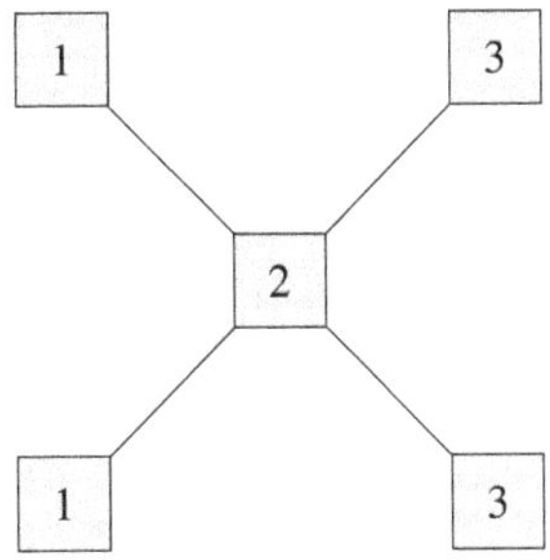

Figure 2.3 Shorthand for the heap E over the graph Γ shown in Figure 2.1

$E =$

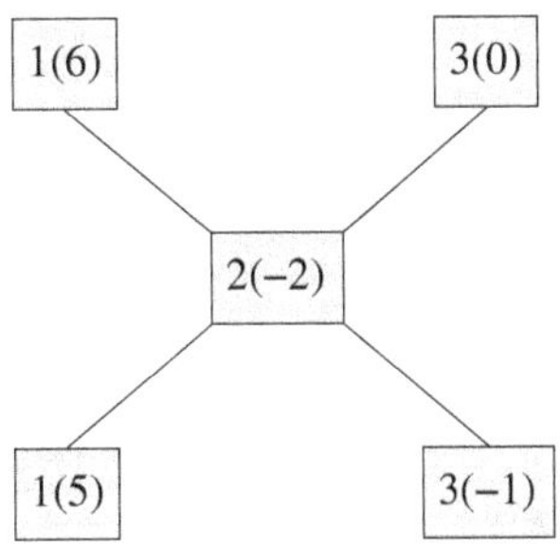

Figure 2.4 An example of the labelling $E(a, z)$ for a heap whose shorthand notation is given in Figure 2.3

rather than x. For this notation to make sense, it must be understood what the graph Γ is.

Example 2.1.3 The shorthand notation for the heap shown in Figure 2.1 is shown in Figure 2.3.

We may be in a situation where a heap is defined in terms of the above shorthand and the graph Γ is understood, but the names of the elements of E are not specified. If E is locally finite, as will be the case for nearly all the heaps that we consider, then the vertex chains of E are isomorphic, as partially ordered sets, to convex subsets of $\mathbb{Z}$ (see Exercise 2.1.8 below). If a is a vertex of Γ, we may therefore mark the elements of $\varepsilon^{-1}(a)$ by consecutive integers in such a way that higher vertices in the partial order $\leq$ are marked by greater integers. We will write $a(z)$ or $E(a, z)$ to refer to the element $\varepsilon^{-1}(a) \in E$ that is marked by the integer z. This labelling system is not canonical, but is fixed once we have decided upon a name for any one element of $\varepsilon^{-1}(a)$.

Example 2.1.4 Let Γ be the graph in Example 2.1.2, and let E be a heap over Γ whose shorthand notation is shown in Figure 2.3. One way in which the elements of E may be labelled using the notation $E(a, z)$ explained above is shown in Figure 2.4. In this case, the names of the elements of the heap E are

$$\{E(1, 5),\ E(1, 6),\ E(2, -2),\ E(3, -1),\ E(3, 0)\}.$$

Lemma 2.1.5 *Let $\varepsilon : E \to \Gamma$ be a heap over a graph, and let $x, y \in E$.*

(i) *We have $x \leq y$ if and only if there is a (possibly trivial) finite sequence*

$$x = z_0 < z_1 < \cdots < z_k = y$$

in E such that for each $0 \leq i < k$, the following two properties hold:
   (a) $z_i < z_{i+1}$;
   (b) *either $\varepsilon(z_i)$ and $\varepsilon(z_{i+1})$ are adjacent vertices of Γ, or $\varepsilon(z_i) = \varepsilon(z_{i+1})$.*

(ii) *If $\varepsilon(x)$ and $\varepsilon(y)$ lie in different connected components of Γ, then x and y are not comparable in E.*

(iii) *If y covers x in E, then either $\varepsilon(x) = \varepsilon(y)$, or $\varepsilon(x)$ and $\varepsilon(y)$ are adjacent vertices of Γ.*

(iv) *If, in addition, E is locally finite, then the z_i above may be chosen so that z_{i+1} covers z_i, for all $0 \leq i < k$.*

Proof To prove (i), we recall that the partial order $\leq$ on E is the minimal partial order extending the total order on each vertex and edge chain; in other words, it is the reflexive, transitive extension of the vertex and edge chain order. This implies that we may choose the z_i such that each pair $\{z_i, z_{i+1}\}$ satisfies $z_i < z_{i+1}$ and belongs to a vertex chain or edge chain. The latter condition means that $\varepsilon(z_i)$ and $\varepsilon(z_{i+1})$ are either equal or adjacent vertices of Γ.

Part (ii) is a consequence of (i). If x and y are comparable, then we may assume without loss of generality that $x \leq y$. The sequence $\varepsilon(z_0), \ldots, \varepsilon(z_k)$ of (i) then produces a path in Γ from $\varepsilon(x)$ to $\varepsilon(y)$, which implies that $\varepsilon(x)$ and $\varepsilon(y)$ lie in the same connected component of Γ.

If y covers x in E as in (iii), we must have $k = 1$ in the sequence of (i), and the assertion follows.

If E is locally finite, the sequence in (i) may be refined if necessary until the relations shown are covering relations. Since E is locally finite, the interval $[x, y]$ is finite, and this refinement must terminate after a finite number of steps. The assertion of (iv) follows. □

Exercise 2.1.6 Verify that **Heap** is a category. Prove that **Heap**(Γ) is a subcategory of **Heap** for each graph Γ. Give an example of a graph Γ such that

(i) **Heap**(Γ) is a full subcategory of **Heap**;
(ii) **Heap**(Γ) is not a full subcategory of **Heap**.

(Hint: you may wish to consider the case of heaps over the singleton graph.)

Exercise 2.1.7 Suppose that F is a *convex* subheap of a heap E via a morphism f that is injective on vertices. Prove that the partial order on F may be identified with the restriction of the partial order on E to the subset $f_E(F)$.

Exercise 2.1.8 Let E be a locally finite heap over a graph Γ. Show that each vertex chain of E is isomorphic, as a partially ordered set, to a nonempty convex subset of $\mathbb{Z}$.

Exercise 2.1.9 Let E and Γ be as in Example 2.1.2. Show that, in the category **Heap**(Γ), E has 24 isomorphism classes of subheaps: one of size 5, five of size 4, eight of size 3, seven of size 2, three of size 1 and one of size 0.

Exercise 2.1.10 Verify that E^* is a heap.

Exercise 2.1.11 Let E be a heap over a graph Γ_1, and let Γ_2 be the singleton graph. Show that there is a natural bijection between (a) morphisms in **Heap** that are injective on vertices from E to heaps over Γ_2 and (b) total orders refining the underlying partial order on E.

Exercise 2.1.12 Construct an explicit example of an uncountable heap E over the singleton graph, and show that it is not locally finite. Give an example of two elements $x, y \in E$ such that we have $x < y$, but such that it is impossible to find a sequence of z_i as in Lemma 2.1.5 for which z_{i+1} covers z_i for all $0 \leq i < k$.

Exercise 2.1.13 Let Γ be a graph, and let $(E, \leq)$ be a partially ordered set. Let us say that a function $\varepsilon : E \to \Gamma$ satisfies Property H if all covering pairs in E have labels that are equal or adjacent in Γ, and incomparable pairs in E have distinct, nonadjacent labels.

(i) Show that any heap satisfies Property H.
(ii) Show that if E is locally finite and $\varepsilon : E \to \Gamma$ satisfies Property H, then ε is a heap.
(iii) Now let Γ be a graph with vertex set $\{a, b, c\}$ in which b is adjacent to a and to c, but a and c are not adjacent to each other. Let $(E, \leq)$ be the set of real numbers under their usual (total) order, and define $\varepsilon : E \to \Gamma$ by letting $\varepsilon(r) = a$ if r is rational, and $\varepsilon(r) = c$ otherwise. Show that $\varepsilon : E \to \Gamma$ satisfies Property H, but that ε is not a heap.

Exercise 2.1.14 Let Γ be a disconnected graph with connected components Γ_1 and Γ_2, and let $\varepsilon_1 : E_1 \to \Gamma_1$ and $\varepsilon_2 : E_2 \to \Gamma_2$ be heaps equipped with partial orders $\leq_1$ and $\leq_2$, respectively. Let E be the disjoint union of E_1 and E_2, equipped with the natural partial order (as defined in Exercise A.1.5). Let $\varepsilon : E \to \Gamma$ be the extension of ε_1 and ε_2 to E.

(i) Show that $\varepsilon : E \to \Gamma$ is a heap.
(ii) Show that E has convex subheaps F_1 and F_2 isomorphic (in **Heap**) to each of E_1 and E_2.
(iii) Show that no element of F_1 is comparable to an element of F_2.

Definition 2.1.15 A *semigroup* is a set equipped with an associative binary operation, and a *monoid* is a semigroup that has an identity element. If S is a set, the *free monoid* on S, denoted by S^*, is the set of all words in the alphabet S (including the empty word), equipped with the binary operation of juxtaposition (which is clearly associative).

Example 2.1.16 Let $S = \{a, b, c\}$, so that the elements $x = cbbac$ and $y = bcb$ are elements of S^*. In this case, the product is given by

$$xy = cbbacbcb.$$

Definition 2.1.17 Let $\varepsilon : E \to \Gamma$ be a heap over a graph, and let S be the set of vertices of Γ. The *commutation monoid*, $\mathrm{Co}(\Gamma)$, of Γ is the quotient of the free monoid S^* by the congruence $\equiv$ generated by the relations $st \equiv ts$ whenever s and t are nonadjacent in Γ. The equivalence class of the word $x \in S^*$ is denoted by $[x]$. The multiplication in $\mathrm{Co}(\Gamma)$ is given by

$$[x][y] := [xy],$$

and is well-defined. Elements of $\mathrm{Co}(\Gamma)$ are called *traces*.

Definition 2.1.18 Let $\varepsilon : E \to \Gamma$ be a finite heap, let S be the set of vertices of Γ, and let $\leq'$ be a refinement of the partial order of E to a total order; recall from Section A.1 that such a refinement exists by Szpilrajn's Theorem. Let $x_1, x_2, \ldots, x_N$ be the set of elements in E, listed so that $x_1 <' x_2 <' \cdots <' x_N$. Denote by $f_{<'}(E)$ the element of S^* given by $\varepsilon(x_1)\varepsilon(x_2)\cdots\varepsilon(x_N)$.

Exercise 2.1.19 Explain why the multiplication in $\mathrm{Co}(\Gamma)$ is well-defined.

We now state one of the most well-known results about heaps. It shows that the isomorphism types of heap over a fixed graph are in natural correspondence with the traces of the associated commutation monoid. In particular, this gives a convenient notation for finite heaps in terms of sequences, and also endows the set of isomorphism types of finite heap over a fixed graph with the structure of a monoid.

Theorem 2.1.20 *Let Γ be a graph and let $\mathrm{Co}(\Gamma)$ be its commutation monoid. Let f_Γ be the map from the set of isomorphism classes of finite heaps over Γ to $\mathrm{Co}(\Gamma)$ given by*

$$f_\Gamma(\overline{E}) = [f_{<'}(E)],$$

where $\overline{E}$ is the isomorphism type of E and $f_{<'}$ is as in Definition 2.1.18. Then f_Γ is (i) independent of the choice of total order $<'$, and (ii) is a bijection. □

Proof A proof may be found in [88, proposition 3.4]. □

Example 2.1.21 Let E be the heap shown in Example 2.1.2, and let $E' = E\backslash\{d\}$ be the subheap of E obtained by removing the element $d \in \varepsilon^{-1}(1)$. Then there are four (respectively, two) possible refinements of E (respectively, E') to a total order. We have

$$f_\Gamma(E) = [13213] = [31213] = [13231] = [31231]$$

and

$$f_\Gamma(E') = [3213] = [3231].$$

2.2 Full heaps over Dynkin diagrams

Most of the heaps of interest in this book are convex subheaps of full heaps. Full heaps are defined in terms of certain matrices, which we now introduce.

Let A be an n by n matrix with integer entries. We call A a *generalized Cartan matrix* if it satisfies the following conditions:

(i) $a_{ii} = 2$ for all $1 \leq i \leq n$,
(ii) $a_{ij} \leq 0$ for $i \neq j$ and
(iii) $a_{ij} = 0 \Leftrightarrow a_{ji} = 0$.

Unless otherwise stated, we will only consider generalized Cartan matrices with entries in the set

$$\{2, 0, -1, -2\};$$

such matrices are sometimes called *doubly laced*. If, furthermore, A has no entries equal to -2, we will call A *simply laced*. A generalized Cartan matrix is called *symmetrizable* if there exists a nonsingular diagonal matrix D and a symmetric matrix B such that $A = DB$. In particular, if A is simply laced, then A is symmetrizable: in this case, A is symmetric, so we may take D to be the identity matrix.

The *Dynkin diagram* $\Gamma = \Gamma(A)$ associated with a generalized Cartan matrix is a directed graph, possibly with multiple edges, and vertices indexed (for now) by the integers 1 up to n. Suppose that $i \neq j$. If $a_{ij}a_{ji} \leq 4$ and $|a_{ij}| \geq |a_{ji}|$, we connect the vertices corresponding to i and j by $|a_{ij}|$ lines; this set of lines is equipped with an arrow pointing towards i if $|a_{ij}| > 1$. (For example, if $a_{ij} = a_{ji} = -2$, this will result in a double edge between i and j equipped with an arrow pointing in each direction.) There are further rules if $a_{ij}a_{ji} > 4$, but we do not need these for our purposes.

The Dynkin diagram (up to re-indexing of the vertices) and the generalized Cartan matrix determine each other, so we may write $A = A(\Gamma)$. If Γ is connected, we call A *indecomposable*. It will be convenient for some purposes to relabel the vertices of the Dynkin diagram.

A *heap over a Dynkin diagram* is by definition a heap over the underlying graph obtained from the Dynkin diagram after replacing all multiple directed edges by single, undirected edges.

Example 2.2.1 An example of a generalized Cartan matrix is

$$A = \begin{pmatrix} 2 & 0 & -1 & 0 \\ 0 & 2 & 1 & 0 \\ -1 & -1 & 2 & -1 \\ 0 & 0 & -2 & 2 \end{pmatrix}.$$

In this case, A is doubly laced, but not simply laced. The corresponding Dynkin diagram $\Gamma = \Gamma(A)$ is as shown in Figure 2.5. We will follow Kac's notation for Dynkin diagrams [37], in which Γ is called $B_3^{(1)}$. In order to retain compatibility with this notation, we renumber the vertices of Γ to produce the Dynkin diagram in Figure 2.6.

Figure 2.7 shows an infinite heap over the Dynkin diagram Γ, with the numbering conventions of Figure 2.6. The heap has the form of a vertically repeating motif, namely the contents of the dashed box, and it follows that the heap is periodic. In this case, the finite subheap enclosed in the dashed box is associated with the trace [203213] by Theorem 2.1.20.

We are now ready to introduce the notion of a full heap. Full heaps are one of the primary objects of study in this book.

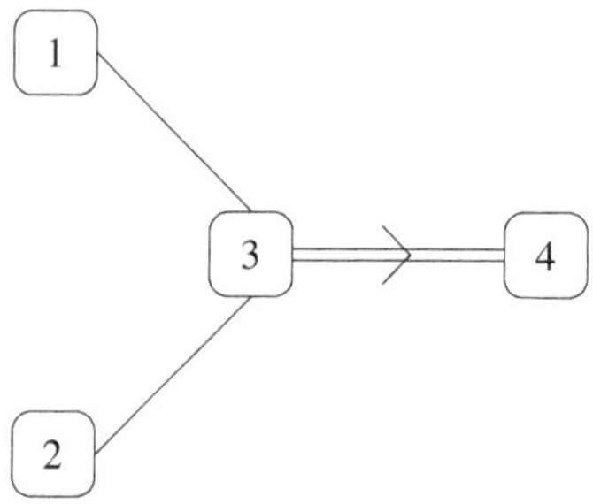

Figure 2.5 The Dynkin diagram corresponding to A in Example 2.2.1

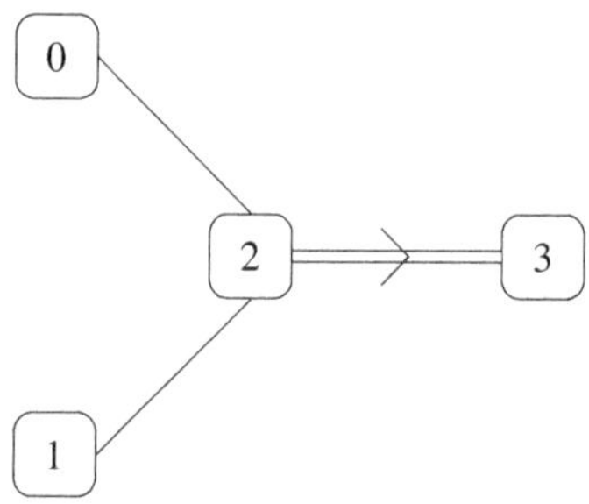

Figure 2.6 Kac's vertex labelling for the Dynkin diagram $B_3^{(1)}$

Definition 2.2.2 Let $\varepsilon : E \to \Gamma$ be a nonempty locally finite heap over a Dynkin diagram corresponding to the generalized Cartan matrix $A = (a_{ij})$. If $x, y \in E$ have the properties that $\varepsilon(x) = \varepsilon(y) = p$ and $(x, y) \cap \varepsilon^{-1}(p) = \emptyset$, we call (x, y) an *open p-interval* and $[x, y]$ a *closed p-interval*. We call E a *full heap over* Γ if the following three conditions are satisfied.

(F1) Every vertex chain of E is isomorphic, as a partially ordered set, to $\mathbb{Z}$.
(F2) For every edge $\{a, b\}$ of Γ and every vertex $x \in E$ with $\varepsilon(x) = a$, there exists $y \in E$ with $\varepsilon(y) = b$ such that either x covers y or y covers x.
(F3) For every closed p-interval $[x, y]$ in E, we have $\sum_{x \leq z \leq y} a_{p,\varepsilon(z)} = 2$.

Remark 2.2.3 Note that condition (F1) implies that every vertex chain of E is nonempty.

Roughly speaking, condition (F2) says that the labelled Hasse diagram of E locally looks like the underlying Dynkin diagram.

An interval $[x, y]$ satisfying condition (F3) may be thought of as "full", and the term "full heap" is intended to convey this idea.

Example 2.2.4 The heap E shown in Figure 2.7 is a full heap over the Dynkin diagram $\Gamma = B_3^{(1)}$.

To see this, first note that it is clear from the picture that E is locally finite, and that each vertex chain is isomorphic to $\mathbb{Z}$. Condition (F1) now follows.

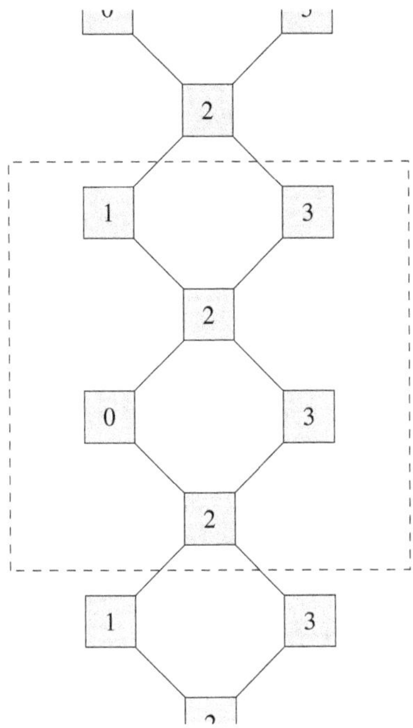

Figure 2.7 A vertically repeating periodic heap over the Dynkin diagram $B_3^{(1)}$

To check condition (F2), we need to verify that in the Hasse diagram of E, we have the following.

(i) Every occurrence of 0 is adjacent to at least one occurrence of 2.
(ii) Every occurrence of 1 is adjacent to at least one occurrence of 2.
(iii) Every occurrence of 3 is adjacent to at least one occurrence of 2.
(iv) Every occurrence of 2 is adjacent to at least one occurrence of each of 0, 1 and 3.

These conditions, which are easy to verify from the picture, arise from the adjacency relations in the graph Γ, treating multiple and directed edges as ordinary edges.

Finally, we need to check condition (F3). If x and y are consecutive elements of $\varepsilon^{-1}(0)$, the open interval (x, y) contains one vertex labelled 1, one vertex labelled 3, and two vertices labelled 2. In the generalized Cartan matrix (with the Kac indexing for rows and columns), we have $a_{0,0} = 2, a_{0,1} = 0, a_{0,2} = -1$ and $a_{0,3} = 0$. Since 0 is adjacent only to 2, only vertices in $[x, y]$ labelled 0 or 2 make a nonzero contribution to the sum in condition (F3). There are two contributions of 2, from x and y themselves, and two contributions of -1, from each of the vertices labelled 2. The sum then totals $2 - 1 - 1 + 2 = 2$, as required. A similar argument deals with the vertex chain $\varepsilon^{-1}(1)$.

Suppose now that x and y are consecutive elements of $\varepsilon^{-1}(2)$. In this case, the open interval (x, y) either contains (a) one occurrence of 1 and one occurrence of 3, or (b) one occurrence of 0 and one occurrence of 3. In the generalized Cartan matrix, we have $a_{2,0} = -1$, $a_{2,1} = -1$, $a_{2,2} = 2$ and $a_{2,3} = -1$. In either of the two cases, this makes a total contribution to the sum of $2 - 1 - 1 + 2 = 2$, as required.

Finally, we suppose that x and y are consecutive elements of $\varepsilon^{-1}(3)$. In this case, (x, y) contains a single element, z, with $\varepsilon(z) = 2$. In the generalized Cartan matrix, we have $a_{3,0} = 0$, $a_{3,1} = 0$, $a_{3,2} = -2$ and $a_{3,3} = 2$. This means that z contributes -2 to the sum, and each of x and y contributes 2, totalling $2 - 2 + 2 = 2$. It follows that condition (F3) is always satisfied, and that E is a full heap.

Exercise 2.2.5 Let $\varepsilon : E \to \Gamma$ be a full heap. Use the fact that E is locally finite to show that condition (F1) in Definition 2.2.2 may be replaced by the condition (F1′): "Every vertex chain of E is nonempty, unbounded above and unbounded below."

Exercise 2.2.6 Show that condition (F2) in Definition 2.2.2 may be replaced by the condition (F2′): "Every edge chain of E is a (not necessarily disjoint) union of covering pairs from E."

Exercise 2.2.7 Show that the requirement in axiom (F3) that $[x, y]$ be a p-interval can be weakened to "$x < y$ and $\varepsilon(x) = \varepsilon(y) = p$". (We have chosen the more restrictive definition because it makes full heaps easier to identify; see Example 2.2.4.)

Exercise 2.2.8 Let $\varepsilon : E \to \Gamma$ be a full heap, and let p and q be adjacent vertices of Γ. Show that there exists a p-interval $[x, y]$ of E containing an element $z \in \varepsilon^{-1}(q)$.

Exercise 2.2.9 Let Γ be the Dynkin diagram consisting of a single vertex and no edges, corresponding to the generalized Cartan matrix (2). Show that there are no full heaps over Γ.

Exercise 2.2.10 Let E be a full heap over a Dynkin diagram Γ. Use Lemma 2.1.5 to show that if y covers x in E, then $\varepsilon(x)$ and $\varepsilon(y)$ must be distinct adjacent vertices of Γ.

Exercise 2.2.11 Show that if E is a full heap, then so is its dual heap E^*. Show also that the full heap E shown in Figure 2.7 is isomorphic to its dual.

Exercise 2.2.12 Let Γ be the Dynkin diagram $B_3^{(1)}$. Show that there is only one isomorphism class of full heaps over Γ in the category **Heap**(Γ).

2.3 Local structure of full heaps

In this section, we prove the Local Structure Theorem for full heaps (Theorem 2.3.15), which is the key ingredient of the applicability of the theory of full heaps. This section may be safely skipped at a first reading, since only the statement of the theorem will be used in the sequel.

As explained in Appendix A.1, any partially ordered set P gives rise to a distributive lattice $J(P)$ whose elements are the (order) ideals of P, and whose meet and join operations are $\cap$ and $\cup$, respectively.

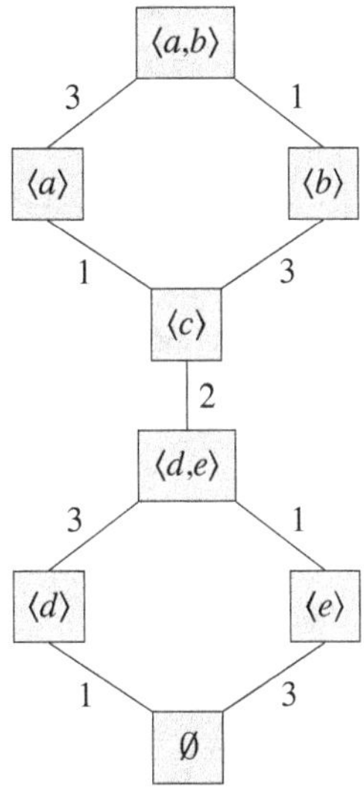

Figure 2.8 The Hasse diagram of the distributive lattice $J(E)$ corresponding to the heap E in Figure 2.1

If the poset P is the underlying poset of a heap over a graph Γ, then the lattice $J(P)$ acquires some additional structure. Lemma A.1.10 shows that if I' covers I in $J(P)$, then $I' \backslash I$ is a singleton, $\{x\}$. In this case, we may label the edge of the Hasse diagram of $J(P)$ corresponding to the covering relation $I < I'$ by $\varepsilon(x) \in \Gamma$. The "local structure" in Theorem 2.3.15 refers to the structure of the covering relations in the partially ordered set $J(P)$.

Example 2.3.1 Let E be the heap shown in Figure 2.1 over the graph Γ. In this case, the Hasse diagram of the distributive lattice $J(E)$ is shown in Figure 2.8. The labels on the edges are the labels $\varepsilon(x)$ described above, and the labels on the vertices are the names of the ideals of E, following the conventions in Appendix A.1.

The next result shows that upper and lower bounds for elements always exist for full heaps over connected graphs, although this is not true in general (see Exercise 2.3.3 below).

Lemma 2.3.2 *Let $\varepsilon : E \to \Gamma$ be a full heap over a connected Dynkin diagram.*

(i) *Suppose that p and q are adjacent vertices in Γ, and that $\alpha \in \varepsilon^{-1}(p)$. Then there exist elements $\beta, \gamma \in \varepsilon^{-1}(q)$ such that $\beta < \alpha < \gamma$.*
(ii) *Suppose that p and q are arbitrary vertices in Γ, and that $\alpha \in \varepsilon^{-1}(p)$. Then there exist elements $\beta, \gamma \in \varepsilon^{-1}(q)$ such that $\beta < \alpha < \gamma$.*
(iii) *If α and β are arbitrary elements of E, then α and β have an upper bound and a lower bound in E.*
(iv) *Any finite subset of E has an upper bound in E, and a lower bound in E.*
(v) *There are no maximal or minimal elements of E.*

Proof By axiom (F1), $\varepsilon^{-1}(q)$ is isomorphic to $\mathbb{Z}$. Every element of $\varepsilon^{-1}(p)$ is comparable to every element of $\varepsilon^{-1}(q)$ by the edge chain requirement of axiom (H1).

Suppose for a contradiction that no element γ exists as described. This implies that every element of $\varepsilon^{-1}(q)$ is less than α. Taking β to be any element of $\varepsilon^{-1}(q)$, it follows that the interval $[\beta, \alpha]$ would be infinite, contradicting the local finiteness of E. This proves the assertion about γ, and the assertion about β follows similarly, completing the proof of (i).

The assertions of (ii) follow by induction on the (finite) length k of a shortest path from p to q in Γ. The case $k = 0$ follows from the fact that $\varepsilon^{-1}(p)$ is a chain. The case $k = 1$ is part (i), and the inductive step follows easily from the conclusions of (i).

Let $\alpha, \beta \in E$, and define $p = \varepsilon(\alpha)$ and $q = \varepsilon(\beta)$. By (ii), there exists $\gamma \in \varepsilon^{-1}(q)$ with $\gamma > \alpha$. Since $\varepsilon^{-1}(q)$ is a chain containing both β and γ, the larger of β and γ will be an upper bound for both α and β. A similar argument shows that α and β have a lower bound in E, which proves (iii). Part (iv) follows from (iii) by induction. Part (v) holds because every element of E lies in some vertex chain, and the vertex chains are unbounded above and below by axiom (F1). □

Exercise 2.3.3 Show that parts (ii) and (iii) of Lemma 2.3.2 fail if Γ is not connected.

Many of the applications of full heaps we discuss in this book rely on the existence of certain chains in the lattice of ideals. In order to explain these, it is convenient to introduce the following notation.

Definition 2.3.4 Let $\varepsilon : E \to \Gamma$ be a heap over a graph, and let $J(E)$ be the lattice of ideals of E. We write $I_1 \prec_p I_2$ or $I_2 \succ_p I_1$ to mean that $I_1, I_2 \in J(E)$ and $I_2 \backslash I_1 = \{\alpha\}$ is a singleton with $\varepsilon(\alpha) = p \in \Gamma$. We define $\sim_{pq}$ to be the minimal equivalence relation on $J(E)$ containing all of the four relations $\prec_p$, $\prec_q$, $\succ_p$ and $\succ_q$. A (p, q)-*string* is by definition an equivalence class of $\sim_{pq}$.

Exercise 2.3.5 Suppose that $\varepsilon : E \to \Gamma$ is a heap over a graph, p is a vertex of Γ and $I \in J(E)$. Show that there is at most one $I' \in J(E)$ satisfying $I \prec_p I'$, and at most one $I' \in J(E)$ satisfying $I \succ_p I'$.

Example 2.3.6 Consider the full heap of Example 2.2.4, labelled as in Figure 2.9.
The ideal $\langle 2(0) \rangle$ lies in the following three (p, q)-strings:

(a) the $(1, 2)$-string $\{\langle 3(-1)\rangle, \langle 1(-1), 3(-1)\rangle, \langle 2(0)\rangle\}$, where

$$\langle 3(-1)\rangle \prec_1 \langle 1(-1), 3(-1)\rangle \prec_2 \langle 2(0)\rangle,$$

(b) the $(0, 2)$-string $\{\langle 1(-1), 3(-1)\rangle, \langle 2(0)\rangle, \langle 0(0)\rangle\}$, where

$$\langle 1(-1), 3(-1)\rangle \prec_2 \langle 2(0)\rangle \prec_0 \langle 0(0)\rangle,$$

and

(c) the $(2, 3)$-string $\{\langle 1(-1)\rangle, \langle 1(-1), 3(-1)\rangle, \langle 2(0)\rangle, \langle 3(0)\rangle\}$ where

$$\langle 1(-1)\rangle \prec_3 \langle 1(-1), 3(-1)\rangle \prec_2 \langle 2(0)\rangle \prec_3 \langle 3(0)\rangle.$$

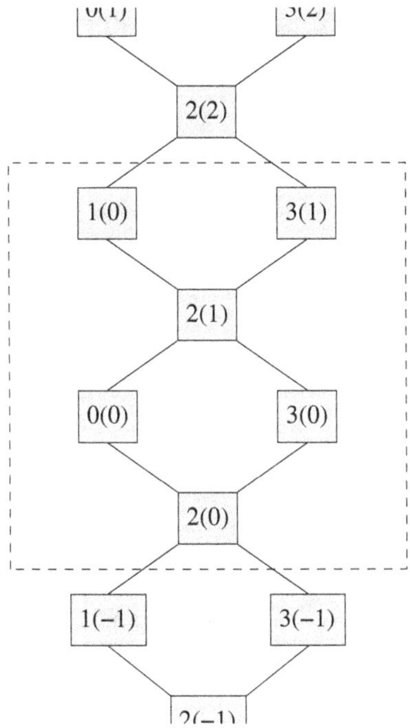

Figure 2.9 Labelling for the full heap over $B_3^{(1)}$ shown in Figure 2.7

Lemma 2.3.7 *Let $\varepsilon : E \to \Gamma$ be a heap over a graph, let $I_1, I_2, I_3 \in J(E)$, and let p and q be adjacent elements of Γ. Then the following situations cannot occur:*

(i) $I_1 \prec_p I_3$ *and* $I_2 \prec_q I_3$*;*
(ii) $I_1 \succ_p I_3$ *and* $I_2 \succ_q I_3$.

Proof For the situation in (i) to occur, there would need to be two maximal elements $\alpha, \beta \in I_3$ with $\varepsilon(\alpha) = p$ and $\varepsilon(\beta) = q$. Since p and q are adjacent, α and β must lie in the chain $\varepsilon^{-1}(\{p, q\})$. This is a contradiction, because the maximal elements of I_3 (or indeed any partially ordered set) form an antichain.

Part (ii) follows from part (i) by duality: replace E by E^* and I_k by $(E \backslash I_k)^*$ for $k \in \{1, 2, 3\}$. □

Lemma 2.3.8 *Let $\varepsilon : E \to \Gamma$ be a full heap over a Dynkin diagram, and let p be a vertex of Γ. Then it is not possible to have $I_1, I_2, I_3 \in J(E)$ with $I_1 \prec_p I_2 \prec_p I_3$.*

Proof Let α (respectively, β) be the unique element of $I_3 \backslash I_2$ (respectively, $I_2 \backslash I_1$). We have $\varepsilon(\alpha) = \varepsilon(\beta) = p$, but the open interval (α, β) is empty. This implies that

$$\sum_{\alpha \leq z \leq \beta} a_{p,\varepsilon(z)} = 4,$$

which contradicts axiom (F3). □

Lemma 2.3.9 *Let $\varepsilon : E \to \Gamma$ be a heap and let $p, q \in \Gamma$ be nonadjacent vertices. Suppose that $I_1, I_2 \in J(E)$ satisfy $I_1 \prec_q I_2$, and that $\alpha \in E$ satisfies $\varepsilon(\alpha) = p$. Then α is a maximal element of I_1 if and only if α is a maximal element of I_2.*

Proof Let β be the unique element of $I_2 \backslash I_1$, so that $\varepsilon(\beta) = q$. If α is maximal in I_2, then it is clearly maximal in I_1, because $I_1 \subset I_2$. Conversely, suppose that α is maximal in I_1. If α is not maximal in I_2, then it must be the case that β covers α, but this contradicts Lemma 2.1.5 (iii). □

Lemma 2.3.10 *Let $\varepsilon : E \to \Gamma$ be a locally finite heap over a graph, and let $p, q \in \Gamma$ be nonadjacent vertices. Suppose that $I_1, I_2 \in J(E)$ satisfy $I_1 \prec_p I_2$.*

(i) *We have $I_1 \prec_q I_3$ for some $I_3 \in J(E)$ if and only if we have $I_2 \prec_q I_4$ for some $I_4 \in J(E)$. If both conditions hold, then we have $I_3 \prec_p I_4$.*
(ii) *We have $I_1 \succ_q I_3$ for some $I_3 \in J(E)$ if and only if we have $I_2 \succ_q I_4$ for some $I_4 \in J(E)$. If both conditions hold, then we have $I_4 \succ_p I_3$.*

Proof Since $I_1 \prec_p I_2$, there is a minimal element $\alpha \in E \backslash I_1$ with $\varepsilon(\alpha) = p$, and α is the unique element of $I_2 \backslash I_1$.

We first prove (i). Suppose that $I_1 \prec_q I_3$ for some $I_3 \in J(E)$. Then $I_4 := I_2 \cup I_3$, which is an ideal of E, satisfies $I_2 \prec_q I_4$ and $I_3 \prec_p I_4$.

For the converse, suppose that $I_2 \prec_q I_4$ for some $I_4 \in J(E)$. By Lemma 2.3.9, the fact that I_2 has a maximal element labelled p means that I_4 has a maximal element labelled p (namely α) and another maximal element $\beta \in I_4 \backslash I_2$ labelled q. It follows that $I_3 := I_4 \backslash \{\alpha\}$ is an ideal of E, and that $I_1 \prec_q I_3 \prec_p I_4$. This completes the proof of (i).

Suppose now that $I_1 \succ_q I_3$ for some I_3, and let β be the unique element of $I_1 \backslash I_3$, so that $\varepsilon(\beta) = q$. By Lemma 2.3.9, the fact that β is a maximal element of I_1 labelled q means that β is also a maximal element of I_2. It follows that $I_4 := I_2 \backslash \{\beta\}$ is an ideal of E and that $I_2 \succ_q I_4 \succ_p I_3$.

Conversely, suppose that $I_2 \succ_q I_4$ for some $I_4 \in J(E)$, and let β be the unique element of $I_2 \backslash I_4$, so that $\varepsilon(\beta) = q$. In this case, I_2 has maximal elements α and β. It follows that $I_3 := I_2 \backslash \{\alpha, \beta\}$ is an ideal of E, and that $I_1 \succ_q I_3$ and $I_4 \succ_p I_3$. This completes the proof of (ii). □

Lemma 2.3.11 *Let $\varepsilon : E \to \Gamma$ be a locally finite heap. Suppose $I \in J(E)$ and $\alpha \in E \backslash I$ satisfy $I \cup \{\alpha\} \notin J(E)$. Then there exists $\beta \in E \backslash I$ such that α covers β in E, and furthermore, the vertices $\varepsilon(\alpha)$ and $\varepsilon(\beta)$ are either equal or adjacent in Γ.*

Proof Because $I \cup \{\alpha\}$ is not an ideal, there must exist $\gamma \in E \backslash I$ with $\gamma < \alpha$. Since E is locally finite, Lemma 2.1.5 (iv) shows that there exists a finite sequence of covering relations

$$\gamma = z_0 < z_1 < \cdots < z_k = \alpha$$

in E, and $k > 0$. Since I is an ideal, none of the z_i can lie in I. We may then take $\beta = z_{k-1}$, and the remaining assertions follow from Lemma 2.1.5 (iii). □

Lemma 2.3.12 *Let $\varepsilon : E \to \Gamma$ be a full heap over a Dynkin diagram Γ with generalized Cartan matrix A. The following are equivalent:*

(i) $a_{pq} = -2$;
(ii) *for every $\alpha \in E$ with $\varepsilon(\alpha) = q$, there is a convex chain $\beta < \alpha < \gamma$ in E with $\varepsilon(\beta) = \varepsilon(\gamma) = p$;*
(iii) *whenever $I_1 \prec_q I_2$ for some $I_1, I_2 \in J(E)$, then there exist unique $I_0, I_3 \in J(E)$ with*

$$I_0 \prec_p I_1 \prec_q I_2 \prec_p I_3.$$

Proof (i) $\Rightarrow$ (ii): Suppose that $a_{pq} = -2$, and let $\alpha \in E$ satisfy $\varepsilon(\alpha) = q$. By Lemma 2.3.2 (i), there exist $\beta, \gamma \in \varepsilon^{-1}(p)$ with $\beta < \alpha < \gamma$. We assume furthermore that β is maximal with this property, and that γ is minimal with this property, and because the edge chain $\varepsilon^{-1}\{p, q\}$ is totally ordered, this implies that $(\beta, \gamma) \cap \varepsilon^{-1}(p)$ is empty.

Suppose for a contradiction that the chain $\beta < \alpha < \gamma$ is not convex, and let $\delta \in (\beta, \gamma)\backslash\{\alpha\}$. By Lemma 2.1.5 (iv), there is a finite sequence of covering relations

$$\delta = z_0 < z_1 < \cdots < z_k = \gamma,$$

so by replacing δ by z_{k-1} if necessary, we may assume that γ covers δ. We have eliminated the possibility that $\varepsilon(\delta) = p$, so Lemma 2.1.5 (iii) shows that $\varepsilon(\delta)$ and $\varepsilon(\gamma)$ are adjacent in Γ. Apply Axiom (F3) with $x = \beta$ and $y = \gamma$ and consider the sum $\sum_{\beta \leq z \leq \gamma} a_{p,\varepsilon(z)}$. (Recall that $a_{p,\varepsilon(z)}$ is nonzero if and only if p and $\varepsilon(z)$ are equal or adjacent in Γ.) There are two summands equal to 2 ($z = \beta$ and $z = \gamma$) and one equal to -2 ($z = \alpha$). The term $z = \delta$ contributes a strictly negative number to the sum and all other terms are nonpositive. It cannot therefore be the case that $\sum_{\beta \leq z \leq \gamma} a_{p,\varepsilon(z)} = 2$. This is the required contradiction.

(ii) $\Rightarrow$ (iii): Assume the hypothesis of (ii) and suppose that $I_1 \prec_q I_2$. Let α be the unique (maximal) element of $I_2\backslash I_1$, and let $\beta, \gamma \in \varepsilon^{-1}(p)$ be such that $\beta < \alpha < \gamma$ is a convex chain. Note that $\beta \in I_1$, and suppose that there exists $\delta \in I_1$ with $\delta > \beta$. Since E is locally finite, we may assume by Lemma 2.1.5 (i) that δ covers β and (by Exercise 2.2.10) that $\varepsilon(\delta)$ and $\varepsilon(\beta)$ are adjacent in Γ. By considering the edge chain $\varepsilon^{-1}\{\varepsilon(\delta), p\}$, we see that $\beta < \delta < \gamma$, which is a contradiction and proves that β is maximal in I_2. It follows that $I_0 = I_1\backslash\{\beta\}$ is an ideal and $I_0 \prec_q I_1$.

We will be done if we can show that $I_3 := I_2 \cap \{\gamma\}$ is an ideal, because it will follow that $I_2 \prec_q I_3$ and the uniqueness assertions are clear. If I_3 is not an ideal, then Lemma 2.3.11 shows that there exists $\delta' \notin I_2$ such that γ covers δ' and that $\varepsilon(\delta')$ and $\varepsilon(\gamma)$ are adjacent. (We cannot have $\varepsilon(\delta') = \varepsilon(\gamma)$ by Exercise 2.2.10.) Considering the edge chain $\varepsilon^{-1}\{\varepsilon(\delta'), p\}$ shows that $\beta < \delta' < \gamma$, which forces $\alpha = \delta'$ and gives a contradiction, because $\alpha \in I_2$.

(iii) $\Rightarrow$ (i): Suppose that

$$I_0 \prec_p I_1 \prec_q I_2 \prec_p I_3,$$

and define the unique elements $\beta \in I_1\backslash I_0$, $\alpha \in I_2\backslash I_1$, and $\gamma \in I_3\backslash I_2$. It follows that $\{\beta, \alpha, \gamma\}$ is convex by Exercise A.1.7, because it is the complement $I_3\backslash I_0$ of two ideals. Since $\beta < \alpha < \gamma$, the interval $[\beta, \gamma]$ contains only three elements. Axiom (F3) now shows that

$$\sum_{\beta \leq z \leq \gamma} a_{p,\varepsilon(z)} = 2 - a_{pq} + 2 = 2,$$

forcing $a_{pq} = -2$ as claimed. □

Lemma 2.3.13 *Let $\varepsilon : E \to \Gamma$ be a full heap over a Dynkin diagram Γ with generalized Cartan matrix A, and let $I \in J(E)$. Suppose that the chain $\varepsilon^{-1}(p) \cap I$ has a maximal element, β. Suppose also that*

$$\sum_{z \in I : \beta < z} a_{p,\varepsilon(z)} \leq -2.$$

Then there exists a unique $\beta' \in E$ with $\varepsilon(\beta') = p$ such that $I \cup \{\beta'\}$ is an ideal.

Proof Note that if β' exists, it must be the minimal element of the chain $\varepsilon^{-1}(p) \cap (E \backslash I)$. We define β' to be this element, so that we have $\varepsilon(\beta') = p$. It remains to show that $I \cup \{\beta'\}$ is an ideal.

Suppose for a contradiction that $I \cup \{\beta'\}$ is not an ideal. By Lemma 2.3.11, there must exist $\gamma \in E \backslash I$ such that β' covers γ in E, and such that $p = \varepsilon(\beta')$ and $q = \varepsilon(\gamma)$ are either adjacent or equal. Now the three elements β', β, γ lie in the chain $\varepsilon^{-1}\{p, q\}$, whether or not we have $p = q$. Since β' covers γ and $\beta < \beta'$, we must have $\beta < \gamma$. Since the open interval (β, β') is disjoint from $\varepsilon^{-1}(p)$, it must be the case that $p \neq q$, and $a_{pq} < 0$.

By axiom (F3), we have

$$\sum_{z \in E : \beta \leq z \leq \beta'} a_{p,\varepsilon(z)} = 2.$$

Since $(\beta, \beta') \cap \varepsilon^{-1}(p) = \emptyset$, the only positive contributions to the sum are the terms $z = \beta$ and $z = \beta'$. By hypothesis, the negative contributions to the sum from elements $z \in I$ total at most -2. However, there is at least one more negative contribution from the term $z = \gamma \notin I$, and this is the required contradiction. □

Lemma 2.3.14 *Let $\varepsilon : E \to \Gamma$ be a full heap over a Dynkin diagram Γ with generalized Cartan matrix A, and suppose that p and q are adjacent vertices satisfying $a_{pq} = -1$. Suppose also that $I_1 \prec_q I_2$ for some $I_1, I_2 \in J(E)$. Then either there exists $I_0 \in J(E)$ with $I_0 \prec_p I_1$, or there exists $I_3 \in J(E)$ with $I_2 \prec_p I_3$, but not both.*

Proof Let α be the unique element of $I_2 \backslash I_1$, so that $\varepsilon(\alpha) = q$. We first note that if I_0 and I_3 were both to exist as described, they would be unique, and Lemma 2.3.12 would then imply that $a_{pq} = -2$, contrary to hypothesis. It therefore remains to show that one of I_0 or I_3 exists as described.

By axiom (F2), there exists $\beta \in E$ with $\varepsilon(\beta) = p$ such that either α covers β or β covers α.

Suppose first that α covers β, so β is maximal in $\varepsilon^{-1}(p) \cap I_2$. If β is maximal in I_1, then we are done by taking $I_0 := I_1 \backslash \{\beta\}$. If β is not maximal in I_1, then let $\gamma \neq \alpha$ be an element of I_1 such that γ covers β. Now α and γ each contribute at most -1 to the sum $\sum_{z \in I_2 : \beta < z} a_{p,\varepsilon(z)}$. Lemma 2.3.13 now shows that there exists $\beta' \in E \backslash I_2$ with $\varepsilon(\beta') = p$ and $I_2 \cup \{\beta'\}$ is an ideal. We are now done by defining $I_3 = I_2 \cup \{\beta'\}$.

Now suppose that β covers α, which means that β is the minimal element of $\varepsilon^{-1}(p)$ that is greater than α (in the chain $\varepsilon^{-1}(\{p, q\})$). Since α is maximal in I_2, we have $\beta \notin I_2$. If $I_2 \cup \{\beta\}$ is an ideal, then we are done by taking $I_3 = I_2 \cup \{\beta\}$, so we may assume that $I_2 \cup \{\beta\}$ is not an ideal. By Lemma 2.3.11, there exists $\gamma \in E \backslash I_2$ such that β covers γ and such that the labels $\varepsilon(\beta)$ and $\varepsilon(\gamma)$ are either equal or adjacent. Since E is a full heap, it must be the case that $\varepsilon(\beta)$ and $\varepsilon(\gamma)$ are adjacent. Now let E^* be the dual heap of E. Lemma 2.3.13 now applies to the ideal $I_1^* = (E \backslash I_1)^*$ of E^*, which contains the elements α, β and γ. We deduce that I_1 contains an element β' with $\varepsilon(\beta') = p$ such that $I_1^* \cup \beta'$ is an ideal. It follows that β' is a maximal element of I_1, and we are done by taking $I_0 = I_1 \backslash \{\beta'\}$. □

The local structure of the lattice $J(E)$ of ideals of a full heap is made clear by the following result.

Theorem 2.3.15 *(Local Structure Theorem for full heaps)*
Let $\varepsilon : E \to \Gamma$ be a full heap over a Dynkin diagram Γ with generalized Cartan matrix A, let I be an ideal of E, and let p and q be distinct vertices of Γ. Suppose that the (p, q)-string $\mathbf{i}$ containing I contains more than one element.

(i) *If $a_{pq} = a_{qp} = -1$, then $\mathbf{i} = \{I_1, I_2, I_3\}$, and precisely one of the following two possibilities holds:*
 (a) $I_1 \prec_p I_2 \prec_q I_3$;
 (b) $I_1 \prec_q I_2 \prec_p I_3$.
(ii) *If $a_{pq} = -2$ and $a_{qp} = -1$, then we have $\mathbf{i} = \{I_1, I_2, I_3, I_4\}$, where $I_1 \prec_p I_2 \prec_q I_3 \prec_p I_4$.*
(iii) *If $a_{pq} = 0$, then precisely one of the following three possibilities holds:*
 (a) *$\mathbf{i} = \{I_1, I_2, I_3, I_4\}$, where both $I_1 \prec_p I_2 \prec_q I_4$ and $I_1 \prec_q I_3 \prec_p I_4$;*
 (b) *$\mathbf{i} = \{I_1, I_2\}$, where $I_1 \prec_p I_2$;*
 (c) *$\mathbf{i} = \{I_1, I_2\}$, where $I_1 \prec_q I_2$.*
(iv) *If $a_{pq} = a_{qp} = -2$, then $\mathbf{i} = \{I_n\}_{n \in \mathbb{Z}}$, where $I_n \prec_p I_{n+1}$ if n is even, and $I_n \prec_q I_{n+1}$ if n is odd.*

Proof We first prove (i). Suppose that $a_{pq} = a_{qp} = -1$. Using the symmetry between p and q in this case, Lemma 2.3.14 shows that I must lie in a sequence satisfying either (a) or (b) (but not both) in the statement of (i). These chains cannot be further extended to the left or the right using an alternating sequence of $\prec_p$ and $\prec_q$, because Lemma 2.3.12 would then imply that either a_{pq} or a_{qp} was equal to -2. Lemmas 2.3.8 and 2.3.7 then show that the chains shown are the complete (p, q)-strings in each case, thus proving (i).

We next prove (iv). Suppose that $a_{pq} = a_{qp} = -2$. By Lemma 2.3.12, it follows that I lies in a sequence satisfying the conditions of (iv). Using Exercise 2.3.5 together with Lemmas 2.3.8 and 2.3.7, we then see that this chain is the entire (p, q)-string, and this proves (iv).

Now suppose that $a_{pq} = -2$ and $a_{qp} = -1$. Since the (p, q)-string containing I contains more than one element, I must be involved in a chain of the form $I' \prec_q I''$, or in a chain of the form $I' \prec_p I''$ (or possibly both). In the former case, Lemma 2.3.12 shows that we can form a chain $I_1 \prec_p I' \prec_q I'' \prec_p I_4$. In the latter case, Lemma 2.3.14 shows that the chain $I' \prec_p I''$ can be extended either to a chain of the form $I_1 \prec_p I_2 \prec_q I_3$, or to a chain of the form $I_2 \prec_q I_3 \prec_p I_4$; furthermore, such a chain will contain I. Applying Lemma 2.3.12 as before to the chain $I_2 \prec_q I_3$, we see that each of these chains extends further to a chain $I_1 \prec_p I_2 \prec_q I_3 \prec_p I_4$ that contains I. This chain of four ideals can be shown to be maximal using the same argument used to deal with case (iv), and this completes the proof of (ii).

Finally, suppose that $a_{pq} = a_{qp} = 0$. If I is involved in a chain of the form $I_1 \prec_p I_2$ but not in a chain of the form $I_1 \prec_q I_2$, then Lemma 2.3.10 shows that we cannot have $I_1 \prec_q I'$, $I_1 \succ_q I'$, $I_2 \prec_q I'$ or $I_2 \succ_q I'$ for any ideal I'. Exercise 2.3.5 and Lemma 2.3.8 now show that the (p, q)-string containing I consists solely of I_1 and I_2, as described in case (iii) (b). Similarly, if I is involved in a chain of the form $I_1 \prec_q I_2$ but not in one of the form $I_1 \prec_p I_2$, then we are in case (iii) (c). The other possibility is that I is involved in a chain of the form $J_1 \prec_p J_2$ and in a chain of the form $J_3 \prec_q J_4$. There are four subcases to consider: $I_1 \prec_p I \prec_q I_4$, $I_1 \prec_q I \prec_p I_4$, $I_3 \succ_q I \prec_p I_2$ and $I_3 \prec_p I \succ_q I_2$. Each of these configurations can be extended to the one described in the statement of case (iii) (a) by using Lemma 2.3.10. Using Exercise 2.3.5 and Lemma 2.3.8, we find that the (p, q)-string containing I consists only of these four ideals, and this completes the proof of (iii). □

Exercise 2.3.16 Supppose that Γ_1 and Γ_2 are graphs with two vertices, and that $\varepsilon_1 : E_1 \to \Gamma_1$ and $\varepsilon_2 : E_2 \to \Gamma_2$ are full heaps. Prove that E_1 and E_2 are isomorphic in **Heap**.

2.4 Quotient heaps

In Section 2.4, we define the notion of a quotient heap. This section may be skipped at a first reading, but the constructions will be needed to construct full heaps in type affine A in Section 6.1. In this section, we will maintain two assumptions about the heap $\varepsilon : E \to \Gamma$:

(1) the quotient poset $(G\backslash E, \preceq)$ exists in the sense of Definition A.2.5, and
(2) for all $p \in E$ and $g \in G$, we have $g(\varepsilon^{-1}(\varepsilon(p))) = \varepsilon^{-1}(\varepsilon(g(p)))$.

Definition 2.4.1 Let $\varepsilon : E \to \Gamma$ be a heap, subject to the assumptions above. The action of G on E induces an action on Γ by $g(\varepsilon(p)) = \varepsilon(g(p))$ for $g \in G$, $p \in E$, and this action is well-defined by assumption (2). Let $G\backslash\Gamma$ be the graph whose vertices are the G-orbits $\{\overline{x} : x \in \Gamma\}$ and whose edges are the pairs

$$\{\{\overline{x}, \overline{y}\} : \{x', y'\} \text{ is an edge in } \Gamma \text{ for some } x' \in \overline{x} \text{ and } y' \in \overline{y}\}.$$

Define $\overline{\varepsilon} : G\backslash E \to G\backslash\Gamma$ by $\overline{\varepsilon}(\overline{p}) := \overline{\varepsilon(p)}$.

Lemma 2.4.2 *The map $\overline{\varepsilon}$ is well-defined.*

Proof For any $h \in G$, we need to show that $\overline{\varepsilon(p)} = \overline{\varepsilon(h(p))}$. By definition, we have

$$\overline{\varepsilon(p)} = \{\varepsilon(g(\varepsilon^{-1}(\varepsilon(p)))) : g \in G\}.$$

By assumption (2), this set is equal to

$$\{\varepsilon(\varepsilon^{-1}(\varepsilon(g(p)))) : g \in G\},$$

or, more simply, $\{\varepsilon(g(p)) : g \in G\}$. A similar argument shows that

$$\overline{\varepsilon(h(p))} = \{\varepsilon(g(h(p))) : g \in G\} = \{\varepsilon(g(p)) : g \in G\},$$

as required. □

Lemma 2.4.3 *The graph $G\backslash\Gamma$ has no loops.*

Proof It is clear from the definition that $G\backslash\Gamma$ is a graph, but the nonexistence of loops is not immediate. Suppose that $\{\overline{x}, \overline{x}\}$ is an edge of $G\backslash\Gamma$. This implies the existence of an edge $\{x, y\}$ of Γ for which $y \in \overline{x}$; note that there is no loss in generality in assuming that one of the endpoints of this edge is x itself. By definition of $\overline{x}$, there exists $g \in G$ such that $y = \varepsilon(g(\varepsilon^{-1}(x)))$. Choose $p \in \varepsilon^{-1}(x)$, so that $y = \varepsilon(g(\varepsilon^{-1}(\varepsilon(p))))$. Assumption (2) now shows that

$$y = \varepsilon(\varepsilon^{-1}(\varepsilon(g(p)))) = \varepsilon(g(p)),$$

so that $g(p)$ is labelled y. Thus p and $g(p)$ are in the edge chain $\varepsilon^{-1}\{x, y\}$ and therefore $p = g(p)$, so $x = y$, contradicting the fact that Γ has no loops. □

Lemma 2.4.4 *The function $\overline{\varepsilon} : G\backslash E \to G\backslash\Gamma$ satisfies Axiom (H1).*

Proof Let $\overline{x} \in G\backslash\Gamma$. Let $\overline{p}, \overline{q} \in \overline{\varepsilon}^{-1}(\overline{x})$, so that

$$\overline{x} = \{\varepsilon(g(p)) : g \in G\} = \{\varepsilon(g(q)) : g \in G\}.$$

Clearly, we have $x \in \overline{x}$, and this means that there exist $h_1, h_2 \in G$ such that $\varepsilon(h_1(p)) = x = \varepsilon(h_2(q))$. Since $\varepsilon : E \to \Gamma$ is a heap, $\varepsilon^{-1}(x)$ is a chain in E, which implies that $h_1(p)$ and $h_2(q)$ are comparable in E. If $h_1(p) \leq h_2(q)$, then we have $p \leq h_1^{-1}h_2(q)$, which means that $\overline{p} \prec \overline{q}$. Similarly, if $h_2(q) \leq h_1(p)$, we have $\overline{q} \preceq \overline{p}$. In any case, $\overline{p}$ and $\overline{q}$ are comparable, and $\overline{\varepsilon}^{-1}(\overline{x})$ is a vertex chain.

Now let $\{\overline{x}, \overline{y}\}$ be an edge of $G\backslash\Gamma$. We already know that $\overline{\varepsilon}^{-1}(\overline{x})$ and $\overline{\varepsilon}^{-1}(\overline{y})$ are chains, so it remains to show that if $\overline{p} \in \overline{\varepsilon}^{-1}(\overline{x})$ and $\overline{q} \in \overline{\varepsilon}^{-1}(\overline{y})$, then $\overline{p}$ is comparable to $\overline{q}$. Since $\{\overline{x}, \overline{y}\}$ is an edge in $G\backslash\Gamma$, we can assume, by picking different representatives if necessary, that $\{x, y\}$ is an edge in Γ. As in the previous paragraph, there exist $h_1, h_2 \in G$ such that $\varepsilon(h_1(p)) = x$ and $\varepsilon(h_2(q)) = y$. Since $\varepsilon^{-1}(\{x, y\})$ is an edge chain of E, it follows that $h_1(p)$ and $h_2(q)$ are comparable, and this implies that $\overline{p}$ and $\overline{q}$ are comparable. We conclude that $\overline{\varepsilon}^{-1}(\{\overline{x}, \overline{y}\})$ is an edge chain, as required. □

Lemma 2.4.5 *The function $\overline{\varepsilon} : G\backslash E \to G\backslash\Gamma$ satisfies Axiom (H2).*

Proof Suppose that $\overline{p} \leq \overline{q}$, meaning that there exists $g \in G$ with $p \leq g(q)$. By Axiom (H2) applied to E, there is a chain

$$p = p_1 \leq p_2 \leq \cdots \leq p_s = g(q)$$

such that for all i, either (a) $\varepsilon(p_i) = \varepsilon(p_{i+1})$ or (b) $\{\varepsilon(p_i), \varepsilon(p_{i+1})\}$ is an edge in Γ. The labels $\overline{p_i}$ and $\overline{p_{i+1}}$ are in the same vertex chain in case (a), and are in the same edge chain in case (b); the assertion now follows. □

Theorem 2.4.6 *Let $\varepsilon : E \to \Gamma$ be a heap for which the quotient poset $(G\backslash E, \preceq)$ exists in the sense of Definition A.2.5. If, for all $p \in E$ and $g \in G$, we have $g(\varepsilon^{-1}(\varepsilon(p))) = \varepsilon^{-1}(\varepsilon(g(p)))$, then, in the notation of Definition 2.4.1, $\overline{\varepsilon} : G\backslash E \to G\backslash\Gamma$ is a heap. Furthermore, there is a morphism $\pi \in$* **Heap** *from $\varepsilon : E \to \Gamma$ to $\overline{\varepsilon} : G\backslash E \to G\backslash\Gamma$ that is surjective on vertices and surjective on edges.*

Proof Lemmas 2.4.2–2.4.5 show that $\overline{\varepsilon} : G\backslash E \to G\backslash\Gamma$ is a heap.

Let $\pi_E : E \to G\backslash E$ be the projection map $\pi_E(p) = \overline{p}$, and let $\pi_\Gamma : \Gamma \to G\backslash\Gamma$ be the projection map $\pi_\Gamma(x) = \overline{x}$. The definition of $\preceq$ shows that π_E is a morphism of posets, and the definition of edges in $G\backslash E$ ensures that π_Γ is a morphism of graphs. The diagram

$$\begin{array}{ccc} E & \xrightarrow{\varepsilon} & \Gamma \\ {\scriptstyle \pi_E}\downarrow & & \downarrow{\scriptstyle \pi_\Gamma} \\ G\backslash E & \xrightarrow{\overline{\varepsilon}} & G\backslash\Gamma \end{array}$$

commutes, because the definitions show that

$$\overline{\varepsilon}(\pi_E(p)) = \overline{\varepsilon(p)} = \pi_\Gamma(\varepsilon(p)),$$

and this shows that $\pi = (\pi_E, \pi_\Gamma)$ is a heap morphism. The maps π_E and π_Γ are clearly surjective, and this completes the proof. □

2.5 Notes and references

1 The notion of heap here extends that of Viennot [88], who considers only finite heaps. However, the category-theoretic definition of a heap given here is new. Our approach here makes some of the subsequent definitions (such as the definition of "subheap") considerably more transparent.

2 Heaps also appear in theoretical computer science, where they are used to model concurrency. We have taken the term "trace" from the computer science literature, although computer scientists refer to heaps as *dependence graphs*. Dependence graphs that are not locally finite play a significant role in the computer science applications, although this will not be the case for the applications in this book. For details on the applications to concurrency, the reader is referred to Diekert and Rozenberg [19].

3 Our approach to generalized Cartan matrices in Section 2.2 follows that of Kac [37, section 1.1]. Kac introduces Dynkin diagrams in [37, section 4.7]. In the case that $a_{ij}a_{ji} > 4$, the corresponding vertices i and j are connected by a bold-faced line equipped with an ordered pair of integers $|a_{ij}|$, $|a_{ji}|$. Some authors use a different convention for Cartan matrices that turns out to be the transpose of ours. This difference applies only to the matrices, and not to any of the other objects we will associate with them.

4 Property H from Exercise 2.1.13 is what Stembridge [79] calls "Property (H1)".

5 The bijection of Theorem 2.1.20 induces a monoid structure on the set of isomorphism classes of finite heaps over a fixed graph. However, this monoid structure may also be defined directly in terms of the heap, albeit in a more complicated way, as in [88, definition 2.7].

6 Full heaps were introduced by the author in [28], and developed further in [29]. Their definition is based on Stembridge's notion of a minuscule heap in [79, section 3]. Our definition of generalized Cartan matrix is the transpose of Stembridge's, but this

distinction only affects the matrices, and not the corresponding heaps. A *minuscule heap* over a Dynkin diagram is a finite heap for which every p-interval (α, β) satisfies either (a) (α, β) contains precisely two elements with labels (q_1, q_2, say) adjacent to p and such that there is no arrow from q_1 (or q_2) to p in the Dynkin diagram, or that (b) (α, β) contains precisely one element with label (q, say) adjacent to p such that there is an arrow from q to p in the Dynkin diagram.

7 Theorem 2.3.15 is new.

8 The (p, q)-strings of Section 2.3 are reminiscent of the $\{s, t\}$-strings appearing in the work of Shi [74]. In turn, these are based on classes of elements appearing in the context of star operations in Coxeter groups, which were introduced by Kazhdan and Lusztig in [40] and further developed by Lusztig [49].

9 The notion of p-interval is due to Hagiwara [32, section 2].

10 The notion of quotient heap is also essentially due to Hagiwara [32]. Quotient heaps will be important in the classification of full heaps over Dynkin diagrams of type affine A in Section 6.1. We remark that it would be very awkward to define quotient heaps without the category theoretic definition given here.

11 Theorem 2.4.6 was proved by McGregor-Dorsey in his Master's thesis [57, theorem 1.2.20].

3

Weyl group actions

One main purpose of the Local Structure Theorem for full heaps (Theorem 2.3.15) is to enable the construction of various familiar algebraic objects using full heaps as a starting point. We explain in Section 3.1 how to associate linear operators with full heaps; these linear operators will act on vector spaces whose bases are indexed by ideals of the heap itself. The most important linear operators for our purposes will be the operators X_p, Y_p, H_p, S_p and $T_{p,q}$, all of which are defined in Section 3.1. In Chapter 3, we will concentrate on the operators S_p and $T_{p,q}$, which will give rise to representations of the associated Weyl group (Theorem 3.2.27) and Hecke algebra (Theorem 3.1.13) respectively. The operators X_p, Y_p and H_p will reappear later in the construction of Lie algebras.

It turns out to be more convenient to consider the action of the linear operators on a subset of the ideals of the full heap. These are the so-called "proper ideals": an ideal is proper if its intersection with each vertex chain of the original heap is both proper and nonempty, considered as a set. Section 3.2 develops the theory of proper ideals, and also introduces the key concepts of the "content" of a finite heap and the "relative content" of a pair of proper ideals. The action of the Weyl group on the root system emerges naturally from these concepts and gives rise to Theorem 3.2.21, which gives a precise connection between the action of the Weyl group on the proper ideals and the action on the root system. In turn, this leads to Theorem 3.2.27, which explains why the well-known faithfulness of the action of the Weyl group on the root system proves that the Weyl group also acts faithfully on the proper ideals. This means that we can think of affine Weyl groups associated with full heaps as certain infinite permutation groups, thus recovering various previously familiar constructions in a uniform way. We can also identify the positive roots of the Weyl group with certain convex subsets of the heap (Theorem 3.2.30), which will turn out to be the key to understanding both the action of the reflections of the Weyl group in Section 5.4 and the action of the Chevalley bases of simple Lie algebras in Chapter 7.

In Section 3.3, we develop the theory of parabolic subheaps, which are of vital importance in many later applications. These will play the role in the theories of finite Weyl groups and finite dimensional Lie algebras that is played by the full heaps in affine types.

3.1 Linear operators and group actions

In order to be able to apply the theory of full heaps to representation theory, it is necessary to associate with each full heap a family of linear operators on a vector space. The definition of these operators relies on the following classification of the ideals of a full heap.

Definition 3.1.1 Let $\varepsilon : E \to \Gamma$ be a full heap and let p be a vertex of Γ. We define subsets $J_p^+(E)$, $J_p^-(E)$ and $J_p^0(E)$ of $J(E)$ as follows:

$$\begin{aligned} J_p^+(E) &= \{I \in J(E) : I \succ_p I' \text{ for some } I' \in J(E)\}; \\ J_p^-(E) &= \{I \in J(E) : I \prec_p I' \text{ for some } I' \in J(E)\}; \\ J_p^0(E) &= J(E) \backslash (J_p^+(E) \cup J_p^-(E)). \end{aligned}$$

Lemma 3.1.2 *Let $\varepsilon : E \to \Gamma$ be a full heap and let p be a vertex of Γ. The subsets $J_p^+(E)$, $J_p^-(E)$ and $J_p^0(E)$ of $J(E)$ appearing in Definition 3.1.1 partition $J(E)$. Furthermore, each $I \in J_p^-(E)$ satisfies $I \prec_p I'$ for a unique I' in $J_p^+(E)$, and conversely each $I \in J_p^+(E)$ satisfies $I \succ_p I'$ for a unique I' in $J_p^-(E)$.*

Proof The only nontrivial thing to check is that $J_p^+(E)$ is disjoint from $J_p^-(E)$, and this follows from Lemma 2.3.8. □

Definition 3.1.3 Let $\varepsilon : E \to \Gamma$ be a full heap, and let p be a vertex of Γ. Let R be a commutative ring with 1, and let $V_{J(E)} = V_{J(E)}(R)$ be the free R-module with basis $\{v_I : I \in J(E)\}$. For each $a, b, c, d, q \in R$, we define the R-linear operator

$$T = \begin{bmatrix} a & c \\ b & d \end{bmatrix}_{q,p}$$

from $V_{J(E)}$ to $V_{J(E)}$ by its effect on basis elements, as follows. If $I_1, I_2 \in J(E)$ are such that $I_1 \prec_p I_2$, then we define

$$T(v_{I_1}) = a v_{I_1} + b v_{I_2}$$

and

$$T(v_{I_2}) = c v_{I_1} + d v_{I_2}.$$

If $I \in J_p^0(E)$, then we define

$$T(v_I) = q v_I.$$

The operator T is defined and well-defined by Lemma 3.1.2.

Definition 3.1.4 Some of the most important operators for our purposes are

$$\begin{aligned} X_p &= \begin{bmatrix} 0 & 0 \\ 1 & 0 \end{bmatrix}_{0,p}, \\ Y_p &= \begin{bmatrix} 0 & 1 \\ 0 & 0 \end{bmatrix}_{0,p}, \\ H_p &= \begin{bmatrix} -1 & 0 \\ 0 & 1 \end{bmatrix}_{0,p}, \end{aligned}$$

$$S_p = \begin{bmatrix} 0 & 1 \\ 1 & 0 \end{bmatrix}_{1,p} \quad \text{and}$$

$$T_{p,q} = \begin{bmatrix} 0 & q \\ 1 & q-1 \end{bmatrix}_{q,p}.$$

(Notice that $S_p = T_{p,1}$.) We have

$$S_p^2 = \begin{bmatrix} 1 & 0 \\ 0 & 1 \end{bmatrix}_{1,p},$$

which is the identity map, id, and

$$X_p^2 = Y_p^2 = \begin{bmatrix} 0 & 0 \\ 0 & 0 \end{bmatrix}_{0,p},$$

which is the zero map.

Exercise 3.1.5 Let $\varepsilon : E \to \Gamma$ be a full heap, let p be a vertex of Γ, and let $I \in J(E)$. Show that

(i) $S_p(v_I) = X_p(v_I)$ if and only if $I \in J_p^-(E)$;
(ii) $S_p(v_I) = Y_p(v_I)$ if and only if $I \in J_p^+(E)$;
(iii) $S_p(v_I) = v_I$ if and only if $I \in J_p^0(E)$.

Lemma 3.1.6 *Using the above notation, we have*

$$T_{s,q}^2 = q \,\mathrm{id} + (q-1)T_{s,q},$$

where id *is the identity map. If, furthermore, q is an invertible element of R, we have*

$$T_{s,q}^{-1} = \begin{bmatrix} q^{-1}-1 & 1 \\ q^{-1} & 0 \end{bmatrix}_{q^{-1},s} = q^{-1}T_{s,q} + (q^{-1}-1)\,\mathrm{id}.$$

Proof As in Definition 3.1.3, it is enough to check the identities on (a) two-dimensional subspaces $\langle v_I, v_{I'}\rangle$ of $V_{J(E)}$ for which $I \prec_p I'$ and (b) one-dimensional subspaces $\langle v_I\rangle$ of $V_{J(E)}$ for which $I \in J_p^0(E)$.

In case (a), $T_{s,q}$ acts via the matrix

$$M = \begin{pmatrix} 0 & q \\ 1 & q-1 \end{pmatrix},$$

and the assertions reduce to the identities $M^2 = qI + (q-1)M$, $M^{-1} = q^{-1}M + (q^{-1}-1)I$ and

$$M^{-1} = \begin{pmatrix} q^{-1}-1 & 1 \\ q^{-1} & 0 \end{pmatrix}.$$

In case (b), $T_{s,q}$ acts via scalar multiplication by q. In this case, the assertions reduce to the observation that $q^2 = q \cdot 1 + (q-1)q$. □

Lemma 3.1.7 *Maintain the above notation, and suppose that s and t are nonadjacent vertices in Γ and that q is an invertible element of R. Then we have*

$$T_{s,q}T_{t,q} = T_{t,q}T_{s,q}.$$

Proof It is enough to check that each side of the equation acts the same on the subspaces $\langle v_I : I \in \mathbf{i} \rangle$ for each (s, t)-string $\mathbf{i}$. If $\mathbf{i}$ contains a single element, then each side acts as the scalar q^2, and we are done. Otherwise, we are in one of the three situations of Theorem 2.3.15 (iii) (with $p = s$ and $q = t$). In case (iii)(b), $T_{t,q}$ acts as the scalar q on both v_{I_1} and v_{I_2}, so both sides of the equation act as $T_{s,q}$, and we are done. Similarly, in case (iii)(c), both sides of the equation act as $T_{t,q}$.

It remains to deal with the situation of case (iii)(a), in which we have $\mathbf{i} = \{I_1, I_2, I_3, I_4\}$, where both $I_1 \prec_s I_2 \prec_t I_4$ and $I_1 \prec_t I_3 \prec_s I_4$. Both sides of the equation agree when acting on I_1, because we have

$$T_{s,q}T_{t,q}(v_{I_1}) = v_{I_4} = T_{t,q}T_{s,q}(v_{I_1}).$$

In the case of I_2, we have

$$T_{t,q}T_{s,q}^{-1}(v_{I_2}) = v_{I_3} = T_{s,q}^{-1}T_{t,q}(v_{I_2}).$$

However, using Lemma 3.1.6, we can express the operator $T_{s,q}^{-1}$ as a linear combination of the identity operator and $T_{s,q}$, from which it follows that the identities

$$T_{s,q}T_{t,q} = T_{t,q}T_{s,q}$$

and

$$T_{t,q}T_{s,q}^{-1} = T_{s,q}^{-1}T_{t,q}$$

are equivalent. It follows that both sides of the equation agree when acting on v_{I_2}.

The case of I_3 is similar, using the equivalent form

$$T_{t,q}^{-1}T_{s,q} = T_{s,q}T_{t,q}^{-1}.$$

For I_4, we consider the equivalent form

$$T_{t,q}^{-1}T_{s,q}^{-1} = T_{s,q}^{-1}T_{t,q}^{-1}.$$

It follows that the two sides of the equation agree on the given four-dimensional subspace, and therefore that they agree in general. □

Lemma 3.1.8 *Maintain the above notation, and suppose that s and t are adjacent vertices in Γ and that q is an invertible element of R. Let $A = (A_{ij})$ be the generalized Cartan matrix associated with Γ.*

(i) *If $a_{st} = a_{ts} = -1$ then we have*

$$T_{s,q}T_{t,q}T_{s,q} = T_{t,q}T_{s,q}T_{t,q}.$$

(ii) *If $a_{st} = -2$ and $a_{ts} = -1$ then we have*

$$T_{s,q}T_{t,q}T_{s,q}T_{t,q} = T_{t,q}T_{s,q}T_{t,q}T_{s,q}.$$

Proof The proof follows the same strategy as Lemma 3.1.7 above. If I lies in an (s, t)-string of size 1, then both sides of the equation act as the scalar q^3 (respectively, q^4) in part (i) (respectively, (ii)), and we are done.

The other possibility for part (i) is the situation of Theorem 2.3.15 (i), in other words, that $\mathbf{i} = \{I_1, I_2, I_3\}$. By symmetry, we may assume that $I_1 \prec_s I_2 \prec_t I_3$. We find using the definitions that

$$\begin{aligned} T_{s,q}T_{t,q}T_{s,q}(v_{I_1}) &= qv_{I_3} = T_{t,q}T_{s,q}T_{t,q}(v_{I_1}), \\ T_{s,q}T_{t,q}T_{s,q}^{-1}(v_{I_2}) &= qv_{I_2} = T_{t,q}^{-1}T_{s,q}T_{t,q}(v_{I_2}), \text{ and} \\ T_{t,q}T_{s,q}^{-1}T_{t,q}^{-1}(v_{I_3}) &= qv_{I_1} = T_{s,q}^{-1}T_{t,q}^{-1}T_{s,q}(v_{I_3}). \end{aligned}$$

Expanding the formulae for the inverse, we find that the three identities

$$\begin{aligned} T_{s,q}T_{t,q}T_{s,q} &= T_{t,q}T_{s,q}T_{t,q}, \\ T_{s,q}T_{t,q}T_{s,q}^{-1} &= T_{t,q}^{-1}T_{s,q}T_{t,q} \text{ and} \\ T_{t,q}T_{s,q}^{-1}T_{t,q}^{-1} &= T_{s,q}^{-1}T_{t,q}^{-1}T_{s,q} \end{aligned}$$

are all equivalent, which completes the proof of (i).

The proof of (ii) uses a similar argument applied to Theorem 2.3.15 (ii). In this case, we have $\mathbf{i} = \{I_1, I_2, I_3, I_4\}$ with $I_1 \prec_s I_2 \prec_t I_3 \prec_s I_4$. We then have

$$\begin{aligned} T_{t,q}T_{s,q}T_{t,q}T_{s,q}(v_{I_1}) &= qv_{I_4} = T_{s,q}T_{t,q}T_{s,q}T_{t,q}(v_{I_1}), \\ T_{t,q}T_{s,q}T_{t,q}T_{s,q}^{-1}(v_{I_2}) &= qv_{I_3} = T_{s,q}^{-1}T_{t,q}T_{s,q}T_{t,q}(v_{I_2}), \\ T_{s,q}T_{t,q}T_{s,q}^{-1}T_{t,q}^{-1}(v_{I_3}) &= qv_{I_2} = T_{t,q}^{-1}T_{s,q}^{-1}T_{t,q}T_{s,q}(v_{I_3}) \text{ and} \\ T_{t,q}T_{s,q}^{-1}T_{t,q}^{-1}T_{s,q}^{-1}(v_{I_4}) &= qv_{I_1} = T_{s,q}^{-1}T_{t,q}^{-1}T_{s,q}^{-1}T_{t,q}(v_{I_4}). \end{aligned}$$

Expanding the formulae for the inverse, we find that the four identities

$$\begin{aligned} T_{t,q}T_{s,q}T_{t,q}T_{s,q} &= T_{s,q}T_{t,q}T_{s,q}T_{t,q}, \\ T_{t,q}T_{s,q}T_{t,q}T_{s,q}^{-1} &= T_{s,q}^{-1}T_{t,q}T_{s,q}T_{t,q}, \\ T_{s,q}T_{t,q}T_{s,q}^{-1}T_{t,q}^{-1} &= T_{t,q}^{-1}T_{s,q}^{-1}T_{t,q}T_{s,q} \text{ and} \\ T_{t,q}T_{s,q}^{-1}T_{t,q}^{-1}T_{s,q}^{-1} &= T_{s,q}^{-1}T_{t,q}^{-1}T_{s,q}^{-1}T_{t,q} \end{aligned}$$

are all equivalent, which completes the proof of (ii). □

Definition 3.1.9 Let Γ be a Dynkin diagram with associated generalized Cartan matrix A. We define the *Weyl group*, $W(\Gamma)$, associated with Γ to be the group with generators $S(\Gamma) = \{s_i : i \in \Gamma\}$ indexed by the vertices of Γ and defining relations

$$\begin{aligned} s_i^2 &= 1 \text{ for all } i \in \Gamma, \\ s_i s_j &= s_j s_i \text{ if } a_{ij} = 0, \\ s_i s_j s_i &= s_j s_i s_j \text{ if } a_{ij} < 0 \text{ and } a_{ij}a_{ji} = 1, \\ s_i s_j s_i s_j &= s_j s_i s_j s_i \text{ if } a_{ij} < 0 \text{ and } a_{ij}a_{ji} = 2. \end{aligned}$$

Note that no relation is added in the case where $a_{ij} < 0$ and $a_{ij}a_{ji} = 4$. The *rank* of the Weyl group $W(\Gamma)$ is defined to be the number $|\Gamma|$.

Lemma 3.1.10 *Let Γ be a Dynkin diagram with associated generalized Cartan matrix A. As a monoid, the Weyl group $W(\Gamma)$ is the quotient of the commutation monoid* $\mathrm{Co}(\Gamma)$ *by the relations*

$$\begin{aligned} s_i^2 &= 1 \text{ for all } i \in \Gamma, \\ s_i s_j s_i &= s_j s_i s_j \text{ if } a_{ij} < 0 \text{ and } a_{ij}a_{ji} = 1, \\ s_i s_j s_i s_j &= s_j s_i s_j s_i \text{ if } a_{ij} < 0 \text{ and } a_{ij}a_{ji} = 2. \end{aligned}$$

We denote the surjective monoid homomorphism from $\mathrm{Co}(\Gamma)$ *to* $W(\Gamma)$ *by* $\pi_{\mathrm{Co}(\Gamma)}$.

Proof Note that $W = W(\Gamma)$ is a quotient of the free monoid S^*, where $S = S(\Gamma)$, because S generates W. The defining relations for $\mathrm{Co}(\Gamma)$ are precisely those of the form

$$s_i s_j = s_j s_i \text{ if } a_{ij} = 0,$$

and these, together with the relations in the statement, form a presentation for W. □

Definition 3.1.11 Let R be a commutative ring with 1. By an (associative) R-algebra, we mean an R-module A equipped with an associative R-bilinear multiplication making A into a ring. If S is a set, then the free R-algebra RS^* generated by S consists of all R-linear combinations of the words S^* in the generators S, including the empty word, and the R-linear multiplication is that induced by juxtaposition of words in S^*. If I is a collection of elements in RS^* and $\langle I \rangle$ is the two-sided ideal of RS^* generated by I, then we refer to the quotient R-algebra $RS^*/\langle I \rangle$ as *the R-algebra A generated by S subject to the defining relations I*. We will often represent an element $x - y \in I$ as an equation $x - y = 0$ or $x = y$.

Definition 3.1.12 Let R be a commutative ring with 1 and let $q \in R$ be invertible. The *Hecke algebra* over R, $\mathcal{H}(\Gamma)$, associated with Γ is the associative algebra with identity T_e generated by the set $\{T_s : s \in S(\Gamma)\}$ subject to the defining relations

$$\begin{aligned} T_{s_i}^2 &= qT_e + (q-1)T_{s_i} \text{ for all } i \in \Gamma, \\ T_{s_i}T_{s_j} &= T_{s_j}T_{s_i} \text{ if } a_{ij} = 0, \\ T_{s_i}T_{s_j}T_{s_i} &= T_{s_j}T_{s_i}T_{s_j} \text{ if } a_{ij} < 0 \text{ and } a_{ij}a_{ji} = 1, \\ T_{s_i}T_{s_j}T_{s_i}T_{s_j} &= T_{s_j}T_{s_i}T_{s_j}T_{s_i} \text{ if } a_{ij} < 0 \text{ and } a_{ij}a_{ji} = 2. \end{aligned}$$

Theorem 3.1.13 *Let k be a field, let $\varepsilon : E \to \Gamma$ be a full heap over a Dynkin diagram Γ, let $\mathcal{H}(\Gamma)$ be the Hecke algebra of Γ over k, and let $W(\Gamma)$ be the Weyl group of Γ.*

(i) *The k-vector space $V_{J(E)}$ of Definition 3.1.3 has a unique unital (left) $\mathcal{H}$-module structure in which T_{s_i} acts via the endomorphism $T_{i,q}$.*
(ii) *The k-vector space $V_{J(E)}$ of Definition 3.1.3 has a unique unital (left) W-module structure in which s_i acts via the endomorphism S_i. This induces a left W-action on $J(E)$.*

Proof Part (i) follows from the definitions and Lemmas 3.1.6, 3.1.7 and 3.1.8.

The first assertion of part (ii) comes from the observations that (a) the relations governing the s_i are obtained from those governing the corresponding T_{s_i} by specializing q to 1, and (b) the operator S_i coincides with $T_{i,1}$. The second assertion of part (ii) follows from the observation that the operator S_i permutes the basis elements of $V_{J(E)}$. □

Remark 3.1.14 Since the action of W on $V_{J(E)}$ described in Theorem 3.1.13 is via permutation of a basis, there is an obvious induced action of W on $J(E)$ itself. We will use this extensively in the sequel.

Example 3.1.15 (Type affine A.) Let Γ be the Dynkin diagram $A_l^{(1)}$ of Figure 3.1 and let $E = \mathrm{FH}(A_l^{(1)}(l))$ be the labelled partially ordered set shown in Figure 3.2.

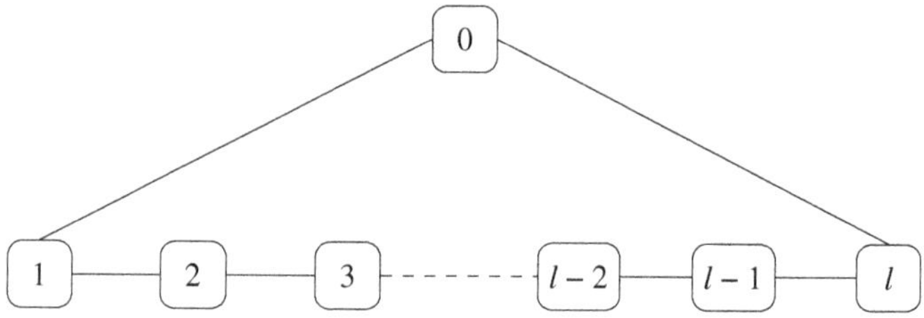

Figure 3.1 Dynkin diagram of type $A_l^{(1)} (l > 1)$

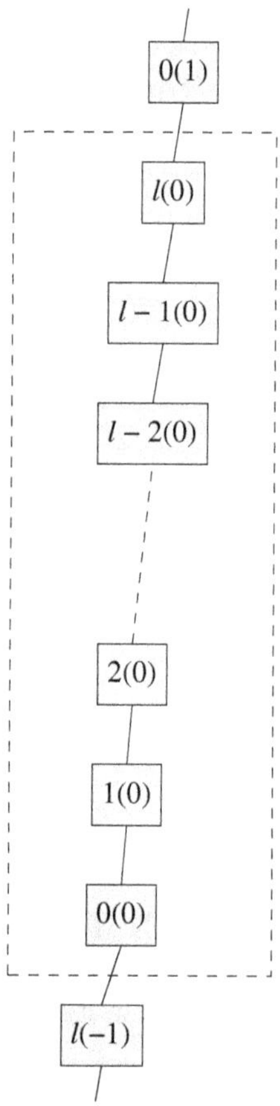

Figure 3.2 The full heap $\mathrm{FH}(A_l^{(1)}(l))$ over the Dynkin diagram of Figure 3.1

Let $J(E)$ denote the set of ideals of E, and define $\mathcal{B} := J(E) \backslash \{\emptyset, E\}$. In this case, every ideal in $\mathcal{B}$ is principal, so we have

$$\mathcal{B} = \{\langle p(y) \rangle : 0 \le p \le l,\ y \in \mathbb{Z}\}.$$

Because $\mathcal{B}$ is totally ordered by inclusion, the map $\zeta : \mathcal{B} \to \mathbb{Z}$ defined by $\zeta(\langle p(y)\rangle) = (l+1)y + p + 1$ is an isomorphism of totally ordered sets (where $\mathbb{Z}$ is ordered in the usual way).

With these identifications, there is a unique action of $W(A_l^{(1)})$ on $\mathbb{Z}$ for which s_i acts as follows:

$$s_i(z) = \begin{cases} z+1 & \text{if } z \equiv i \bmod (l+1), \\ z-1 & \text{if } z \equiv i+1 \bmod (l+1), \\ z & \text{otherwise.} \end{cases}$$

If q is an invertible element of a commutative ring R with identity and V is the free R-module with basis $\{v_z : z \in \mathbb{Z}\}$, then T_{s_i} acts on V as follows:

$$T_{s_i}(v_z) = \begin{cases} v_{z+1} & \text{if } z \equiv i \bmod (l+1), \\ qv_{z-1} + (q-1)v_z & \text{if } z \equiv i+1 \bmod (l+1), \\ qv_z & \text{otherwise.} \end{cases}$$

There is a full heap $\varepsilon : E \to \Gamma$, where E is as shown in Figure 3.2; we denote this by $\mathrm{FH}(A_l^{(1)}(l))$, for reasons that will become clear in Chapter 6. The dashed box has the same meaning as in Example 2.2.1, and corresponds to the trace

$$[0123\cdots(l-2)(l-1)l].$$

Exercise 3.1.16 (Type affine A.) Verify that the heap $\varepsilon : E = \mathrm{FH}(A_l^{(1)}(l)) \to \Gamma$ in Example 3.1.15 is a full heap.

Exercise 3.1.17 (Type affine C.)

(i) Prove that the heap $\varepsilon : E \to \Gamma$, where $E = \mathrm{FH}(C_l^{(1)}(l))$ is as shown in Figure 3.3 and Γ is as shown in Figure 3.4, is full. Find the trace associated with the repeating motif shown in the dashed box.

(ii) Let $J(E)$ denote the set of ideals of E, and define $\mathcal{B} := J(E)\backslash\{\emptyset, E\}$. Show that every ideal in $\mathcal{B}$ is principal, that

$$\mathcal{B} = \{\langle p(y)\rangle : 0 \le p \le l,\ y \in \mathbb{Z}\},$$

and that $\mathcal{B}$ is totally ordered by inclusion.

(iii) Show that there is an isomorphism $\zeta : \mathcal{B} \to \mathbb{Z}$ of totally ordered sets given by

$$\zeta(\langle p(y)\rangle) = \begin{cases} 2ly + 1 & \text{if } p = 0, \\ 2ly + l + 1 & \text{if } p = l, \\ ly + p + 1 & \text{if } p \notin \{0, l\} \text{ and } y \text{ is even}, \\ l(y+1) - p + 1 & \text{if } p \notin \{0, l\} \text{ and } y \text{ is odd}. \end{cases}$$

(iv) Show that, under the identifications of (iii), there is a unique action of $W(C_l^{(1)})$ on $\mathbb{Z}$ for which the action of s_j on $\mathbb{Z}$ is given by

$$s_j(z) = \begin{cases} z+1 & \text{if } z \equiv \pm j \bmod 2l, \\ z-1 & \text{if } z \equiv (\pm j)+1 \bmod 2l, \\ z & \text{otherwise.} \end{cases}$$

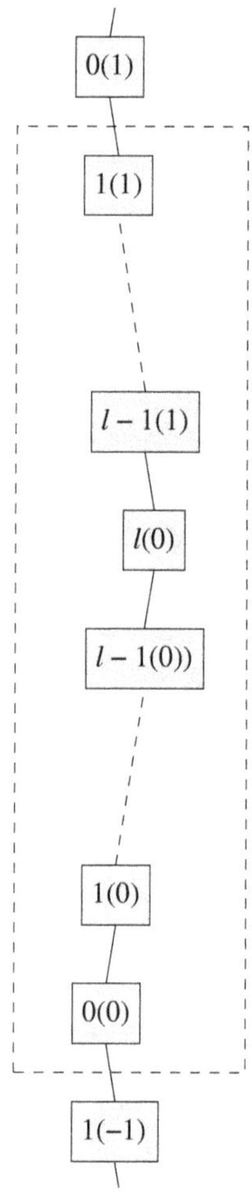

Figure 3.3 The full heap $\mathrm{FH}(C_l^{(1)}(1))$ over the Dynkin diagram of Figure 3.4

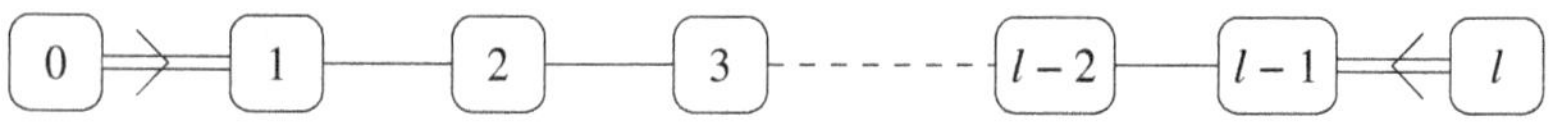

Figure 3.4 Dynkin diagram of type $C_l^{(1)} (l > 1)$

(v) Let V be as in Example 3.1.15. Describe the action of $T_{s_i} \in \mathcal{H}(C_l^{(1)})$ on V.

Exercise 3.1.18 (Type affine B.)

(i) Prove that the heap $\varepsilon : E \to \Gamma$, where $E = \mathrm{FH}(A_{2l-1}^{(2)}(l))$ is as shown in Figure 3.5 and Γ is as shown in Figure 3.6, is full. Find the trace associated with the repeating motif shown in the dashed box.

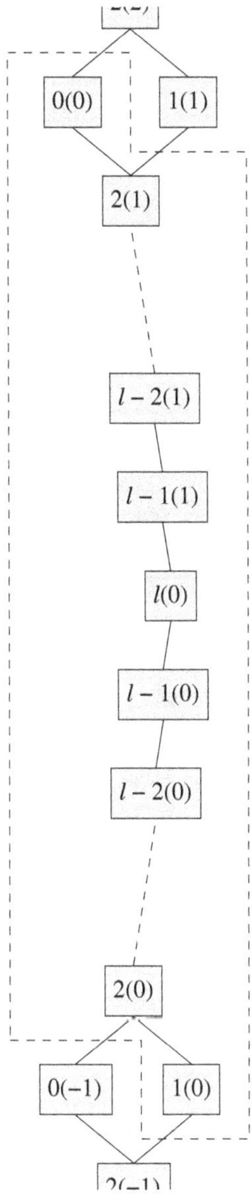

Figure 3.5 The full heap $\mathrm{FH}(A^{(2)}_{2l-1}(l))$ over the Dynkin diagram of type $A^{(2)}_{2l-1}$

(ii) Let $J(E)$ denote the set of ideals of E, and define $\mathcal{B} := J(E)\backslash\{\emptyset, E\}$. Show that every ideal in $\mathcal{B}$ is principal, except those of the form $\langle 0(y), 1(y+1)\rangle$. Show also that the inclusion partial order on $\mathcal{B}$ can be refined to a total order by stipulating that, for all $y \in \mathbb{Z}$, $\langle 0(y)\rangle < \langle 1(y+1)\rangle$.

(iii) With the refinement in (ii), show that there is an isomorphism $\zeta : \mathcal{B} \to \mathbb{Z}$ of totally ordered sets given by

$$\zeta(\langle 0(y-1), 1(y)\rangle) = 2ly + 2$$

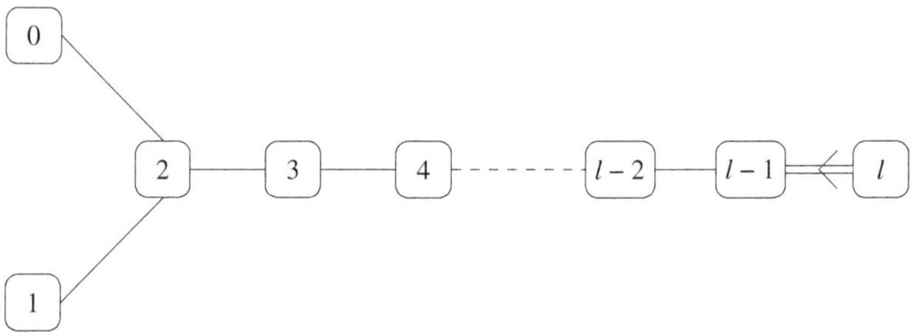

Figure 3.6 Dynkin diagram of type $A^{(2)}_{2l-1}$

and

$$\zeta(\langle p(y)\rangle) = \begin{cases} 2l(y+1) & \text{if } p = 0, \\ 2ly + 1 & \text{if } p = 1, \\ ly + p + 1 & \text{if } 1 < p < l \text{ and } y \text{ is even}, \\ l(y+1) - p + 1 & \text{if } 1 < p < l \text{ and } y \text{ is odd}, \\ 2ly + l + 1 & \text{if } p = l. \end{cases}$$

(iv) Show that, under the identifications of (iii), there is a unique action of $W(A^{(2)}_{2l-1}) = W(B^{(1)}_l)$ on $\mathbb{Z}$ for which the action of s_j for $j \notin \{0, 1\}$ is given by

$$s_j(z) = \begin{cases} z+1 & \text{if } z \equiv \pm j \bmod 2l, \\ z-1 & \text{if } z \equiv (\pm j) + 1 \bmod 2l, \\ z & \text{otherwise.} \end{cases}$$

Show also that the action of s_0 is given by

$$s_0(z) = \begin{cases} z+1 & \text{if } z \equiv \pm 1 \bmod 2l, \\ z-1 & \text{if } z \equiv 0 \text{ or } z \equiv 2 \bmod 2l, \\ z & \text{otherwise,} \end{cases}$$

and that the action of s_1 is given by

$$s_1(z) = \begin{cases} z+2 & \text{if } z \equiv -1 \text{ or } z \equiv 0 \bmod 2l, \\ z-2 & \text{if } z \equiv 1 \text{ or } z \equiv 2 \bmod 2l, \\ z & \text{otherwise.} \end{cases}$$

(v) Let V be as in Example 3.1.15. Describe the action of $T_{s_i} \in \mathcal{H}(A^{(2)}_{2l-1}) = \mathcal{H}(B^{(1)}_l)$ on V.

Exercise 3.1.19 (Type affine D.)

(i) Prove that the heap $\varepsilon : E \to \Gamma$, where $E = \mathrm{FH}(D^{(1)}_l(l))$ is as shown in Figure 3.7 and Γ is as shown in Figure 3.8, is full. Find the trace associated with the repeating motif shown in the dashed box.

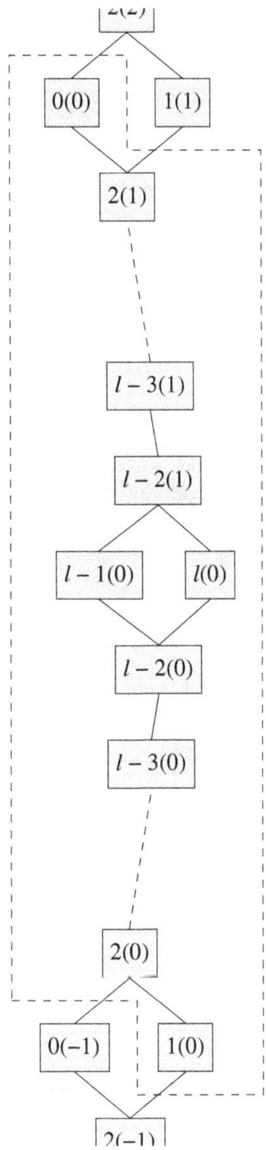

Figure 3.7 The full heap $\text{FH}(D_l^{(1)}(1))$ over the Dynkin diagram of type $D_l^{(1)}$

(ii) Let $J(E)$ denote the set of ideals of E, and define $\mathcal{B} := J(E)\backslash\{\emptyset, E\}$. Show that every ideal in $\mathcal{B}$ is principal, except those of the form $\langle 0(y), 1(y+1)\rangle$ or $\langle l-1(y), l(y)\rangle$, where $y \in \mathbb{Z}$. Show also that the inclusion partial order on $\mathcal{B}$ can be refined to a total order by stipulating that, for all $y \in \mathbb{Z}$, $\langle l-1(y)\rangle < \langle l(y)\rangle$ and $\langle 0(y)\rangle < \langle 1(y+1)\rangle$.

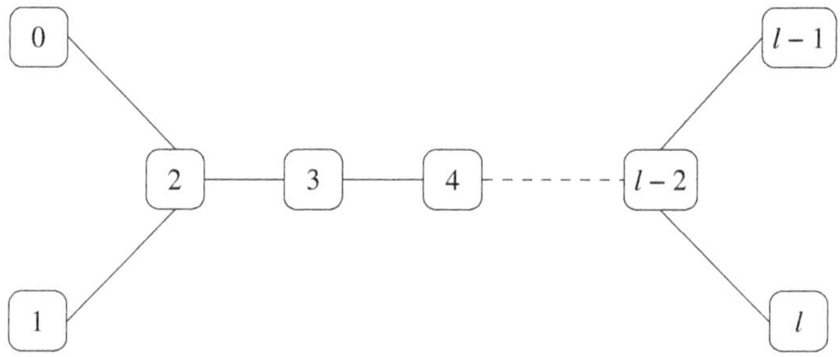

Figure 3.8 Dynkin diagram of type $D_l^{(1)}$

(iii) With the refinement in (ii), show that there is an isomorphism $\zeta : \mathcal{B} \to \mathbb{Z}$ of totally ordered sets given by

$$\zeta(\langle 0(y-1), 1(y)\rangle) = 2ly + 2,$$
$$\zeta(\langle l-1(y), l(y)\rangle) = 2ly + l + 2$$

and

$$\zeta(\langle p(y)\rangle) = \begin{cases} 2l(y+1) & \text{if } p = 0, \\ 2ly + p & \text{if } p = 1, \\ 2ly + p + 1 & \text{if } p \in \{l-1, l\}, \\ ly + p + 1 & \text{if } 1 < p < l-1 \text{ and } y \text{ is even}, \\ l(y+1) - p + 1 & \text{if } 1 < p < l-1 \text{ and } y \text{ is odd}. \end{cases}$$

(iv) Show that, under the identifications of (iii), there is a unique action of $W(D_l^{(1)})$ on $\mathbb{Z}$ for which the action of s_j on $\mathbb{Z}$ for $j \notin \{0, 1, l-1, l\}$ is given by

$$s_j(z) = \begin{cases} z+1 & \text{if } z \equiv \pm j \bmod 2l, \\ z-1 & \text{if } z \equiv (\pm j) + 1 \bmod 2l, \\ z & \text{otherwise.} \end{cases}$$

Show also that the action of s_j for $j \in \{0, l-1\}$ is given by

$$s_j(z) = \begin{cases} z+1 & \text{if } z \equiv j - 1 + c(j) \text{ or } z \equiv j + 1 + c(j) \bmod 2l, \\ z-1 & \text{if } z \equiv j + c(j) \text{ or } z \equiv j + 2 + c(j) \bmod 2l, \\ z & \text{otherwise,} \end{cases}$$

where we define $c(0) = 0$ and $c(l-1) = 1$. Finally, show that the action of s_j for $j \in \{1, l\}$ is given by

$$s_j(z) = \begin{cases} z+2 & \text{if } z \equiv j - 2 + c(j) \text{ or } z \equiv j - 1 + c(j) \bmod 2l, \\ z-2 & \text{if } z \equiv j + c(j) \text{ or } z \equiv j + 1 + c(j) \bmod 2l, \\ z & \text{otherwise,} \end{cases}$$

where we define $c(1) = 0$ and $c(l) = 1$.

(v) Describe the action of $T_{s_i} \in \mathcal{H}(D_l^{(1)})$ on V.

3.2 Proper ideals

As can be seen from the examples of full heaps in Example 3.1.15 and Exercises 3.1.17, 3.1.18 and 3.1.19, the ideals $\emptyset$ and E in $J(E)$ are not very interesting from the point of view of linear operators. For this reason, we usually restrict our attention to the so-called "proper ideals" in $J(E)$. For heaps over finite connected graphs, these turn out to be all elements of $J(E)$ except for $\emptyset$ and E itself (Lemma 3.2.4 (v)), but the general definition is as follows.

Definition 3.2.1 Let $\varepsilon : E \to \Gamma$ be a heap over a graph. If F is an ideal of E such that for each vertex p of Γ we have $\emptyset \neq F \cap \varepsilon^{-1}(p) \neq \varepsilon^{-1}(p)$, then we call F a *proper ideal* of E. We denote the set of proper ideals of E by $\mathcal{B}(E)$, or by $\mathcal{B}$ if the context is clear.

For R a commutative ring with 1, we denote by $V_{\mathcal{B}} = V_{\mathcal{B}(E)} = V_{\mathcal{B}(E)}(R)$ the free R-module with basis $\{v_I : I \in \mathcal{B}(E)\}$.

Example 3.2.2 If E is a nonempty heap over any graph, then $\emptyset$ and E itself are ideals of E, but not proper ideals.

If E is one of the four full heaps discussed in Example 3.1.15 and Exercises 3.1.17, 3.1.18 and 3.1.19, then $\mathcal{B}(E) = J(E)\backslash\{\emptyset, E\}$.

Exercise 3.2.3 Suppose Γ is a disconnected graph with connected components Γ_1 and Γ_2 and that $\varepsilon : E \to \Gamma$ is a heap with the property that $E_1 = \varepsilon^{-1}(\Gamma_1)$ and $E_2 = \varepsilon^{-1}(\Gamma_2)$ are both nonempty. Regard E_1 and E_2 as heaps over Γ_1 and Γ_2 in the obvious way. (In other words, the heaps E, E_1 and E_2 are related as in Exercise 2.1.14.)

(i) Show that there is a bijection between the set of all proper ideals of E and the set of all pairs (I, I'), where I is a proper ideal of E_1 and I' is a proper ideal of E_2.
(ii) Formulate an analogous statement in the general case where Γ has $k > 1$ connected components.
(iii) Construct an explicit example of a heap E over a graph Γ with the property that E has at least three ideals that are not proper.
(iv) Show that if E_1 and E_2 are full heaps over Dynkin diagrams Γ_1 and Γ_2 respectively, then E is a full heap over Γ. Deduce that there are infinitely many isomorphism classes of full heaps in **Heap**.

We now collect some of the basic properties of proper ideals for later use.

Lemma 3.2.4 *Let $\varepsilon : E \to \Gamma$ be a full heap over a Dynkin diagram, let F and F' be proper ideals of E and let J be an ideal of E.*

(i) *The ideal J is proper if and only if for all vertices $p \in \Gamma$, we have*

$$J \cap \varepsilon^{-1}(p) = E_p(N) := \{E(p, t) : t < N\}$$

for some integer N depending on J and p. (Here, $E(p, t)$ refers to the labelling convention for elements of E described in Section 2.1.)
(ii) *The subheaps $F \cap F'$ and $F \cup F'$ are proper ideals of E.*
(iii) *If $F \subseteq J$ and $J\backslash F$ is finite, then J is a proper ideal of E.*

(iv) *If* $J \subseteq F$ *and* $F \backslash J$ *is finite, then* J *is a proper ideal of* E.
(v) *If* Γ *is finite and connected, then* J *is a proper ideal if and only if* $\emptyset \neq J \neq E$.
(vi) *If* Γ *is finite and* $F \subseteq F'$, *then* $F' \backslash F$ *is finite.*
(vii) *If* Γ *is finite and connected and* L *is a finite convex subheap of* E, *then there is a proper ideal* L' *of* E *such that* $L \subset L'$ *and* $L' \backslash L$ *is a proper ideal of* E.
(viii) *If* p *and* q *are adjacent vertices of* Γ *then we cannot have both* $F \cap \varepsilon^{-1}(p) \subsetneq F' \cap \varepsilon^{-1}(p)$ *and* $F \cap \varepsilon^{-1}(q) \supsetneq F' \cap \varepsilon^{-1}(q)$.

Proof Part (i) is immediate from the definitions.

Part (ii) follows from (i) and the fact that, for a fixed vertex p, the collection of chains $E_p(N)$ is closed under finite intersections and finite unions.

For part (iii), fix $p \in \Gamma$ and choose N so that $F \cap \varepsilon^{-1}(p) = E_p(N)$. Since $J \backslash F$ is finite,

$$(J \cap \varepsilon^{-1}(p)) \backslash (F \cap \varepsilon^{-1}(p))$$

must be finite. However, since J is an ideal, $J \cap \varepsilon^{-1}(p)$ must be downward closed (meaning that if $E(p, y) \in J \cap \varepsilon^{-1}(p)$ and $x < y$, then $E(p, x) \in J \cap \varepsilon^{-1}(p)$). This shows that $J \cap \varepsilon^{-1}(p) = E_p(N')$ for some $N' \geq N$.

Part (iv) follows by an argument similar to that used to prove part (iii), *mutatis mutandis*.

If J is a proper ideal, then it follows easily from (i) that $\emptyset \neq J \neq E$, as required for (v), so suppose that $\emptyset \neq J \neq E$ for an ideal J. Let $\alpha \in E \backslash J$ and let $p = \varepsilon(\alpha)$.

Suppose first that $J \cap \varepsilon^{-1}(p) \neq \emptyset$. Since $J \cap \varepsilon^{-1}(p)$ is an ideal of $\varepsilon^{-1}(p)$ and $\alpha \notin J$, we have $J \cap \varepsilon^{-1}(p) = E_p(N)$. Let q be any vertex of Γ. By Lemma 2.3.2 (ii), the definition of full heap ensures that there exists an element $\beta > \alpha$ with $\varepsilon(\beta) = q$, and since $\alpha \notin J$, we must have $\beta \notin J$. On the other hand, Lemma 2.3.2 (ii) also proves the existence of an element $\beta' < E(p, N-1)$ with $\varepsilon(\beta') = q$, and the fact that J is an ideal means that $\beta' \in J$. Combining these observations, we see that $J \cap \varepsilon^{-1}(q) = E_q(N')$ for some integer N'.

The other possibility is that $J \cap \varepsilon^{-1}(p) = \emptyset$. By reversing the argument of the above paragraph, we find that $J \cap \varepsilon^{-1}(q) = \emptyset$ for all vertices q, in other words, that $J = \emptyset$, contrary to the hypothesis of (v).

Under the assumptions of (vi), the sets $(F' \backslash F) \cap \varepsilon^{-1}(p)$ are all finite by (i), which means that $F' \backslash F$ is also finite because Γ is.

For part (vii), define $L' = \{\alpha \in E : \alpha \leq \beta \text{ for some } \beta \in L\}$. It is easily checked that L' is a nonempty ideal of E and that $L' \backslash L$ is an ideal of E. Since L is bounded above by Lemma 2.3.2 (iv), it follows that L' is bounded above so $L' \neq E$, L' is a proper ideal by (v) and $L' \backslash L$ is a proper ideal by (iii).

Part (viii) is proved by considering the edge chain $\varepsilon^{-1}(\{p, q\})$. The first condition in the statement implies that $F \cap \varepsilon^{-1}(\{p, q\}) \subsetneq F' \cap \varepsilon^{-1}(\{p, q\})$ and the second condition implies that $F \cap \varepsilon^{-1}(\{p, q\}) \supsetneq F' \cap \varepsilon^{-1}(\{p, q\})$, which are mutually contradictory conclusions. □

Exercise 3.2.5 Suppose that $\varepsilon : E \to \Gamma$ is a full heap and I is a proper ideal of E. Show that if $I \prec_p I'$ or $I \succ_p I'$ for some $p \in \Gamma$, then I' is also a proper ideal. Deduce that the R-linear operators

$$T = \begin{bmatrix} a & c \\ b & d \end{bmatrix}_{q,p}$$

of Definition 3.1.3 act on the subspace $V_{\mathcal{B}}(R)$ of $V_{J(E)}(R)$.

Exercise 3.2.6 Show that the finiteness assumption of Lemma 3.2.4 (v) cannot be dropped. (Hint: use Exercise 3.2.3.)

Exercise 3.2.7 If $\varepsilon : E \to \Gamma$ is a full heap, show that $\mathcal{B}(E)$ is a distributive lattice under the usual operations of intersection and union.

Definition 3.2.8 Let $\varepsilon : E \to \Gamma$ be a full heap. We define V_Γ to be the $\mathbb{R}$-vector space with basis $\{\alpha_p : p \in \Gamma\}$. The *support* of an element $\alpha = \sum_{p \in \Gamma} \lambda_p \alpha_p$ of V_Γ is the subset of Γ given by

$$\{p \in \Gamma : \lambda_p \neq 0\}.$$

We define a partial order $\leq$ on V_Γ by the condition that $\lambda \leq \mu$ if and only if we have $c_p \geq 0$ for all p, where $\mu - \lambda = \sum_{p \in \Gamma} c_p \alpha_p$.

If F is a finite subheap of E, we define the *content* of F to be

$$\chi(F) = \sum_{p \in \Gamma} c_p \alpha_p,$$

where $c_p = |\varepsilon^{-1}(p) \cap F|$ is the number of vertices of F with label p. The *support* of F is by definition the support of $\chi(F)$.

If $\alpha \in V_\Gamma$ then we define $\mathcal{L}_\alpha = \mathcal{L}_\alpha(E)$ to be the set of all finite convex subheaps L of E with $\chi(L) = \alpha$. We may abbreviate $\mathcal{L}_{\alpha_i}$ to $\mathcal{L}_i$.

Let I and I' be proper ideals of E. Using the labelling convention described in Section 2.1, we define $t_p(I)$ (respectively, $t_p(I')$) to be the largest integer t for which $E(p, t)$ is an element of I (respectively, I'); note that these integers exist by the definition of proper ideal. We define the *relative content*, $\chi(I, I') \in V_\Gamma$, as

$$\chi(I, I') = \sum_{p \in \Gamma} (t_p(I') - t_p(I))\alpha_p.$$

We say that the pair (I, I') is *relatively trivial* if either $I = I'$ or both $I \subset I'$ and $I' \backslash I$ is a trivial heap. Equivalently, (I, I') is relatively trivial if and only if $\chi(I, I')$ is a (possibly trivial) sum of elements α_p corresponding to nonadjacent vertices of Γ.

Example 3.2.9 Let $\varepsilon : E \to \Gamma$ be the full heap of Exercise 3.1.19, shown in Figure 3.7. Let $I = \langle E(l-2, 0) \rangle$ and let $I' = \langle E(l-1, 0), E(l, 0) \rangle$. Then $\chi(I, I') = \alpha_{l-1} + \alpha_l$, which implies that the pair (I, I') is relatively trivial.

Exercise 3.2.10 Show that the definition of $\chi(I, I')$ is independent of the labellings chosen for the vertex chains of E.

Exercise 3.2.11 Suppose that $I, I' \in \mathcal{B}(E)$ satisfy $I \subset I'$, and let $F = I' \backslash I$. Show that $\chi(F) = \chi(I, I')$.

Proposition 3.2.12 *Let $\varepsilon : E \to \Gamma$ be a full heap over a finite graph, and let F and F' be proper ideals.*

(i) *If $F \subseteq F'$ then there exist finite sequences*

$$Y_{j_1} Y_{j_2} \cdots Y_{j_l} v_{F'} = v_F$$

and

$$X_{i_1} X_{i_2} \cdots X_{i_k} v_F = v_{F'},$$

where X_p and Y_p are the linear operators defined in Definition 3.1.4, and where each operator X_p and Y_p occurring has the property that p lies in the support of $\chi(F, F')$.

(ii) *There exists a finite sequence*

$$F = F_0, F_1, \ldots, F_r = F'$$

such that for all $0 \leq i < r$, there exists p_i in the support of $\chi(F, F')$ such that $S_{p_i}(F_i) = F_{i+1}$. In particular, for each $0 \leq i < r$, either $F_i \subset F_{i+1}$ and $F_{i+1} \backslash F_i$ is a singleton, or $F_{i+1} \subset F_i$ and $F_i \backslash F_{i+1}$ is a singleton.

(iii) *There exists an element $w \in W(\Gamma)$ such that $w(F) = F'$, and hence $W(\Gamma)$ acts transitively on $\mathcal{B}(E)$.*

Proof By Lemma 3.2.4 (vi), $F' \backslash F$ is finite, and we prove (i) by induction on $k = |F' \backslash F|$. The case $k = 0$ is trivial. For the inductive step, let α (respectively, β) be a minimal (respectively, maximal) element of $F' \backslash F$. Let $p = \varepsilon(\alpha)$ and $q = \varepsilon(\beta)$, and note that p and q both lie in the support of $\chi(F' \backslash F) = \chi(F, F')$. By Lemma 3.2.4 (iv), $Y_q(F')$ is a proper ideal of E, and by construction, $Y_q(F')$ contains F. Furthermore, $|Y_q(F') \backslash F| = k - 1$, and the assertion regarding the Y operators follows, with $Y_{j_l} = Y_q$. The assertion regarding the X operators follows similarly by considering $X_p(F)$, which is a proper ideal by Lemma 3.2.4 (iii), and then taking $X_{i_k} = X_p$.

To prove (ii) and (iii), we first note that $F \cap F'$ is a proper ideal, by Lemma 3.2.4 (ii). Applying part (i) twice, once to the pair $F \cap F' \subseteq F$ and once to the pair $F \cap F' \subseteq F'$, we then have

$$v_{F'} = X_{i_1} X_{i_2} \cdots X_{i_k} Y_{j_1} Y_{j_2} \cdots Y_{j_l} v_F,$$

where

$$v_{F \cap F'} = Y_{j_1} Y_{j_2} \cdots Y_{j_l} v_F.$$

Using the definitions (or the result of Exercise 3.1.5), we then have

$$F' = S_{i_1} S_{i_2} \cdots S_{i_k} S_{j_1} S_{j_2} \cdots S_{j_l} F,$$

from which the assertions of (ii) and (iii) follow. □

Exercise 3.2.13 Prove that $W(\Gamma)$ does not act transitively on $J(E)$.

Lemma 3.2.14 *Let $\varepsilon : E \to \Gamma$ be a full heap and let E^* be its dual heap. The map $*_{\mathcal{B}(E)} : \mathcal{B}(E) \to \mathcal{B}(E^*)$ given by*

$$*_{\mathcal{B}(E)}(I) = *(E \backslash I)$$

is an isomorphism of W-sets.

Proof The proof that $*_{\mathcal{B}(E)}(I)$ is a proper ideal of E^* whenever I is a proper ideal of E is left as a routine exercise. Note that $*_{\mathcal{B}(E^*)}$ is the inverse of $*_{\mathcal{B}(E)}$, so both maps are bijections.

Now let p be a vertex of Γ and let $I, I' \in \mathcal{B}(E)$ satisfy $I \prec_p I'$. It follows from the definitions that we have $*_{\mathcal{B}(E)}(I) \succ_p *_{\mathcal{B}(E)}(I')$ in the heap E^*. The definition of S_p then ensures that $S_p(I) = I'$ if and only if $S_p(*_{\mathcal{B}(E)}(I)) = *_{\mathcal{B}(E)}(I')$. The claim follows from the fact that the generators s_p of W act via the permutations S_p. $\square$

Exercise 3.2.15 Show that, in the notation of Lemma 3.2.14, $*_{\mathcal{B}(E)}(I)$ is a proper ideal of E^* whenever I is a proper ideal of E.

Exercise 3.2.16 Let $\varepsilon : E \to \Gamma$ be the full heap of Example 3.1.15. Show that E is not isomorphic to its dual heap E^*, but that the sets $\mathcal{B}(E)$ and $\mathcal{B}(E^*)$ are isomorphic as W-sets.

Lemma 3.2.17 *Let $\varepsilon : E \to \Gamma$ be a full heap corresponding to the generalized Cartan matrix A, and let $I, I', I'' \in \mathcal{B}(E)$. Then we have*

$$\chi(I, I'') = \chi(I, I') + \chi(I', I'').$$

Proof This is immediate from the observation that

$$t_p(I'') - t_p(I) = (t_p(I'') - t_p(I')) + (t_p(I') + t_p(I)). \quad \square$$

Lemma 3.2.18 *Let $\varepsilon : E \to \Gamma$ be a full heap over a finite connected graph and let $p \in \Gamma$. Then there exist proper ideals I and I' of E for which $\chi(I, I') = \alpha_p$.*

Proof Let $\alpha = E(p, t)$ be an element of the vertex chain $\varepsilon^{-1}(p)$, and define $I' = \langle \alpha \rangle$. Since $E(p, t+1) \notin I'$, I' is a proper ideal by Lemma 3.2.4 (v). Since α is maximal in I', the set $I = I' \backslash \{\alpha\}$ is an ideal of E, which is proper by Lemma 3.2.4 (iv). We then have $\chi(I, I') = \alpha_p$, as required. $\square$

Lemma 3.2.19 *Let $\varepsilon : E \to \Gamma$ be a full heap corresponding to the generalized Cartan matrix A, and let $I, I' \in \mathcal{B}(E)$ be such that $\chi(I, I') = \alpha_p$ for some $p \in \Gamma$. Then for every $q \in \Gamma$, we have*

$$\chi(S_q(I), S_q(I')) = \alpha_p - a_{qp}\alpha_q.$$

Proof The hypothesis $\chi(I, I') = \alpha_p$ is equivalent to the condition $I \prec_p I'$. The proof proceeds by a case analysis based on the Local Structure Theorem.

If $a_{qp} = -1$, then parts (i) and (ii) of Theorem 2.3.15 shows that either there exists I'' with $I'' \prec_q I \prec_p I'$, or there exists I'' with $I \prec_p I' \prec_q I''$, but not both. In the former case we have $S_q(I) = I''$ and $S_q(I') = I'$, and in the latter case we have $S_q(I) = I$ and $S_q(I') = I''$. In both cases, we have $\chi(S_q(I), S_q(I')) = \alpha_p + \alpha_q$, as required.

The case $a_{p,q} = -2$ follows similar lines; in this case, parts (ii) and (iv) of Theorem 2.3.15 show that there exists a chain of proper ideals

$$I_1 \prec_q I \prec_p I' \prec_q I_2,$$

which implies that $S_q(I) = I_1$ and $S_q(I') = I_2$. Since $\chi(I_1, I_2) = \alpha_p + 2\alpha_q$, the claim follows.

If $a_{pq} = 0$, then Theorem 2.3.15 (iii) applies to the (p, q)-string containing I and I'. Of the three possibilities, case (c) is impossible, case (b) implies that $S_q(I) = I$ and $S_q(I') = I'$, and case (a) implies that $S_q(I) \prec_p S_q(I')$. In any case, we must have $\chi(S_q(I), S_q(I')) = \alpha_p$, as required. □

Exercise 3.2.20 Show that the formula

$$s_q(\alpha_p) = \alpha_p - a_{qp}\alpha_q,$$

when extended linearly, defines an action of W on V_Γ; in other words, check that the defining relations of W are respected. (See also Theorem 4.2.3 below.)

Theorem 3.2.21 *Let $\varepsilon : E \to \Gamma$ be a full heap over a finite connected graph with Weyl group W, and let W act on V_Γ as above (Exercise 3.2.20). For any $w \in W$ and any $I, I' \in \mathcal{B}(E)$, we have $\chi(w(I), w(I')) = w(\chi(I, I'))$.*

Proof Since the group W is generated by involutions $s_i \in S(\Gamma)$, any element $w \in W$ can be written as a product $s_{i_1} s_{i_2} \cdots s_{i_l}$ of elements of $S(\Gamma)$.

The proof is by induction on l, and the case $l = 0$ (meaning $w = 1$) holds trivially. The general case is immediate from the case $l = 1$, so we may assume from now on that $w = s_p \in S(\Gamma)$.

Let I, I' be as in the statement, and let $I = I_0, I_1, \ldots, I_r = I'$ be the sequence defined in Proposition 3.2.12 (ii). The proof is by induction on r; the case $r = 0$ is trivial, and the case $r = 1$ is Lemma 3.2.19. For the inductive step, Lemma 3.2.17 shows that

$$\chi(I, I') = \sum_{i=0}^{r-1} \chi(I_i, I_{i+1}).$$

Using the additivity of the action of s_p, the case $r = 1$, and Lemma 3.2.17 again, we then obtain

$$\begin{aligned} s_p(\chi(I, I')) &= \sum_{i=0}^{r-1} s_p(\chi(I_i, I_{i+1})) \\ &= \sum_{i=0}^{r-1} \chi(s_p(I_i), s_p(I_{i+1})) \\ &= \chi(s_p(I), s_p(I')), \end{aligned}$$

as required. □

Definition 3.2.22 We define a *real root* of W to be an element of V_Γ of the form $w(\alpha_i)$ for $i \in \Gamma$ and $w \in W$. The set of real roots is denoted by Φ^{Re}. We call the elements α_i *fundamental roots*. The set of fundamental roots is denoted by Δ.

Remark 3.2.23 It is immediate from Definition 3.2.22 that the set of real roots of W is permuted by the action of W.

Exercise 3.2.24 Let $\varepsilon : E \to \Gamma$ be a full heap over a finite connected graph with Weyl group W. Show that the real roots of W are precisely those elements of V_Γ of the form $\chi(w(I), w(I'))$, where $I, I' \in \mathcal{B}$, $I \subset I'$, and $I' \backslash I$ is a singleton. Show also that this definition only depends on W, whichever full heap E is chosen.

Definition 3.2.25 Let $\varepsilon : E \to \Gamma$ be a full heap over a finite graph and let $\alpha \in V_\Gamma$. Then α is said to be *representable in* E if and only if there exist proper ideals $I, I' \in \mathcal{B}(E)$ such that $\chi(I, I') = \alpha$.

Example 3.2.26 Lemma 3.2.18 says that the fundamental roots are representable in E.

Theorem 3.2.27 *Maintain the notation of Theorem 3.2.21.*

(i) *For each* $1 \neq w \in W$, *there exists a fundamental root* α_i *for which* $w(\alpha_i) \neq \alpha_i$.
(ii) *If* $w \neq 1$, *then there exists* $I \in \mathcal{B}(E)$ *such that* $w(I) \neq I$; *in other words,* W *acts faithfully on* $\mathcal{B}(E)$.

Proof Part (i) is a standard result in Lie theory and follows, for example, from [37, lemma 3.11 (a), (3.12.1)].

To prove (ii), let $w \in W$ be a nonidentity element, and let α_i be a fundamental root satisfying the hypotheses of (i). By Lemma 3.2.18, there exist $I, I' \in \mathcal{B}(E)$ with $\chi(I, I') = \alpha_i$. By Theorem 3.2.21, we have

$$\chi(w(I), w(I')) = w(\chi(I, I')) = w(\alpha_i) \neq \alpha_i = \chi(I, I').$$

Since $\chi(w(I), w(I')) \neq \chi(I, I')$, we must have either $w(I) \neq I$, or $w(I') \neq I'$, or both. In any case, there exists an element of $\mathcal{B}(E)$ that is moved by the action of w, and (ii) follows. □

Lemma 3.2.28 *Let* $\varepsilon : E \to \Gamma$ *be a full heap over a finite graph.*

(i) *If* F *is a finite convex subheap of* E, *then there exist proper ideals* $I, I' \in \mathcal{B}(E)$ *such that* $I \subset I'$ *and* $I' \backslash I = F$.
(ii) *If* $I, I' \in \mathcal{B}(E)$ *are proper ideals such that* $I \subset I'$, *then* $I' \backslash I$ *is a finite convex subheap of* E.

Proof Part (i) is Lemma 3.2.4 (vii).

Now assume the hypotheses of (ii). The set $F = I' \backslash I$ is convex because I and I' are convex and I is an ideal of I'. Lemma 3.2.4 (vi) now shows that F is finite, which completes the proof. □

In certain situations, representability can also be characterized in terms of the content of finite convex subheaps, as follows.

Lemma 3.2.29 *Let* $\varepsilon : E \to \Gamma$ *be a full heap over a finite graph and let* $\alpha = \sum_{i \in \Gamma} \lambda_i \alpha_i \in V_\Gamma$ *be such that* $\lambda_i \geq 0$ *for all* i. *Then* α *is representable in* E *if and only if there exists a finite convex subheap* F *of* E *with* $\chi(F) = \alpha$.

Proof Suppose that α is representable in E, so there exist $I, I' \in \mathcal{B}(E)$ such that $\chi(I, I') = \alpha$. The hypotheses of positivity show that $I \subset I'$. By Lemma 3.2.28 (ii),

$F = I'\backslash I$ is a finite convex subset of E, and we have $\chi(F) = \chi(I, I') = \alpha$ by construction.

Conversely, suppose that α is representable in E by the finite convex subheap F. By Lemma 3.2.28 (i), there exist $I, I' \in \mathcal{B}(E)$ with $I \subset I'$ and $I'\backslash I = F$, and it follows from the definitions that $\chi(I, I') = \chi(F) = \alpha$. □

Theorem 3.2.30 *If $\varepsilon : E \to \Gamma$ is a full heap over a Dynkin diagram, then any real root α associated with Γ is representable in E.*

Proof If $\alpha = \alpha_p$ is a fundamental root, Lemma 3.2.18 shows that there are proper ideals $I \subset I'$ of E with $\chi(I, I') = \alpha_p$. If α is an arbitrary real root, then we can write $\alpha = w(\alpha_p)$ for some $w \in W$ and some fundamental root α_p. By Theorem 3.2.21, we have

$$\chi(w(I), w(I')) = w(\alpha_p) = \alpha.$$

The conclusion follows from the fact that $w(I)$ and $w(I')$ are proper ideals. □

3.3 Parabolic subheaps

Definition 3.3.1 Let W be a Weyl group associated with a Dynkin diagram Γ, and let $S = S(\Gamma)$ be its standard set of generating involutions. Let S' be a subset of S. We define the *parabolic subgroup*, $W_{S'}$, associated with S' to be the subgroup of W generated by S'.

For future applications, we will be interested in restricting the action of W on the set of proper ideals $\mathcal{B}(E)$ of a full heap E to a parabolic subgroup. We showed in Proposition 3.2.12 that W acts transitively on $\mathcal{B}(E)$, and the next result describes the orbits of $W_{S'}$ acting on $\mathcal{B}(E)$.

Lemma 3.3.2 *Let $\varepsilon : E \to \Gamma$ be a full heap over a finite graph, let $\mathcal{B}(E)$ be the set of proper ideals of E, and let W be the associated Weyl group with generators $S(\Gamma)$. Let S' be a subset of $S(\Gamma)$, and let $W_{S'}$ be the associated parabolic subgroup. Then two elements $I, I \in \mathcal{B}(E)$ are conjugate under the action of $W_{S'}$ (i.e., there exists $w \in W_{S'}$ such that $w(I) = I'$) if and only if $\chi(I, I')$ lies in the span of the set $\{\alpha_p : p \in S'\}$.*

Proof Suppose that $w \in W_{S'}$ satisfies $w(I) = I'$. It follows from the definition of the action of W that for any $s_p \in S(\Gamma)$ and $I \in \mathcal{B}(E)$, $\chi(I, s_p(I))$ is a scalar multiple of α_p. Since w is a product of elements of the set $\{s_p : p \in S'\}$, it follows by induction from Lemma 3.2.17 that $\chi(I, I') = \chi(I, w(I))$ is a linear combination of the elements $\{\alpha_p : p \in S'\}$.

Conversely, suppose that $\chi(I, I')$ lies in the span of the set $\{\alpha_p : p \in S'\}$. The assertions about supports in Proposition 3.2.12 show that the $w \in W$ appearing in Proposition 3.2.12 (iii) can be chosen to be in the subgroup $W_{S'}$, and this completes the proof. □

Lemma 3.3.3 *Let $\varepsilon : E \to \Gamma$ be a full heap over a finite connected graph, let $\mathcal{B}(E)$ be the set of proper ideals of E, and let W be the associated Weyl group with generators $S(\Gamma)$. Let S' be a proper subset of $S(\Gamma)$, and let $W_{S'}$ be the associated parabolic* *subgroup. Let $I \in \mathcal{B}(E)$ and let $[I]_{S'}$ be the orbit of I under the action of $W_{S'}$. Then the set $[I]_{S'}$ has the following properties:*

(i) *it is closed under finite intersections and unions;*
(ii) *it is finite;*
(iii) *it contains a unique maximal element $I^+_{S'}$ and a unique minimal element $I^-_{S'}$ (in the sense of set-theoretic containment); and*
(iv) *it consists precisely of the proper ideals*

$$\{J \in \mathcal{B}(E) : I^-_{S'} \subseteq J \subseteq I^+_{S'}\}.$$

(v) *The set*

$$F_{S'} = I^+_{S'} \backslash I^-_{S'}$$

is a finite convex subset of E. There is an inclusion-preserving bijection b from the ideals of $F_{S'}$ to the orbit $[I]_{S'}$ given by

$$b(J) = J \cup I^-_{S'}.$$

(vi) *Every minimal element of $E \backslash I^+_{S'}$ and every maximal element of $I^-_{S'}$ has a label in the set $S \backslash S'$.*
(vii) *If $S \backslash S' = \{p\}$ is a singleton, then there exists $z \in \mathbb{Z}$ such that*

$$I^-_{S'} = \langle E(p, z) \rangle$$

and

$$E \backslash I^+_{S'} = \{x \in E : x \geq E(p, z + 1)\}.$$

Proof Let p be an element of $S \backslash S'$; such a p exists by hypothesis. By Lemma 3.3.2, any two elements J and J' in $[I]_{S'}$ have the property that α_p appears with coefficient zero in $\chi(J, J')$. It follows that

$$J \cap \varepsilon^{-1}(p) = J' \cap \varepsilon^{-1}(p) = (J \cup J') \cap \varepsilon^{-1}(p) = (J \cap J') \cap \varepsilon^{-1}(p).$$

Since this is true for any $p \in S \backslash S'$, Lemma 3.3.2 shows that $J \cup J' \in [I]_{S'}$ and $J \cap J' \in [I]_{S'}$, and assertion (i) follows by induction.

Given $p \in S \backslash S'$, we can define $t_p \in \mathbb{Z}$ to be the largest integer for which $E(p, t) \in J$ for $J \in [I]_{S'}$; the reasoning above shows that this definition is independent of the choice of J. Let q be any vertex of Γ that is different from p. By Lemma 2.3.2 (ii), there exist elements $\alpha = E(q, m)$ and $\beta = E(q, m')$ of $\varepsilon^{-1}(q)$ with the properties that $\alpha > E(p, t + 1)$ and $\beta < E(p, t)$. If $J \in [I]_{S'}$, it follows that the largest integer $t_q(J)$ for which $E(p, t_q(J)) \in J$ satisfies $m' \leq t_q(J) \leq m$. Since S is finite, the set

$$\{(q, t_q(J)) : q \in S, J \in [I]_{S'}\}$$

must be finite. However, the ideal J may be reconstructed from a knowledge of the set $\{(q, t_q(J)) : q \in S\}$, so it must be the case that $[I]_{S'}$ is finite, which proves (ii).

Part (iii) is immediate from (i) and (ii): take

$$I^+_{S'} = \bigcup_{J \in [I]_{S'}} J$$

and

$$I_{S'}^- = \bigcap_{J \in [I]_{S'}} J.$$

The construction of $I_{S'}^+$ and $I_{S'}^-$ in (iii) proves that any $J \in [I]_{S'}$ must satisfy

$$I_{S'}^- \subseteq J \subseteq I_{S'}^+.$$

Conversely, if we have $I_{S'}^- \subseteq J \subseteq I_{S'}^+$ for some $J \in \mathcal{B}(E)$, it follows (by arguing as above) that for any $p \in S \backslash S'$, we have

$$t_p = t_p(I_{S'}^-) = t_p(I_{S'}^+)$$

but also

$$t_p(I_{S'}^-) \leq t_p(J) \leq t_p(I_{S'}^+).$$

The conclusion of (iv) now follows from Lemma 3.3.2.

Because the ideals $I_{S'}^-$ and $I_{S'}^+$ satisfy $I_{S'}^- \subset I_{S'}^+$, the set $F_{S'}$ in the statement of (v) must be convex; furthermore, it is finite by Lemma 3.2.4 (vi). The fact that the two ideals are nested also implies that there is an inclusion-preserving bijection between (a) the ideals of $F_{S'}$ and (b) the ideals of $I_{S'}^+$ that contain $I_{S'}^-$. The latter set agrees with $[I]_{S'}$ by part (iv), and this completes the proof of (v).

Suppose that α is a maximal element of $I_{S'}^-$, and let $p = \varepsilon(\alpha)$. Since $S_p(I_{S'}^-) \prec_p I_{S'}^-$ and $I_{S'}^-$ is minimal in its $W_{S'}$-orbit, we must have $p \notin S'$. This proves one of the two assertions of (vi); the other follows by a dual argument.

Let $E(p, z)$ be the maximal element of $\varepsilon^{-1}(p) \cap I_{S'}^-$, and let $E(p, z+1)$ be minimal element of $\varepsilon^{-1}(p) \cap (E \backslash I_{S'}^+)$. If $S \backslash S'$ is a singleton, part (vi) shows that $E(p, z)$ is the unique maximal element of $I_{S'}^-$, and that $E(p, z+1)$ is the unique minimal element of $E \backslash I_{S'}^+$. The assertions of (vii) now follow. □

Exercise 3.3.4 Fill in the details of the assertions of Lemma 3.3.3 (v) regarding the convexity of $F_{S'}$ and the inclusion-preserving bijection.

Definition 3.3.5 The convex subheaps $F_{S'}$ of $\varepsilon : E \to \Gamma$ appearing in Lemma 3.3.3 are known as *parabolic subheaps* of E with respect to S'.

Lemma 3.3.6 *Let $\varepsilon : E \to \Gamma$ be a full heap, let $S' \subset S(\Gamma)$, and let $[I]_{S'}$ be a corresponding orbit in $\mathcal{B}(E)$. If $p, q \in S'$, then $[I]_{S'}$ is a union of (p, q)-strings.*

Proof It follows from the definition of (p, q)-string that the (p, q)-string containing I is simply an orbit of the form $[I]_{\{p,q\}}$. The assertion now follows immediately. □

Remark 3.3.7 Lemma 3.3.3 (v) and the preceding result show that we may identify the (p, q)-strings of $F_{S'}$ with the (p, q)-strings of $[I]_{S'}^+$ that do not lie in $[I]_{S'}^-$. We will often make this identification without comment; for example, we may speak of the action of $W_{S'}$ on the ideals of $F_{S'}$.

Proposition 3.3.8 *Let $\varepsilon : E \to \Gamma$ be a full heap, let $S' \subset S(\Gamma)$, and let $F_{S'}$ be a parabolic subheap of E corresponding to S'. If $F_{S'} \cap \varepsilon^{-1}(p)$ is nonempty for each*

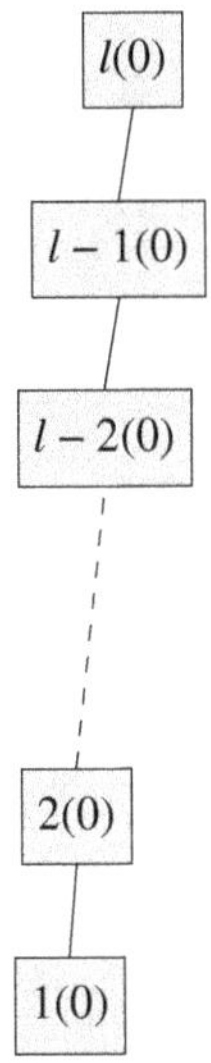

Figure 3.9 A parabolic subheap corresponding to the symmetric group S_{l+1}

$p \in S'$, then the parabolic subgroup $W_{S'}$ acts faithfully and transitively on the ideals $J(F_{S'})$ of $F_{S'}$ (under the identifications of Remark 3.3.7).

Proof The assertion about transitivity follows by construction of $F_{S'}$, so it remains to establish that the action is faithful. Let $w \in W_{S'}$. It is enough to show that there exists an ideal $J' \in J(F_{S'})$ for which $w(J') \neq J'$.

By Theorem 3.2.27 (i), there exists a fundamental root α_p with $p \in S'$ for which $w(\alpha_p) \neq \alpha_p$. By hypothesis, there exists $\gamma \in F_{S'}$ such that $\varepsilon(\gamma) = p$. Let I and I' be the ideals of $F_{S'}$ given by $I' = \langle \gamma \rangle$ and $I = I' \backslash \{\gamma\}$, respectively. Since $\chi(I, I') = \alpha_p$ and $w(\alpha_p) \neq w(\alpha_p)$, Theorem 3.2.21 shows that

$$\chi(w(I), w(I')) \neq \chi(I, I').$$

It follows that we have either $w(I) \neq I$, or $w(I') \neq I'$, or both, and this completes the proof. □

Example 3.3.9 Consider the full heap $\varepsilon : E \to \Gamma$ of Figure 3.2, where $S(\Gamma) = \{s_0, s_1, \ldots, s_l\}$. Let $S' = S \backslash \{s_0\}$. If I is the proper ideal $\langle 0(-1) \rangle$ of E, then the orbit $[I]_{S'}$ consists of the $l + 1$ proper ideals

$$I, \langle 1(0) \rangle, \langle 2(0) \rangle, \ldots, \langle l - 1(0) \rangle, \langle l(0) \rangle,$$

which we will identify with the integers $1, 2, \ldots, l + 1$, respectively. The parabolic subheap $F_{S'}$, which is shown in Figure 3.9, consists of the l elements

$$1(0), 2(0), \ldots, l - 1(0), l(0).$$

Since the support of $F_{S'}$ is the whole of S', we know that $W_{S'}$ acts faithfully on $J(F_{S'})$ by Proposition 3.3.8. Identifying ideals with integers as above, we find that the action of s_i (for $1 \leq i \leq l$) is to interchange the integers i and $i + 1$, and to fix the

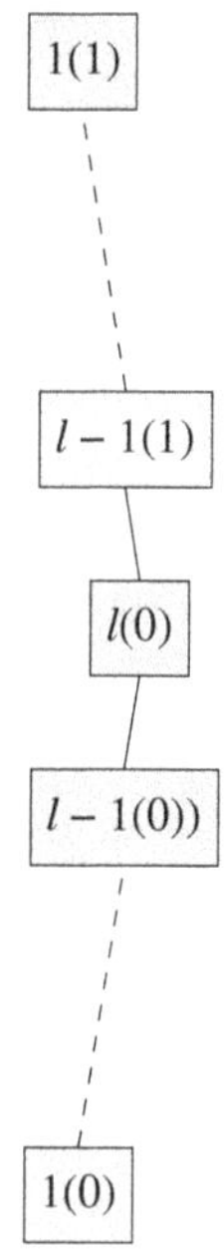

Figure 3.10 A parabolic subheap corresponding to a finite Weyl group of type C_l

other integers. It is well known that the set of transpositions

$$\{(i, i+1) : 1 \leq i \leq l\}$$

generate the symmetric group S_{l+1}. A consequence of this is the well-known fact that $W_{S'}$ is isomorphic to S_{l+1}. Notice that in this example, all the parabolic subheaps $F_{S'}$ corresponding to S' are isomorphic.

Exercise 3.3.10 Consider the full heap $\varepsilon : E \to \Gamma$ of Figure 3.3, where $S(\Gamma) = \{s_0, s_1, \ldots, s_l\}$. Let $S' = S \backslash \{s_0\}$.

(i) Show that the heap $F_{S'}$ shown in Figure 3.10 is a parabolic subheap of E corresponding to S'.

(ii) Show that $F_{S'}$ has $2l$ ideals and that they are totally ordered by inclusion. We index the ideals by integers so that

$$\emptyset = I_1 \subset I_2 \subset \cdots \subset I_l \subset I_{-l} \subset I_{-(l-1)} \subset \cdots \subset I_{-1} = F_{S'}.$$

(iii) List the proper ideals of E corresponding to the ideals of $F_{S'}$ listed in (ii).

(iv) Show that the Weyl group $W_{S'}$ acts faithfully on the symbols $\{\pm 1, \pm 2, \ldots, \pm l\}$ as follows:

  (a) each generator s_i for $1 \leq i < l$ exchanges $\pm i$ and $\pm(i+1)$ and fixes the other $2l - 4$ symbols;

  (b) the generator s_l exchanges the symbols $+l$ and $-l$, and fixes the other $2l - 2$ symbols.

(v) Are all the parabolic subheaps $F_{S'}$ associated with S' isomorphic in **Heap**(Γ)?

Exercise 3.3.11 Use the result of Exercise 3.3.10 (iv) to show that the Weyl group of type B_l (or C_l) is isomorphic to the wreath product $\mathbb{Z}_2 \wr S_l$, in other words, the semidirect product

$$\underbrace{(\mathbb{Z}_2 \times \mathbb{Z}_2 \times \cdots \times \mathbb{Z}_2)}_{l} \rtimes S_l.$$

Deduce that the Weyl group $W(B_l) \cong W(C_l)$ has order $2^l l!$.

Exercise 3.3.12 Show that the faithful action of $W_{S'}$ described in Exercise 3.3.10 may also be constructed from the full heap of Figure 3.5.

Exercise 3.3.13 Consider the full heap $\varepsilon : E \to \Gamma$ of Exercise 3.3.10, but this time let $S'' = S \backslash \{s_i\}$ for some $0 < i < l$.

(i) Show that not all the parabolic subheaps $F_{S''}$ associated with S'' are isomorphic in **Heap**(Γ).
(ii) Show that $W_{S''}$ does not act faithfully on any of the parabolic subheaps $F_{S''}$, and explain why this does not contradict Proposition 3.3.8.
(iii) Show that it is possible (for certain values of i and l) for two parabolic subheaps $F_{S''}$ and $F'_{S''}$ to be isomorphic in **Heap** but nonisomorphic in **Heap**(Γ).

Exercise 3.3.14 Consider the full heap $\varepsilon : E \to \Gamma$ of Figure 3.7, where $S(\Gamma) = \{s_0, s_1, \ldots, s_l\}$. Let $S' = S \backslash \{s_0\}$.

(i) Show that the heap $F_{S'}$ shown in Figure 3.11 is a parabolic subheap of E corresponding to S'.
(ii) Show that $F_{S'}$ has $2l$ ideals but that they are not totally ordered by inclusion.
(iii) List the proper ideals of E corresponding to the ideals of $F_{S'}$ listed in (ii).
(iv) Show (by finding a suitable indexing set for the ideals of $F_{S'}$) that the Weyl group $W_{S'}$ acts faithfully on the symbols $\{\pm 1, \pm 2, \ldots, \pm l\}$ as follows:
   (a) each generator s_i for $1 \leq i < l$ exchanges $\pm i$ and $\pm(i+1)$ and fixes the other $2l - 4$ symbols;
   (b) the generator s_l sends $\pm(l-1)$ to $\mp l$ and fixes the other $2l - 4$ symbols.
(v) Are all the parabolic subheaps $F_{S'}$ associated with S' isomorphic in **Heap**(Γ)?

Exercise 3.3.15 Use the result of Exercise 3.3.14 (iv) to show that the Weyl group of type D_l is isomorphic to the subgroup of the semidirect product

$$\underbrace{(\mathbb{Z}_2 \times \mathbb{Z}_2 \times \cdots \times \mathbb{Z}_2)}_{l} \rtimes S_l$$

consisting of those elements effecting an even number of sign changes. Deduce that the Weyl group $W(D_l)$ has order $2^{l-1} l!$.

3.4 Notes and references

1 The Hecke algebra action defined in Section 3.1 is new to the best of our knowledge.
2 The permutation representations of affine Weyl groups constructed in Example 3.1.15 and Exercise 3.1.16 (in type A), Exercise 3.1.18 (in type B), Exercise 3.1.17 (in type C)

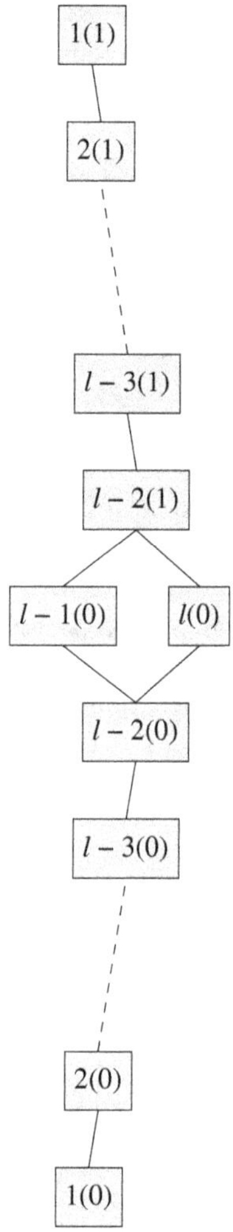

Figure 3.11 A parabolic subheap corresponding to a finite Weyl group of type D_l

and Exercise 3.1.19 (in type D) are considered by Björner and Brenti [4, p293] to be folklore. The $q = 1$ cases of these representations were constructed in terms of full heaps in Green [29, section 5]. In type affine A, the permutation representation first appeared in work of Lusztig [48], although without a proof of faithfulness, and the representation was further developed by Shi [72] and Björner and Brenti [3]. In type affine C, the permutation representation first appeared in work of Bédard [1], again without a proof of faithfulness, and was further studied by Shi [73].

3 A unified treatment (with proofs) of affine Weyl groups of types A, B, C and D as permutations of $\mathbb{Z}$ is given in H. Eriksson's thesis [24]. These explicit constructions of affine Weyl groups may easily be reformulated so as to express the affine Weyl groups in terms of semidirect products of familiar groups. Conversely, the representations of these

groups as permutations of $\mathbb{Z}$ can easily be produced from their realizations as semidirect products, although neither construction is obvious from the usual definition of affine Weyl groups via generators and relations.

4 Although the definition of parabolic subheap is new, the concept is closely related to Stembridge's notion of a minuscule heap [79], and it appears in special cases in the work of Wildberger [90].

4

Lie theory

In Chapter 4, we recall some important concepts from the theory of Lie algebras and Weyl groups; these will play an important role in subsequent chapters. Although it would be possible to develop some of these concepts in the heap framework, there are already good introductory texts dealing with this material. We will therefore omit most of the proofs, which are elementary (although not trivial).

In Section 4.1, we recall the definition of the derived algebra of an affine Kac–Moody algebra in terms of generators and relations. We then use the Local Structure Theorem (Theorem 2.3.15) to obtain a representation of the derived algebra of an affine Kac–Moody algebra from a full heap; this is proved in Theorem 4.1.6.

In Sections 4.2 and 4.3, we review some key definitions and results from the theories of Lie algebras and Weyl groups, respectively. These sections are not self-contained and do not contain proofs: for details, we provide references to the books of Kac [37] and Carter [11] for results on Lie algebras, and to Humphreys' book [36] for results on Weyl groups.

Finally, Section 4.4 introduces the notion of a strongly orthogonal set. We will call a set S of mutually orthogonal positive roots "strongly orthogonal" if each root $s \in S$ is orthogonal to every other positive root that is orthogonal to all members of $S\backslash\{s\}$; see Definition 4.4.1 for the full definition. The strongly orthogonal pairs in type D_n (with $n > 4$) and the strongly orthogonal triples in type E_7 turn out to have interesting combinatorial properties that will play an important role in later applications, especially in the study of Steiner complexes in Section 9.5.

4.1 Representations of Lie algebras from heaps

Definition 4.1.1 In Definition 3.1.11, we defined RS^*, the free R-algebra generated by S. If $R = k$ is a field, then we can make kS^* into a Lie algebra using the Lie bracket of Example 1.1.1. We define $\mathfrak{FL}\langle S\rangle$, the *free Lie algebra on* S, to be the smallest Lie subalgebra of kS^* that contains the elements of S. If I is a collection of elements in $\mathfrak{FL}\langle S\rangle$ and $\langle I\rangle$ is the ideal of $\mathfrak{FL}\langle S\rangle$ generated by I, then we refer to the Lie algebra $\mathfrak{g} = \mathfrak{FL}\langle S\rangle/\langle I\rangle$ as *the Lie algebra generated by* S *subject to the defining relations* I. We will often represent an element $x - y \in I$ as an equation $x - y = 0$ or $x = y$.

Example 4.1.2 A typical element of $\mathfrak{F}\mathfrak{L}\langle S\rangle$ is a k-linear combination of elements such as

$$[[s_1, [s_2, [s_3, s_2]]], [[s_4, s_3], s_1]],$$

where the s_i are elements of S. The Lie bracket is typically nonassociative, so the brackets in the expression cannot be suppressed.

Definition 4.1.3 Let A be a symmetrizable generalized Cartan matrix and Γ its associated Dynkin diagram. The *derived Kac–Moody algebra* $\mathfrak{g} = \mathfrak{g}'(A)$ corresponding to A is the Lie algebra over $\mathbb{C}$ generated by the elements $\{e_i, f_i, h_i : i \in \Gamma\}$ subject to the defining relations

$$[h_i, h_j] = 0, \tag{4.1}$$
$$[h_i, e_j] = a_{ij}e_j, \tag{4.2}$$
$$[h_i, f_j] = -a_{ij}f_j, \tag{4.3}$$
$$[e_i, f_j] = \delta_{ij}h_i, \tag{4.4}$$
$$\underbrace{[e_i, [e_i, \cdots [e_i}_{1-a_{ij}\text{ times}}, e_j]\cdots]] = 0, \tag{4.5}$$
$$\underbrace{[f_i, [f_i, \cdots [f_i}_{1-a_{ij}\text{ times}}, f_j]\cdots]] = 0, \tag{4.6}$$

where δ is the Kronecker delta, meaning that $\delta_{ij} = 1$ if $i = j$, and 0 otherwise.

As its name suggests, the algebra $\mathfrak{g}'(A)$ is the derived algebra of a larger Lie algebra $\mathfrak{g}(A)$, and this larger algebra is the Kac–Moody algebra corresponding to A.

Recall from Section 3.1 that we constructed representations of Weyl groups and their Hecke algebras in terms of the linear operators S_i acting on a vector space associated with a full heap. We will now describe the analogous construction for Lie algebras.

Recall from Exercise 3.2.5 that the linear operators of Definition 3.1.3 may be restricted to the space $V_{\mathcal{B}(E)}$ associated with a full heap E.

Lemma 4.1.4 *Let $\varepsilon : E \to \Gamma$ be a full heap over a Dynkin diagram. For each $p \in \Gamma$, let H_p, X_p and Y_p be the linear operators on $V_{\mathcal{B}}$ defined in Definition 3.1.4. The operators satisfy the following relations, where $p, q \in \Gamma$ and $\circ$ denotes composition of maps:*

$$H_p \circ X_q = X_q \circ H_p + a_{p,q}X_q, \tag{4.7}$$
$$H_p \circ X_p = X_p = -X_p \circ H_p, \tag{4.8}$$
$$H_p \circ Y_q = Y_q \circ H_p - a_{p,q}Y_q, \tag{4.9}$$
$$H_p \circ Y_p = -Y_p = -Y_p \circ H_p, \tag{4.10}$$
$$X_p \circ Y_q = Y_q \circ X_p = 0 \textit{ if } a_{p,q} < 0, \tag{4.11}$$
$$X_p \circ Y_q = Y_q \circ X_p \textit{ if } a_{p,q} = 0, \tag{4.12}$$
$$X_p \circ X_p = 0, \tag{4.13}$$

$$X_p \circ X_q = X_q \circ X_p \text{ if } a_{p,q} = 0, \tag{4.14}$$
$$X_p \circ X_q \circ X_p = 0 \text{ if } a_{p,q} = -1, \tag{4.15}$$
$$Y_p \circ Y_p = 0, \tag{4.16}$$
$$Y_p \circ Y_q = Y_q \circ Y_p \text{ if } a_{p,q} = 0, \tag{4.17}$$
$$Y_p \circ Y_q \circ Y_p = 0 \text{ if } a_{p,q} = -1. \tag{4.18}$$

Proof Observe first that if v_I is a basis element of $\mathcal{B}(E)$, then $H_p(v_I)$ is a scalar multiple of v_I.

Let us first consider Equation (4.7), and consider the action of each of the operators mentioned on a basis vector v_I. If $X_q(v_I) = 0$ then the observation of the previous paragraph shows that both sides of the equation act as zero. If $X_q(v_I) \neq 0$ then it is enough to restrict our attention to the subspace V_c spanned by the basis vectors $v_{I'}$ for which $I \sim_{pq} I'$ (in the sense of Definition 2.3.4). The assertions are now consequences of the Local Structure Theorem, as follows.

If $a_{pq} = -1$, then part (i) of the theorem applies. If we are in subcase (a) of Theorem 2.3.15 with $I_1 \prec_p I_2 = I \prec_q I_3$, then we have $X_q(v_I) = v_{I_3}$, $H_p \circ X_q(v_I) = 0$, $H_p(v_I) = v_I$ and $X_q \circ H_p(v_I) = v_{I_3}$, completing the proof in this case. The consideration of subcase (b) is similar, except that we have $I = I_1 \prec_q I_2 \prec_p I_3$, $H_p \circ X_q(v_I) = -v_{I_2}$, $H_p(v_I) = 0$ and $X_q \circ H_p(v_I) = 0$.

If $a_{pq} = -2$, then part (ii) or (iv) of the theorem applies, and shows that we have $I_1 \prec_p I_2 = I \prec_q I_3 \prec_p I_4$. In this case, we have $H_p \circ X_q(v_I) = -v_{I_3}$ and $X_q \circ H_p(v_I) = v_{I_3}$, from which the equation follows.

If $a_{pq} = 0$ then part (iii) of the theorem applies. As before, the only nontrivial case is the case $X_q(v_I) = v_{I'} \neq 0$, which excludes subcase (b). In subcase (c), the equation follows from the fact that $H_p(v_I) = H_p(v_{I'}) = 0$, and in subcase (a) the equation follows from the fact that $H_p(v_I) = cv_I$ and $H_p(v_{I'}) = cv_{I'}$ for the same constant $c \in \{\pm 1\}$. This completes the proof of Equation (4.7).

Equation (4.8) follows by a similar but easier argument. As before, the only nontrivial case to consider is the case where $X_p(v_I) \neq 0$, in which case $H_p \circ X_p(v_I) = X_p(v_I)$, $H_p(v_I) = -v_I$ and $X_p \circ H_p(v_I) = -X_p(v_I)$.

Equation (4.9) (respectively, (4.10)) is proved by an argument similar to that used to prove Equation (4.7) (respectively, (4.8)).

For Equation (4.11), we note that for $X_p \circ Y_q(v_I) \neq 0$, we would need $I' \in \mathcal{B}(E)$ for which both $I' \prec_q I$ and $I'' \succ_p I'$ for some $I'' \in \mathcal{B}(E)$. This contradicts Lemma 2.3.7, and the equation follows.

Equation (4.12) is proved by considering the various subcases of part (iii) of the Local Structure Theorem, and similar lines of argument establish Equations (4.14) and (4.17).

Equation (4.13) holds because we cannot have $I_1 \prec_p I_2 \prec_p I_3$ for any $I_1, I_2, I_3 \in \mathcal{B}(E)$, and the same argument establishes Equation (4.16).

For Equation (4.15), we can reduce as in previous arguments to the case where $X_p \circ X_q \circ X_p(v_I) \neq 0$, and this can only happen if we have four ideals satisfying $I_1 \prec_p I_2 \prec_q I_3 \prec_p I_4$. By the Local Structure Theorem, this can only happen if $a_{pq} = -2$; this is a contradiction, which establishes the equation. The same argument can be used to establish Equation (4.18). □

Definition 4.1.5 Let $\varepsilon : E \to \Gamma$ be a full heap over a Dynkin diagram and let Γ' be a subgraph of Γ. Let $\mathcal{O}_{\Gamma'}$ be the set of $W_{\Gamma'}$-orbits of $\mathcal{B}(E)$ under the action of the Weyl group. For each orbit $[I] \in \mathcal{O}_{\Gamma'}$, let $V_{\Gamma',[I]}$ be the subspace of $V_{\mathcal{B}(E)}$ spanned by the set $\{v_J : J \in [I]\}$.

Theorem 4.1.6 *Let $\varepsilon : E \to \Gamma$ be a full heap over a Dynkin diagram, and let $\mathfrak{g}'(A)$ be the associated derived Kac–Moody algebra.*

(i) *The space $V_{\mathcal{B}(E)}$ has the structure of a $\mathfrak{g}'(A)$-module in which the action of the generator e_i (respectively, f_i, h_i) is given by the linear operator X_i (respectively, Y_i, H_i).*
(ii) *Let Γ' be a proper subgraph of Γ and let A' (respectively, $\mathfrak{g}'(A')$) be the generalized Cartan matrix (respectively, derived Kac–Moody algebra) associated with Γ'. Then the spaces $V_{\Gamma',[I]}$ of Definition 4.1.5 become modules for $\mathfrak{g}'(A')$ by restricting the action of $\mathfrak{g}'(A)$ defined in (i).*

Proof To prove (i), we need to check that the defining relations in Definition 4.1.3 are satisfied.

For the identity (4.1), we need to show that $H_i \circ H_j = H_j \circ H_i$ for all $i, j \in \Gamma$. This holds because the operators H_i are simultaneously diagonalizable with respect to the basis $\{v_I : I \in \mathcal{B}(E)\}$ and, as such, they commute with each other.

To prove the identity (4.2), we need to show that $H_i \circ X_j - X_j \circ H_i = a_{ij} X_j$, which follows from (4.7) and (4.8). Similarly, identity (4.3) (respectively, (4.4)) follows from (4.9) and (4.10) (respectively, (4.11) and (4.12)).

If $a_{ij} = 0$, establishing (4.5) is equivalent to showing that

$$X_i \circ X_j = X_j \circ X_i,$$

which follows from (4.14). If $a_{ij} = -1$, we need to show that

$$X_i \circ X_i \circ X_j + X_i \circ X_j \circ X_i + X_j \circ X_i \circ X_i = 0.$$

In this case, the first and third terms are zero by (4.13), and the second term is zero by (4.15). The only other possibility is that $a_{ij} = -2$, in which case we need to show that

$$X_i \circ X_i \circ X_i \circ X_j + X_i \circ X_i \circ X_j \circ X_i + X_i \circ X_j \circ X_i \circ X_i + X_j \circ X_i \circ X_i \circ X_i = 0,$$

in which case all four terms on the left are zero by (4.13).

The consideration of (4.6) uses a similar argument, this time using (4.16), (4.17) and (4.18). This completes the proof of (i).

The assertions of (ii) are immediate from Definition 4.1.3 and Lemma 3.3.6, which shows that the orbit $[I]$ is a union of (p, q)-strings. □

Remark 4.1.7 It is often easier to think of the module $V_{\Gamma',[I]}$ of Theorem 4.1.6 (ii) in terms of parabolic subheaps via the bijection described in Lemma 3.3.3 (v). More precisely, let S' be the set of vertices of Γ', let $[I]^+_{S'}$ and $[I]^-_{S'}$ be the maximal and minimal elements of the orbit $[I]$ as in Lemma 3.3.3 (iii), and let $F_{S'} = [I]^+_{S'} \backslash [I]^-_{S'}$ be the corresponding parabolic subheap. The inclusion-preserving bijection means that we may regard the basis $\{v_J : J \in [I]\}$ as being indexed by the set of all ideals of $F_{S'}$. For this reason, it is often convenient to denote the module $V_{\Gamma',[I]}$ by $V_{F_{S'}}$.

It turns out that we can use the ideals of the finite heap $F_{S'}$ as the starting point for describing the representations, as in the following examples.

Example 4.1.8 (Compare with Example 1.3.3.) Let E be the dual of the full heap of Figure 3.2, where Γ has type $A_n^{(1)}$ and $S(\Gamma) = \{s_0, s_1, \ldots, s_l\}$. Let $S' = S\backslash\{s_0\}$, and let $F_{S'}$ be the corresponding parabolic subheap, shown in Figure 1.1; notice that this is the dual of the parabolic subheap of Figure 3.9. Recall from Section 1.3 that $F_{S'}$ has l elements. Because $F_{S'}$ is totally ordered, the set of ideals $J(F_{S'})$ is also totally ordered and has $l + 1$ elements:

$$F_{S'} = v_1 \supset v_2 \supset \cdots \supset v_{l+1} = \emptyset.$$

We can then regard the basis for the module $V_{\Gamma',[I]}$ as the set $\{v_i : 1 \leq i \leq l + 1\}$. The endomorphisms of the module can then be regarded as $(l + 1) \times (l + 1)$ matrices.

With this notation, we have

$$e_i(v_j) = \begin{cases} v_{j-1} & \text{if } j = i + 1, \\ 0 & \text{otherwise.} \end{cases}$$

This information may be expressed more concisely by identifying e_i with the matrix unit $E_{i,i+1}$, that is, the matrix with entry 1 in the $(i, i + 1)$-position, and zeros elsewhere. Applying similar arguments, we find that the generator f_i corresponds to the matrix unit $E_{i+1,i}$, and the generator h_i corresponds to $E_{i,i} - E_{i+1,i+1}$.

Notice that all these matrices have trace zero, and thus (if we work over the field $k = \mathbb{C}$) lie in the Lie algebra $\mathfrak{sl}_{l+1}(\mathbb{C})$. It turns out to be true (although this is not obvious) that the operators e_i, f_i and h_i generate the whole of the Lie algebra $\mathfrak{sl}_{l+1}(\mathbb{C})$. The representation just constructed from the parabolic subheap is known as the *natural representation* (or *defining representation*) of $\mathfrak{sl}_{l+1}(\mathbb{C})$.

Exercise 4.1.9 (Compare with Example 1.3.4.) Consider the full heap $\varepsilon : E \to \Gamma$ of Figure 3.3, where Γ has type $C_n^{(1)}$ and $S(\Gamma) = \{s_0, s_1, \ldots, s_l\}$. Let $S' = S\backslash\{s_0\}$, and let $F_{S'}$ be the corresponding parabolic subheap, shown in Figure 3.10. Recall from Example 3.3.10 that $F_{S'}$ has $2l - 1$ elements. Because $F_{S'}$ is totally ordered, the set of ideals $J(F_{S'})$ is also totally ordered and has $2l$ elements:

$$\emptyset = I_1 \subset I_2 \subset \cdots \subset I_l \subset I_{-l} \subset I_{-(l-1)} \subset \cdots \subset I_{-1} = F_{S'}.$$

Label these ideals respectively by the symbols

$$l, l - 1, \ldots, 1, \overline{1}, \overline{2}, \ldots, \overline{l},$$

and regard the basis for the module $V_{\Gamma',[I]}$ as the set $\{v_1, v_2, \ldots, v_l, v_{\overline{l}}, v_{\overline{l-1}}, \ldots, v_{\overline{1}}\}$. The endomorphisms of the module can then be regarded as $2l \times 2l$ matrices.

(i) Show that, for $1 \leq i < l$, the actions of the generators e_i, f_i and h_i on the module are given by the matrices

$$E_{i,i+1} + E_{\overline{i+1},\overline{i}} \quad E_{i+1,i} + E_{\overline{i},\overline{i+1}} \quad \text{and} \quad E_{i,i} - E_{i+1,i+1} - E_{\overline{i},\overline{i}} + E_{\overline{i+1},\overline{i+1}},$$

respectively.

(ii) Show that, the actions of the generators e_l, f_l and h_l on the module are given by the matrices

$$E_{l,\bar{l}}, \quad E_{\bar{l},l} \quad \text{and} \quad E_{l,l} - E_{\bar{l},\bar{l}},$$

respectively.

Exercise 4.1.10 (Compare with Example 1.3.5.) Consider the full heap $\varepsilon : E \to \Gamma$ of Figure 3.7, where Γ has type $D_n^{(1)}$ and $S(\Gamma) = \{s_0, s_1, \ldots, s_l\}$. Let $S' = S \backslash \{s_0\}$, and let $F_{S'}$ be the corresponding parabolic subheap, shown in Figure 3.11. Recall from Example 3.3.14 that $F_{S'}$ has $2l - 1$ elements. The set of ideals $J(F_{S'})$ has $2l$ elements, but neither it nor $F_{S'}$ is a totally ordered set.

(i) Show that the heap $F_{S'}$ is self-dual, and that any ideal $I \in J(F_{S'})$ is either one of the ideals in the sequence

$$J' = \emptyset \subset \langle 1(0) \rangle \subset \langle 2(0) \rangle \subset \cdots \subset \langle l - 1(0) \rangle,$$

or is of the form $(F_{S'} \backslash I)^*$ for a unique $I \in J'$ (we will call this the ideal dual to I).

Now label the elements of J' in order by the symbols $\bar{1}, \bar{2}, \ldots, \bar{l}$, where $\bar{1}$ corresponds to the empty set, and write i for the dual of the ideal labelled by $\bar{i}$. We now regard the basis for the module $V_{\Gamma',[l]}$ as the set $\{v_1, v_2, \ldots, v_l, v_{\bar{l}}, v_{\overline{l-1}}, \ldots, v_{\bar{1}}\}$, and we can identify the endomorphisms of the module with $2l \times 2l$ matrices.

(ii) Show that, for $1 \leq i < l$, the actions of the generators e_i, f_i and h_i on the module are given by the matrices

$$E_{i,i+1} + E_{\overline{i+1},\bar{i}}, \quad E_{i+1,i} + E_{\bar{i},\overline{i+1}} \quad \text{and} \quad E_{i,i} - E_{i+1,i+1} - E_{\bar{i},\bar{i}} + E_{\overline{i+1},\overline{i+1}},$$

respectively.

(iii) Show that, the actions of the generators e_l, f_l and h_l on the module are given by the matrices

$$E_{l-1,\bar{l}} + E_{l,\overline{l-1}}, \quad E_{\bar{l},l-1} + E_{\overline{l-1},l} \quad \text{and} \quad E_{l-1,l-1} + E_{l,l} - E_{\overline{l-1},\overline{l-1}} - E_{\bar{l},\bar{l}},$$

respectively.

4.2 Review of Lie theory

Theorem 3.2.21 shows that there is a close relationship between full heaps and the roots of a Weyl group. In this section, we will summarize some of the basic properties of root systems of Weyl groups; these will turn out to have interesting applications in the context of full heaps.

Definition 4.2.1 Let A be a $n \times n$ generalized Cartan matrix, and let $u = (u_1, u_2, \ldots, u_n)^\top$ be a column vector in $\mathbb{R}^n$. We say $u \geq 0$ if $u_i \geq 0$ for each i, and we say $u > 0$ if $u_i > 0$ for each i. (If we identify V_Γ with $\mathbb{R}^n$, it should be understood that V_Γ is equipped with the basis $\Delta = \{\alpha_1, \ldots, \alpha_n\}$ of *fundamental roots*.)

We say A has *finite type* if the following three conditions hold:

(i) $\det A \neq 0$;
(ii) there exists $u > 0$ such that $Au > 0$;
(iii) if $Au \geq 0$ then either $u > 0$ or $u = 0$.

We say A has *affine type* if the following three conditions hold:

(i) A has corank 1 (i.e., the rank of A is $n - 1$);
(ii) there exists $u > 0$ such that $Au = 0$;
(iii) if $Au \geq 0$ then $Au = 0$.

We say A has *indefinite type* if the following two conditions hold:

(i) there exists $u > 0$ such that $Au < 0$;
(ii) if $Au \geq 0$ and $u \geq 0$ then $u = 0$.

Theorem 4.2.2 *Let A be an indecomposable generalized Cartan matrix. Then exactly one of the following three possibilities holds:*

(i) *A has finite type;*
(ii) *A has affine type;*
(iii) *A has indefinite type.*

Proof This result is stated in Carter [11, theorem 15.1] and proved in [11, section 15]. □

A list of Dynkin diagrams of finite and affine type may be found in Appendix B.

Theorem 4.2.3 *Let W be the Weyl group associated with the indecomposable generalized Cartan matrix A, and let V_Γ be the associated real vector space as in Definition 3.2.8.*

(i) *The formula*

$$s_q(\alpha_p) = \alpha_p - A_{qp}\alpha_q$$

extends linearly to an action of W on V_Γ. An element of V_Γ of the form $w(\alpha_i)$ for $i \in \Gamma$ and $w \in W$ is called a real root of W.

(ii) *There is a nonzero symmetric bilinear form $(,)$ on V_Γ for which*

$$A_{ij} = 2\frac{(\alpha_i, \alpha_j)}{(\alpha_i, \alpha_i)}$$

for all $1 \leq i, j \leq n$. Furthermore, the form $(,)$ is W-invariant, meaning that

$$(wx, wy) = (x, y)$$

for all $x, y \in V_\Gamma$ and $w \in W$.

(iii) *For all $1 \leq i, j \leq n$, we have*

$$\frac{(\alpha_j, \alpha_j)}{(\alpha_i, \alpha_i)} = \frac{A_{ij}}{A_{ji}}.$$

If A is simply laced, then all real roots have the same length.

(iv) *If A is of finite type, then W is finite, every root is real, and the radical of the form* (,), *i.e., the set*

$$\{v \in \mathbb{R}^n : (x, v) = 0 \text{ for all } x \in \mathbb{R}^n\},$$

is zero. In this case, the form (,) *is positive definite and can be identified with (a nonzero multiple of) the usual scalar product on $\mathbb{R}^n$. Under these identifications, the action of s_q is reflection in the hyperplane orthogonal to α_q.*

(v) *If A is of affine type, then the radical of the form* (,) *is one-dimensional and contains a unique vector $\delta > 0$ whose entries are pairwise coprime positive integers. In particular, we have $s_i(\delta) = \delta$ for all $i \in \Gamma$.*

(vi) *If α is a real root, then $k\alpha$ is also a root if and only if $k = \pm 1$.*

Proof Part (i) follows from Kac [37, section 3.7], using the identifications of [37, (2.3.5)]. The W-invariance of (,) is proved in [37, proposition 3.9], and the formula for A_{ij} appears in [37, section 2.3]. This proves (ii). Part (iii) follows from (ii) and the fact that if A is simply laced, then A is symmetric.

The fact that W is finite in (iv) follows from [37, proposition 4.9], and the fact that every root is real follows from [37, theorem 5.6 (a)]. Let D be the diagonal matrix for which

$$D_{ii} = \frac{(\alpha_i, \alpha_i)}{2},$$

so that DA is the symmetric matrix $B = ((\alpha_i, \alpha_j))$. Now $(x, y) = x^\top B y$, and the radical of (,) is the null space of B, which is the same as the null space of A. If A is of finite type, this null space is trivial since $\det A \neq 0$ by definition, which proves the second assertion of (iv). By Humphreys [36, theorem 6.4], there is (up to nonzero scalar multiples) a unique nondegenerate W-invariant symmetric bilinear form on V_Γ, and it is positive definite, so that it can be identified with the usual scalar product. Using (ii), the formula in (i) then becomes the usual formula for the reflection in Euclidean space orthogonal to α_q, and this completes the proof of (iv).

If A is of affine type, it has corank 1 and its null space has dimension 1. The conclusion of (v) now follows from [37, theorem 4.8 (c)].

Part (vi) is proved in [37, proposition 5.1(b)]. □

Corollary 4.2.4 *Any two forms $(,)_1$ and $(,)_2$ satisfying the hypotheses of Theorem 4.2.3 with respect to the same matrix A are nonzero scalar multiples of each other.*

Proof This follows from Theorem 4.2.3 (iii) and the fact that A is indecomposable (so that Γ is connected). □

Definition 4.2.5 The set of roots associated with a Weyl group W is called the *root system* of W and denoted by Φ. It follows from Theorem 4.2.3 (iv) that Φ consists of the real roots of W if the associated generalized Cartan matrix has finite type.

If A is an $n \times n$ generalized Cartan matrix of finite type, the associated Lie algebra $\mathfrak{g}$ has a *Cartan decomposition* (or *root decomposition*) of the form

$$\mathfrak{g} = \mathfrak{h} \oplus \bigoplus_{\alpha \in \Phi} \mathfrak{g}_\alpha.$$

Here, $\mathfrak{h}$ is a certain n-dimensional abelian subalgebra of $\mathfrak{g}$, called the *Cartan subalgebra*. For each positive or negative root $\alpha \in \Phi$, $\mathfrak{g}_\alpha$ is a one-dimensional $\mathfrak{h}$-submodule of $\mathfrak{g}$. An immediate consequence of this is the following.

Proposition 4.2.6 *Let A be an $n \times n$ generalized Cartan matrix of finite type, let $\mathfrak{g}$ and Φ be the corresponding Lie algebra and root system, respectively. Then we have*

$$\dim \mathfrak{g} = n + |\Phi|. \qquad \square$$

Definition 4.2.7 If α is a real root of a Weyl group, then we define

$$\alpha^\vee = \frac{2\alpha}{(\alpha, \alpha)}.$$

The the element $\alpha^\vee \in V_\Gamma$ is called the *coroot* corresponding to α. We say that $\alpha^\vee$ is a *fundamental coroot* if α is a fundamental root.

Exercise 4.2.8 If α and $\alpha^\vee$ are as in Definition 4.2.7, show that we have

$$\alpha = \frac{2\alpha^\vee}{(\alpha^\vee, \alpha^\vee)}.$$

Exercise 4.2.9 Show that if A is a generalized Cartan matrix, then we have

$$A_{ij} = (\alpha_i^\vee, \alpha_j).$$

Proposition 4.2.10 *Let A be a generalized Cartan matrix of finite type; let Γ be the associated Dynkin diagram, W the associated Weyl group and Φ the associated root system.*

(i) *The transpose, A^{T} of A is also a generalized Cartan matrix of finite type, and has the same Weyl group of A.*
(ii) *The Dynkin diagram of A^{T} is obtained by reversing the directions of any arrows in Γ.*
(iii) *The root system, $\Phi^\vee$, of A^{T} is obtained by replacing each root $\alpha \in \Phi$ by its corresponding coroot $\alpha^\vee$.*

Proof See Carter [11, section 8.8]. $\square$

Definition 4.2.11 The root systems Φ and $\Phi^\vee$ in Proposition 4.2.10 are known as *dual root systems*. (By Exercise 4.2.8, the dual of a dual root system is the original root system.)

Remark 4.2.12 The root systems of types A_l, D_l, E_6, E_7, E_8, F_4 and G_2 are each self-dual, and the root systems B_l and C_l are dual to each other.

Proposition 4.2.13 *Let A be a generalized Cartan matrix with associated Weyl group W; let Φ^{Re} be the set of real roots of W.*

(i) *Let $\alpha = \sum_{i=1}^{n} \lambda_i \alpha_i$ be a real root of W. Then $\alpha \neq 0$, and the coefficients λ_i are either all nonnegative or all nonpositive.*
(ii) *If A has finite type, then the set*

$$L = \{(\alpha, \alpha) : \alpha \in \Phi^{\text{Re}}\}$$

has cardinality at most 2.

Proof For part (i), see [11, section 16.3].

If A has finite type, then any root α is real by Theorem 4.2.3 (iv). Since $(,)$ is W-invariant (Theorem 4.2.3 (ii)), it is enough to prove the assertion when α is a fundamental root, and this follows using a case by case check, as in [11, proposition 17.9]. □

Definition 4.2.14 If A is of affine type with δ as in Theorem 4.2.3 (v), the elements of the set

$$\Phi^{\text{Im}} := \{k\delta : k \in \mathbb{Z}, k \neq 0\}$$

are called *imaginary roots* of A [11, theorem 16.27 (ii)]. It follows from Theorem 4.2.3 (iv) that Φ consists of the real roots of W if the associated generalized Cartan matrix has finite type. The root system of an affine Weyl group is then given by

$$\Phi = \Phi^{\text{Re}} \dot{\cup} \Phi^{\text{Im}}.$$

If A is of finite type, there are no imaginary roots. If A is of indefinite type, there is also a notion of imaginary roots, but we do not require it here; a characterization of such roots in terms of the Weyl group may be found in [11, theorem 16.24].

Let $\alpha = \sum_{i=1}^{n} \lambda_i \alpha_i$ be a root of W (real or imaginary). We call a root α of W *positive* (respectively, *negative*) if $\lambda_i \geq 0$ (respectively, $\lambda_i \leq 0$) for all i. Proposition 4.2.13 (i) shows that every real root is either positive or negative, but not both. It is a general fact [11, section 16.3] that if α is any root (real or imaginary), then $-\alpha$ is also a root. In particular, for affine types, the positive imaginary roots are those of form $k\delta$ for strictly positive integers k, and the negative imaginary roots are those of the form $k\delta$ for strictly negative integers k. We define the *height* of α to be the number $\sum_{i=1}^{n} \lambda_i$. The height of the lowest positive imaginary root δ is called the *Coxeter number* and is denoted by h. The values of h for the various Weyl groups are given in Appendix B.

We call the quantity (α, α) appearing in Proposition 4.2.13 (ii) the *length* of α. If A has finite type and the set L in Proposition 4.2.13 (ii) has cardinality 2, write $L = \{m, M\}$ where $m < M$. We call a root α *short* (respectively, *long*) if it has length m (respectively, M).

Two elements $x, y \in V_\Gamma$ are said to be *orthogonal* if we have $(x, y) = (y, x) = 0$.

Remark 4.2.15 The notation for Dynkin diagrams has the property that a single arrow on an edge points from a longer fundamental root to a shorter fundamental root.

4.3 Review of Weyl groups

Let A be a generalized Cartan matrix of finite type and let W be the associated Weyl group, acting on V_Γ by Euclidean reflections

$$s_\alpha(v) = v - 2\frac{v.\alpha}{\alpha.\alpha}\alpha = v - (v, \alpha^\vee)\alpha,$$

as in Theorem 4.2.3. These generating reflections s_q are known as *fundamental reflections*.

Every element of a group may be written as a word in its generators and their inverses. Since the generators s_q have order 2, every $w \in W$ may be written in the form

$$w = s_{p_1} s_{p_2} \cdots s_{p_k},$$

where the p_i are (not necessarily distinct) generators. The minimal integer k for a given w is known as the *length* of w, and is denoted by $\ell(w)$. If we have $k = \ell(w)$, then the word $s_{p_1} \cdots s_{p_k}$ is called a *reduced expression* for w.

If $s_{p_1} \cdots s_{p_k}$ is any word in S^*, then we define a *subexpression* of the word to be a word of the form

$$s_{p_{i_1}} s_{p_{i_2}} \cdots s_{p_{i_l}},$$

where $1 \leq i_1 < i_2 < \cdots < i_l \leq k$. If the integers $i_1, i_2, \ldots, i_l$ are consecutive, then we call the subexpression a *subword*.

Recall that if $a_{ij} = a_{ji} = 0$, we may use the defining relations of W to replace a subword $s_i s_j$ in a reduced expression by $s_j s_i$. If $a_{ij} a_{ji} = 1$, then we may use the defining relations of W to replace a (consecutive) subword $s_i s_j s_i$ in a reduced expression by $s_j s_i s_j$. Similarly, if $a_{ij} a_{ji} = 2$, then we may use the defining relations of W to replace a subword $s_i s_j s_i s_j$ in a reduced expression by $s_j s_i s_j s_i$. We call these three types of transformation *braid relations*; the first type of relation listed is called a *short* braid relation, and the other two are known as *long* braid relations. It is immediate from the definitions that the result of applying a braid relation to a reduced expression for w results in another reduced expression for w.

Theorem 4.3.1 (Matsumoto's Theorem) *If $\mathbf{u}_1$ and $\mathbf{u}_2$ are two reduced expressions for the same element w of an arbitrary Weyl group, then $\mathbf{u}_1$ may be transformed into $\mathbf{u}_2$ by a finite sequence of braid relations.*

Proof See Geck and Pfeiffer [26, theorem 1.2.2]. □

Remark 4.3.2 In the set-up of Matsumoto's Theorem, it is obvious that one may transform $\mathbf{u}_1$ to $\mathbf{u}_2$ by applying a sequence of defining relations. The surprising part of the theorem is that it is never necessary to use the transformation $1 \to s_i^2$ in order to convert $\mathbf{u}_1$ into $\mathbf{u}_2$.

Definition 4.3.3 Let W be a Weyl group and let $\mathbf{u} = s_{i_1} \cdots s_{i_l}$ be a reduced expression for $w \in W$. If $u \in W$ has the property that some subexpression of $\mathbf{u}$ is equal (as an element of W) to u, then we define $u \leq w$. The resulting relation $\leq$ is a partial order on W, known as the *Bruhat order*.

Remark 4.3.4 It is not obvious from Definition 4.3.3 that the relation $\leq$ is well-defined, since it appears to depend on the reduced expression chosen. The soundness of the definition follows from Humphreys [36, theorem 5.10].

Definition 4.3.5 Let W be a Weyl group, and let $u, w \in W$. We say that $u \leq_{\mathrm{L}} w$ if $\ell(u) + \ell(u^{-1}w) = \ell(w)$. The resulting relation $\leq_{\mathrm{L}}$ is a partial order on W, known as the *(left) weak Bruhat order*.

The verification that $\leq_{\mathrm{L}}$ is a partial order is left as Exercise 4.3.6.

Exercise 4.3.6 Let W be a Weyl group and let $u, w \in W$. Let us say that $u \leq_{\mathrm{R}} w$ if $\ell(wu^{-1}) + \ell(u) = \ell(w)$.

(i) Show that $\leq_{\mathrm{L}}$ is a partial order on W.
(ii) Show that $u \leq_{\mathrm{R}} w$ is a partial order. (The relation $\leq_{\mathrm{R}}$ is called the *(right) weak Bruhat order*.)
(iii) Show that if either $u \leq_{\mathrm{L}} w$ or $u \leq_{\mathrm{R}} w$, then $u \leq w$ in the Bruhat order.
(iv) Give an example of a u and a w for which $u \leq_{\mathrm{L}} w$ but $u \not\leq_{\mathrm{R}} w$.

Exercise 4.3.7 Suppose that W has type A_3 (with the numbering of generators in Appendix B), and that $u = s_2 s_1 s_2$ and $w = s_2 s_1 s_3 s_2$. (It can be shown that these two expressions are both reduced.)

(i) There are 14 elements $x \in W$ such that $x \leq w$; find them.
(ii) Find the six elements $x \in W$ such that $x \leq_{\mathrm{L}} w$ and six elements $x' \in W$ such that $x' \leq_{\mathrm{R}} w$.
(iii) Show that $u \leq w$, but that neither $u \leq_{\mathrm{L}} w$ nor $u \leq_{\mathrm{R}} w$.

Let W be a Weyl group with generating set S, let $I \subset S$, and let W_I be the parabolic subgroup of W generated by I. Define the sets

$$W^I = \{w \in W : \ell(ws) > \ell(w) \text{ for all } s \in I\}$$

and

$${}^I W = \{w \in W : \ell(sw) > \ell(w) \text{ for all } s \in I\}.$$

Proposition 4.3.8 *Let $w \in W$, and $I \subset S$ be as above. Let $s_i \in S$ correspond to the fundamental root α_i. The length function has the following properties:*

(i) $\ell(w^{-1}) = \ell(w)$;
(ii) $\ell(ws_i) = \ell(w) \pm 1$ *and* $\ell(s_i w) = \ell(w) \pm 1$;
(iii) $\ell(ws_i) > \ell(w)$ *if and only if* $w(\alpha_i) > 0$, *and* $\ell(s_i w) > \ell(w)$ *if and only if* $w^{-1}(\alpha_i) > 0$.
(iv) *There is a unique* $u \in W_I$ *and* $v \in W^I$ *such that* $w = uv$, *and we have* $\ell(w) = \ell(u) + \ell(v)$.
(v) *There is a unique* $u \in W_I$ *and* $v \in {}^I W$ *such that* $w = vu$, *and we have* $\ell(w) = \ell(u) + \ell(v)$.

Proof Parts (i) and (ii) are proved in [36, section 5.2], and part (iii) follows from (i) and [36, proposition 5.7]. Parts (iv) and (v) are proved in [36, proposition 1.10 (c)]. □

Definition 4.3.9 The elements v appearing in Proposition 4.3.8 (iv) (respectively, (v)) are known as the *right* (respectively, *left*) *distinguished coset representatives* of the corresponding elements w.

For the rest of Section 4.3, we will assume that the generalized Cartan matrix A has finite type unless otherwise stated.

The fact that $(,)$ is nondegenerate means that the form $(,)$ induces an identification between V_Γ and ${V_\Gamma}^*$ sending $v \in V_\Gamma$ to the linear map ϕ_v, where $\phi_v(v') = (v, v')$. Let $\{\omega_i : 1 \leq i \leq n\}$ be the dual basis in ${V_\Gamma}^*$ to the basis $\{\alpha_i^\vee : 1 \leq i \leq n\}$ of V_Γ consisting of the fundamental coroots. The elements of ${V_\Gamma}^*$ are known as *weights*, and the elements $\omega_i \in {V_\Gamma}^*$ are known as *fundamental weights*. A weight $\sum_{i=1}^n \lambda_i \omega_i$ is called *dominant* (respectively, *integral*) if the scalars λ_i are all nonnegative (respectively, integers).

Suppose that $\mathfrak{g}$ is a simple Lie algebra over $\mathbb{C}$, given by the presentation in Definition 4.1.3. Let $\mathfrak{h}$ be the subalgebra of $\mathfrak{g}$ spanned by the elements $\{h_i : i \in \Gamma\}$; we identify the element h_i with the fundamental coroot $\alpha_i^\vee \in V_\Gamma$. Let $\mathfrak{h}^* = \text{Hom}(\mathfrak{h}, \mathbb{C})$ be the dual vector space of $\mathfrak{h}$, and let $\{\omega_i : i \in \Gamma\}$ be the basis of $\mathfrak{h}^*$ dual to $\{h_i : i \in \Gamma\}$. Using the identification between the h_i and the coroots, we see that the ω_i are nothing other than the fundamental weights. We will write h_α and $h_{\alpha^\vee}$ for the elements of $\mathfrak{h}$ corresponding to the root α and the coroot $\alpha^\vee$, respectively.

Let V be a $\mathfrak{g}$-module. An element $v \in V$ is called a *weight vector* of *weight* $\lambda \in \mathfrak{h}^*$ if for all $h \in \mathfrak{h}$, we have $h.v = \lambda(h)v$. The set of all weight vectors of fixed weight λ is a subspace of V called the λ-*weight space* of V. If the weight vector v is annihilated by the action of all of the elements e_i (respectively, all of the elements f_i), then we call v a *highest weight vector* (respectively, a *lowest weight vector*).

Let Φ be the set of roots of W; these are all real by Theorem 4.2.3 (iv). For $\alpha \in \Phi$, let L_α be the hyperplane in V_Γ orthogonal to α, i.e.,

$$L_\alpha = \{v \in V_\Gamma : s_\alpha(v) = v\} = \{v \in V_\Gamma : (\alpha, v) = 0\}.$$

A connected component C of the complement

$$V_\Gamma - \bigcup_{\alpha \in \Phi} L_\alpha$$

is called a *chamber* of V_Γ; note that the chambers are open sets. If $\partial(C)$ is the boundary of the closure $\overline{C}$ of C and L_α is a hyperplane for which $L_\alpha \cap \partial(C)$ is not contained in any proper subspace of L_α, then we say that L_α is a *bounding hyperplane* of C. The set

$$C = \{v \in V_\Gamma : (\alpha_i, v) > 0\}$$

is a chamber, known as the *fundamental chamber*. The bounding hyperplanes of C are precisely the L_α for which α is a fundamental root.

Theorem 4.3.10 *Let A be a generalized Cartan matrix of finite type with Weyl group W acting on V_Γ.*

(i) *Given any two chambers C and C' of V_Γ, there is a unique element $w \in W$ such that $w(C) = C'$.*

(ii) *The number of chambers of V_Γ is equal to the order of W.*

(iii) *If C is a chamber in V_Γ then its closure $\overline{C}$ is a fundamental domain for the action of W on V_Γ; in other words, it contains precisely one element from each W-orbit of V_Γ.*

(iv) *If C is the fundamental chamber of V_Γ and $v \in \overline{C}$, then the stabilizer of v in W is generated by the fundamental reflections that it contains.*

(v) *A vector $v \in V_\Gamma$ lies in the fundamental chamber if and only if $v = \sum_{i=1}^n \lambda_i \omega_i$ with $\lambda_i > 0$ for all i, where we identify V_Γ with its dual via the form $(,)$.*

(vi) *A vector $v \in V_\Gamma$ lies in the closure of the fundamental chamber if and only if $v = \sum_{i=1}^n \lambda_i \omega_i$ with $\lambda_i \geq 0$ for all i.*

(vii) *A vector $v \in V_\Gamma$ lies in the closure of the fundamental chamber if and only if $s_i(v) \leq v$ for all i; in other words, the vector $s_i(v) - v$ is a nonpositive multiple of the fundamental root α_i.*

(viii) *If there is only one possible root length, then there is a unique root*

$$\theta = \sum_{i=1}^n a_i \alpha_i$$

in the closure of the fundamental chamber. If $\alpha = \sum_{i=1}^n \lambda_i \alpha_i$ is an arbitrary root, we have $\lambda_i \leq a_i$ for all i.

(ix) *If there are two possible root lengths $m < M$, then there are precisely two roots*

$$\theta_l = \sum_{i=1}^n a_i \alpha_i$$

and

$$\theta_s = \sum_{i=1}^n c_i \alpha_i$$

in the closure of the fundamental chamber. If $\alpha = \sum_{i=1}^n \lambda_i \alpha_i$ is an arbitrary root, we have $\lambda_i \leq a_i$ for all i; in particular, we have $c_i \leq a_i$ for all i.

Proof Parts (i)–(iii) are proved in Carter [11, proposition 12.7]. Part (iv) is proved in Kac [37, proposition 3.12 (a)]. Parts (v) and (vi) are proved in [11, proposition 12.8]. Part (vii) follows from (vi) and the formula for the action of s_i given in Theorem 4.2.3. Parts (viii) and (ix) are proved in [11, proposition 12.9]. □

Corollary 4.3.11 *Maintain the notation of Theorem 4.3.10. Let $v \in V_\Gamma$, and let $v' \in V_\Gamma$ be the unique element of the fundamental chamber that is W-conjugate to v. Then there is a sequence*

$$v = v_0, v_1, \ldots, v_k = v'$$

such that for all $0 < i \leq k$, $v_i - v_{i-1}$ is a positive multiple of a fundamental root.

Proof Suppose that $0 \leq j < k$. By Theorem 4.3.10 (vii), v_j does not lie in the fundamental chamber, and there exists s_i such that $s_i(v_j) - v_j$ is a positive multiple of α_i. We define $v_{j+1} = s_i(v_j)$ and proceed by induction. □

Definition 4.3.12 The root θ appearing in Theorem 4.3.10 (viii) is called the *highest root*; it has strictly greater height than any other root. The roots θ_l, θ_s appearing in Theorem 4.3.10 (ix) are called the *highest (long) root* and *highest short root*, respectively.

Proposition 4.3.13 *Let A be a generalized Cartan matrix, and let W be the associated Weyl group.*

(i) *The group W is finite if and only if A has finite type.*
(ii) *If W is finite, then there is a unique element w_0 of W of maximal length.*

Proof Part (i) is proved in Kac [37, proposition 4.9], and part (ii) is proved in Humphreys [36, section 1.8]. □

Proposition 4.3.14 *Let Γ be a Dynkin diagram of finite type, let W be the associated Weyl group, and let h be the associated Coxeter number.*

(i) *There exists a unique partition $\Gamma = \Gamma_1 \dot\cup \Gamma_2$ of the vertices of Γ into two subsets of mutually nonadjacent vertices.*
(ii) *For $i \in \{1, 2\}$, let $w_i = \prod_{j \in \Gamma_i} s_j$. Then the products*

$$\underbrace{(w_1 w_2 w_1 w_2 \cdots)}_{h \text{ factors}}$$

and

$$\underbrace{(w_2 w_1 w_2 w_1 \cdots)}_{h \text{ factors}}$$

are both reduced expressions for the longest element w_0 of W.

Proof It is possible to give a conceptual proof of this using the theory of [36, section 3.17], but instead we give a quick ad hoc sketch.

If the Coxeter number h is even, the result is [36, exercise 3.19.2]; the definition of h here and that of [36] agree by [36, theorem 3.20].

From the table in [36, section 3.7], the only case where a finite Weyl group can have an odd Coxeter number is in the case A_{2n}, where the Coxeter number is $2n + 1$ and the group can be identified with the symmetric group S_{2n+1}. As a permutation, the elements in the statement are both equal to the permutation exchanging i with $2n + 2 - i$ for all $1 \leq i \leq 2n + 1$, which is the longest element in type A_{2n}. (The latter well-known fact follows from the interpretation of length in Weyl groups of type A in terms of inversions; see Björner and Brenti [4, proposition 1.5.2].) □

Example 4.3.15 The Weyl group of type E_7 has Coxeter number $h = 18$. Two reduced expressions for the longest element are

$$((s_1 s_3 s_5)(s_2 s_4 s_6 s_7))^9$$

and

$$((s_2 s_4 s_6 s_7)(s_1 s_3 s_5))^9,$$

in the notation of Appendix B.

Exercise 4.3.16 Repeat Example 4.3.15 for Weyl groups of other finite types, using Proposition 4.3.14.

Proposition 4.3.17 *Let W be a finite Weyl group with connected Dynkin diagram.*

(i) *If W has type A_n (with $n \geq 2$), D_n (with n odd) or E_6, then no element of W acts as the scalar -1 on the roots and the centre, $Z(W)$, is trivial.*
(ii) *If W has a type not listed in (i), then the longest element w_0 acts as the scalar -1 on the roots and $Z(W) = \{1, w_0\}$.*

Proof The statements about the scalar -1 are a special case of a result proved in Humphreys [36, section 3.19], and the statements about the centre follow from [36, exercise 6.3.1]. □

We now return to consideration of generalized Cartan matrices of affine type. It is significant for the theory that in the Dynkin diagrams of such matrices there is a distinguished vertex labelled by 0.

Proposition 4.3.18 *Let A be a generalized Cartan matrix of affine type, and let A^0 be the submatrix obtained by deleting the row and column of A corresponding to the vertex labelled 0 in the Dynkin diagram.*

(i) *The matrix A^0 is of finite type, and every indecomposable generalized Cartan matrix of finite type arises in this way.*
(ii) *The root δ associated with A in Theorem 4.2.3 (v) and the highest root θ associated with A^0 in Theorems 4.3.10 (viii) and (ix) are related by*

$$\delta = \theta + \lambda_0 \alpha_0.$$

We have $\lambda_0 = 1$ if and only if A does not have type $A_{2l}^{(2)}$ for any $l \geq 1$ (in which case we have $\lambda_0 = 2$).
(iii) *We have*

$$s_0(\theta) = \theta + 2\lambda_0\alpha_0 = \delta + \lambda_0\alpha_0;$$

in particular, if A does not have type $A_{2l}^{(2)}$ for any $l \geq 1$, then we have

$$s_0(\theta) = \theta + 2\alpha_0 = \delta + \alpha_0.$$

(iv) *The fundamental roots orthogonal to θ_l are precisely those that are not adjacent to the vertex α_0 in the affine Dynkin diagram, and the stabilizer of θ_l in the finite Weyl group is generated by the fundamental reflections corresponding to these fundamental roots.*

Proof Part (i) follows from Carter [11, section 17.2]. The first assertion of (ii) follows from Kac [37, theorem 4.8 (c), remark 4.9], and the second assertion is proved in [11, proposition 17.2 (ii)].

Recall from Theorem 4.2.3 (v) that $(\delta, \alpha_0) = 0$. Now (ii) implies that

$$(\theta, \alpha_0) = (\delta - \lambda_0\alpha_0, \alpha_0) = -\lambda_0(\alpha_0, \alpha_0).$$

The assertions of (iii) now follow from the formula for $s_0(\theta)$ in terms of the bilinear form.

To prove (iv), recall that the highest root θ lies in the closure of the fundamental chamber $\overline{C}$. Theorem 4.3.10 (iv) then shows that the stabilizer of θ is generated by the fundamental reflections s_i stabilizing θ, and these correspond to the fundamental roots α_i orthogonal to θ. By part (ii) and the fact (Theorem 4.2.3 (v)) that δ is orthogonal to all roots, this shows that the aforementioned fundamental roots α_i are precisely those that are orthogonal to α_0, proving (iv). □

4.4 Strongly orthogonal sets

Theorem 4.3.10 turns out to be very useful in analysing the combinatorics of sets of orthogonal roots associated with a simple Lie algebra.

Definition 4.4.1 Let $S = \{\alpha_1, \alpha_2, \ldots, \alpha_k\}$ be a set of $k \geq 2$ roots associated with a simple Lie algebra $\mathfrak{g}$ of finite type. We say that the set S is a *strongly orthogonal* k-tuple of roots if

(i) the elements of S are mutually orthogonal roots, and
(ii) for every $1 \leq i \leq k$, any root orthogonal to each element of $S \backslash \{\alpha_i\}$ is either equal to $\pm\alpha_i$ or orthogonal to α_i.

A strongly orthogonal set S is said to be *positive* if it consists entirely of positive roots.

Example 4.4.2 Let $l > 4$, and consider the root system of type D_l. The pair $\{\alpha_1, \alpha_3\}$ is not a strongly orthogonal pair, because α_3 is not orthogonal to α_4, despite the fact that the latter vector is orthogonal to α_1.

Proposition 4.4.3 *Every positive root in type D_l lies in a unique strongly orthogonal pair if $l > 4$. There are two orbits of unordered pairs of orthogonal roots in type D_l, and all the strongly orthogonal pairs lie in the same orbit.*

Proof By Theorem 4.3.10 (iii) and (viii), every root is W-conjugate to the highest root θ_l, so it suffices to prove the result for this root. By the formula for a reflection, if α is a root, we have $s_\alpha(\theta_l) = \theta_l$ if and only if $(\alpha, \theta_l) = 0$.

By Proposition 4.3.18 (iv), the fundamental roots orthogonal to θ_l are precisely those that are not adjacent to the vertex α_0 in the corresponding affine Dynkin diagram, $D_l^{(1)}$. According to the numbering in Appendix B, these roots are all the roots $\alpha_1, \ldots, \alpha_l$ other than α_2. This gives a subdiagram of type $\Gamma' = A_1 \dot{\cup} D_{l-2}$, where we interpret D_3 as meaning A_3. There are two orbits of roots associated with Γ': a pair $\{\pm\alpha_1\}$ associated with the component of type A_1, and a single orbit of other roots, associated with the component of type D_{l-2}.

The positive root α_1 of the pair of type A_1 is orthogonal to all the other positive roots associated with Γ'. It follows that $\{\alpha_1, \theta_l\}$ is the unique positive strongly orthogonal pair containing θ_l. The proof follows because W acts transitively on the roots by Theorem 4.2.3 (iii). □

Exercise 4.4.4 Consider a root system of type D_4.

(i) Show that there are exactly eight roots orthogonal to any given one.
(ii) Show that every positive root lies in a strongly orthogonal pair, but that the pair is not unique.

Proposition 4.4.5 *Let $W = W(E_7)$ be the Weyl group with root system of type E_7.*

(i) *There is only one W-orbit of (ordered) pairs of orthogonal roots, and there are no strongly orthogonal pairs.*
(ii) *There are two W-orbits of (ordered) triples of orthogonal roots. Any pair of orthogonal positive roots can be extended uniquely to a strongly orthogonal (ordered) triple, and all strongly orthogonal triples are W-conjugate.*

Proof Let $\{\alpha, \beta\}$ be a pair of orthogonal roots. As in the proof of Proposition 4.4.3, we may assume that β is equal to the highest root, θ_l. The root α is then associated with the subgraph of type $\Gamma' = D_6$ obtained by omitting the fundamental root α_1. Since Γ' is simply laced and connected and has more than one vertex, there are no strongly orthogonal pairs, but all roots orthogonal to θ_l are W-conjugate. This completes the proof of (i).

Let $\{\alpha, \beta, \gamma\}$ be a triple of orthogonal roots. We may assume as before that β is the highest root θ_l of type E_7, and furthermore, that α is the highest root of type D_6 (after omitting the root α_1). As in Proposition 4.4.3, we may take γ to be the positive root for which $\{\alpha, \gamma\}$ is a strongly orthogonal pair of type D_6; that is, $\gamma = \alpha_6$.

Any root orthogonal to β lies in the subsystem of type D_6. It follows that a positive root that is orthogonal to β and α is equal to or orthogonal to γ by the definition of strongly orthogonal pair. A similar argument shows that any positive root orthogonal to β and to γ is equal to or orthogonal to α.

By part (i) and the proof of Proposition 4.4.3, the set of roots of type E_7 orthogonal to a given orthogonal pair forms a root system of type $A_1 \dot\cup D_4$. A similar argument shows that the set of roots in the type D_6 subsystem orthogonal to both α and γ forms a root system of type D_4. It follows that, with two exceptions, all roots in type E_7 that are orthogonal to both α and γ lie in the type D_6 subsystem, and in this case, they are orthogonal to β by definition. The two exceptions are the roots $\pm\beta$. This shows that any positive root orthogonal to α and to γ is equal to or orthogonal to β. This shows that the triple $\{\alpha, \beta, \gamma\}$ is strongly orthogonal.

The above construction shows that in type E_7 there exist orthogonal triples of positive roots that are strongly orthogonal, and also (by making a different choice of γ) that there exist orthogonal triples of positive roots that are not strongly orthogonal. There must therefore be at least two orbits of triples of orthogonal roots.

To show that there are exactly two orbits, suppose that $\{\alpha, \beta, \gamma\}$ and $\{\alpha', \beta', \gamma'\}$ are two triples of orthogonal roots. By part (i), we may assume that $\alpha' = \alpha$ and $\beta' = \beta$. There are then at most two inequivalent choices for γ or γ', because the set of all possible choices forms a root system of type $A_1 \dot\cup D_4$, which has two orbits of roots. Choosing γ from the A_1 subsystem gives a strongly orthogonal triple, and choosing γ' from the D_4 subsystem does not. □

Exercise 4.4.6 Show that Proposition 4.4.3 and Proposition 4.4.5 remain true if the word "unordered" is removed throughout.

Exercise 4.4.7 Show that there is only one $W(E_6)$-orbit of ordered pairs of orthogonal roots in type E_6, and that the same is true with the word "ordered" removed.

4.5 Notes and references

1 The presentation for the derived Kac–Moody algebra given in Definition 4.1.3 is given in [37, (0.3.1)] and proved in Kac [37, theorem 9.11]. By removing the appropriate generators and relations, this gives rise to a presentation for the associated simple Lie algebra over $\mathbb{C}$; this latter presentation is known as the *Serre presentation*.

2 Theorem 4.1.6 was first proved in Green [28, theorem 3.1], which deals with the untwisted affine case. It is possible to enlarge the Lie algebra by adding a derivation that acts on the ideals of the full heap; this gives a representation of the whole Kac–Moody algebra [28, theorem 8.3]. This gives a "level zero" representation of the Kac–Moody algebra, meaning that the one-dimensional centre acts trivially. Our main interest in this book is in finite-dimensional simple Lie algebras, so we do not pursue this here.

3 With the exception of Proposition 4.3.18, the results of Section 4.3 work in the more general context of finite Coxeter groups. A *Coxeter group* is a group W with a distinguished set of generating involutions S with presentation

$$\langle s_1, \ldots, s_n \mid (s_i s_j)^{m_{ij}} = 1 \rangle,$$

where $m_{ij} := m(s_i, s_j) = 1$ if and only if $s_i = s_j$. This presentation may be concisely described using a "Coxeter graph", which plays the role of the Dynkin diagram in this more general situation. A Coxeter group is called "irreducible" if this graph is connected, or equivalently, if the generating set S cannot be partitioned into nonempty subsets S_1 and S_2 such that each element of S_1 commutes with each element of S_2. A finite Weyl group is a finite Coxeter group for which the numbers m_{ij} all come from the set $\{1, 2, 3, 4, 6\}$. There are finite irreducible Coxeter groups that are not Weyl groups, namely those of types H_3, H_4 and $I_2(m)$. These are given by the presentations

$$\begin{aligned} W(H_3) &= \langle s_1, s_2, s_3 : s_1^2 = s_2^2 = s_3^2 = (s_1 s_2)^5 = (s_2 s_3)^3 = (s_1 s_3)^2 = 1 \rangle, \\ W(H_4) &= \langle s_1, s_2, s_3, s_4 : s_1^2 = s_2^2 = s_3^2 = s_4^2 \\ &= (s_1 s_2)^5 = (s_2 s_3)^3 = (s_3 s_4)^3 = (s_1 s_3)^2 = (s_1 s_4)^2 = (s_2 s_4)^2 = 1 \rangle, \\ W(I_2(m)) &= \langle s_1, s_2 : s_1^2 = s_2^2 = (s_1 s_2)^m = 1 \rangle. \end{aligned}$$

The group $W(I_2(m))$ is isomorphic (as an abstract group) to the dihedral group with $2m$ elements. The group $W(H_3)$ is a group of order 120 that is isomorphic to the full symmetry group of the icosahedron or dodecahedron. The group $W(H_4)$ is a group of order 14400 that is isomorphic to the full symmetry group of a four-dimensional analogue of either the icosahedron or the dodecahedron. A good introduction to Coxeter groups is Humphreys' book [36]; in particular, see [36, section 2.13] for more details of these groups of symmetries.

4 Recall that we are following Kac's book [37] for our definition of generalized Cartan matrices. Some other authors (for example, Stembridge [79]) define A_{ij} to be $(\alpha_i, \alpha_j^\vee)$ rather than $(\alpha_i^\vee, \alpha_j)$ (see Exercise 4.2.9), meaning that their generalized Cartan matrices are the transposes of ours.

5

Minuscule representations

In Chapter 5, we will explain how to use full heaps to construct minuscule representations of simple Lie algebras. It will turn out in Section 6.6 that every minuscule representation can be constructed in this way.

We start by recalling the usual construction of simple Lie algebras over $\mathbb{C}$ via generators and relations (Theorem 5.1.1) and the construction of their finite dimensional irreducible representations (Theorem 5.1.2). Certain of these representations are the eponymous minuscule representations; these are classified in Theorem 5.1.5.

Theorem 5.2.14 shows how to construct finite dimensional irreducible modules V_F from certain finite subheaps F of full heaps. The modules are isomorphic if and only if the heaps are isomorphic, and the weight vectors of the modules are easily described in terms of the heap structure.

Section 5.3 proves two very useful results. Theorem 5.3.13 shows that any full heap over an affine Dynkin diagram is periodic, and that the periodicity is intimately related to the lowest positive imaginary root of the associated affine root system. Theorem 5.3.16, which we call the "Trivialization Theorem", is a seemingly technical result about the action of the Weyl group on pairs of proper ideals in a full heap. However, it will have many applications throughout this book.

Section 5.4 discusses the reflections in affine Weyl groups, and Theorem 5.4.13 explains how one can explicitly describe the action of an arbitrary reflection on the proper ideals of a full heap.

Theorem 5.5.6 explains how the isomorphism types of full heaps over a fixed Dynkin diagram are in canonical bijection with the minuscule representations of the corresponding simple Lie algebra. This will play a key role in the classification of full heaps in Chapter 6.

Section 5.6 uses the properties of heaps to construct nondegenerate symplectic and orthogonal invariant bilinear forms on certain minuscule modules. We will revisit the topic of invariant forms in Section 10.5.

5.1 Highest weight modules

Theorem 5.1.1 *Let A be an indecomposable generalized Cartan matrix.*

(i) *If A is of finite type then the Lie algebra $\mathfrak{g}'(A)$ over $\mathbb{C}$ as defined in Definition 4.1.3 is nontrivial, simple and finite-dimensional. Furthermore, every nontrivial simple finite-dimensional Lie algebra over $\mathbb{C}$ arises in this way.*

(ii) *If A is of affine type and is associated with the Dynkin diagram Γ, then every proper subgraph Γ' of Γ is a disjoint union of Dynkin diagrams of finite type. Furthermore, the subgraph Γ^0 obtained from Γ by omitting the vertex corresponding to α_0 (and no other vertices) is a (connected) diagram of finite type.*

Proof It follows from Kac [37, proposition 4.9 (v)] that the nontrivial simple finite-dimensional Lie algebras are precisely the (as yet undefined) Kac–Moody algebras $\mathfrak{g}(A)$ for all possible finite type generalized Cartan matrices A. The assertion of (i) follows from the fact (Exercise 1.1.6) that the derived algebra $\mathfrak{g}'$ of a nontrivial simple Lie algebra $\mathfrak{g}$ is equal to $\mathfrak{g}$ itself.

The first assertion of (ii) follows from [37, proposition 4.7 (c)], and the second assertion is proved in [37, section 4.8]. □

For the rest of Section 5.1, we suppose that $\mathfrak{g}$ is a simple Lie algebra over $\mathbb{C}$, given by the presentation in Definition 4.1.3. The following well-known, but nontrivial, result classifies the finite dimensional irreducible modules of simple Lie algebras over $\mathbb{C}$.

Theorem 5.1.2 *Let $\mathfrak{g}$ be a simple Lie algebra over $\mathbb{C}$.*

(i) *If λ is a dominant integral weight, then up to isomorphism there is a unique finite-dimensional irreducible $\mathfrak{g}$-module $L(\lambda)$ of the form $\mathfrak{g}.v_\lambda$, where v_λ is of weight λ and is the unique nonzero highest weight vector of $L(\lambda)$. The modules $L(\lambda)$ are pairwise nonisomorphic and exhaust all finite-dimensional irreducible modules of $\mathfrak{g}$.*
(ii) *Suppose that V is a finite-dimensional $\mathfrak{g}$-module containing a nonzero highest weight vector v_λ, where λ is a dominant integral weight, and that $\dim(V) = \dim(L(\lambda))$. Then $V \cong L(\lambda)$.*

Proof Part (i) is a special case of Carter [11, theorem 10.21]. For part (ii), it follows from the proof of [11, proposition 10.13] that any $\mathfrak{g}$-module $\mathfrak{g}.v_\lambda$ generated by a highest weight vector of weight λ is a quotient of the Verma module $M(\lambda)$. The Verma module has a unique maximal submodule, $J(\lambda)$ (see [11, theorem 10.9]) and we have $L(\lambda) = M(\lambda)/J(\lambda)$ by definition. It follows that $\mathfrak{g}.v_\lambda$ has a quotient module isomorphic to $L(\lambda)$. The assumption about dimensions allows this only if $V \cong L(\lambda)$ (and $\mathfrak{g}.v_\lambda = V$). □

Definition 5.1.3 If λ is a fundamental weight, the corresponding $\mathfrak{g}$-module $L(\lambda)$ is called a *fundamental module*. If, furthermore, λ has the property that

$$(\lambda, \alpha^\vee) = 2\frac{(\lambda, \alpha)}{(\alpha, \alpha)} \leq 1$$

for all (real) positive roots α, then λ and its associated module and representation are said to be *minuscule*. If $\lambda = \omega_p$ and the Dynkin diagram of $\mathfrak{g}$ has type X_l, then we will also write $L(X_l, \omega_p)$ or $L(\mathfrak{g}, \omega_p)$ for $L(\omega_p)$.

Lemma 5.1.4 *Let A be a generalized Cartan matrix of finite type.*

(i) *If A has only one root length and highest root θ, then $\theta^\vee$ is greater than any other coroot of A with respect to the usual partial order, $\leq$.*

(ii) *If A has two root lengths and highest short root θ_s, then $\theta_s^\vee$ is greater than any other coroot of A with respect to the usual partial order, $\leq$.*

Proof This follows by combining Theorem 4.3.10 (viii) and (ix) with Proposition 4.2.10. □

Theorem 5.1.5 *The following is a complete list of minuscule fundamental weights for simple Lie algebras, using the numbering scheme of Appendix B.*

Type	$\{i : \omega_i$ *is minuscule*$\}$
A_l	$1, 2, \ldots, l$
B_l	l
C_l	1
D_l	$1, l-1, l$
E_6	$1, 5$
E_7	6
E_8	*none*
F_2	*none*
G_2	*none*

Proof Let ω_i be a fundamental weight for a simple Lie algebra $\mathfrak{g}$ associated with a generalized Cartan matrix A of finite type.

By Theorem 4.2.3 (iv), we may assume that the form $(,)$ agrees with the usual Euclidean inner product. If α is a negative root, then $(\omega_i, \alpha^\vee) < 0$.

Assume now that α is a positive root. Let $\theta^\vee$ be the highest coroot of A, as described in Lemma 5.1.4. We have $(\omega_i, \alpha^\vee) \leq (\omega_i, \theta^\vee)$. It follows that a necessary and sufficient condition for ω_i to be minuscule is for $(\omega_i, \theta^\vee) \leq 1$. In other words, the coefficient c of $\alpha_i^\vee$ in the expansion of $\theta^\vee$ as a linear combination of simple coroots satisfies $c \leq 1$ if and only if ω_i is minuscule. Now let θ' be the highest (long) root in the dual root system of A; it follows that c is also the coefficient of α_i in θ'.

Using an exhaustive check, we find that the dual root system of a root system of type B_l is of type C_l, and vice versa; in all other types, the root system is self-dual. The coefficients of α_i in θ' may be read off from the tables in Appendix B using Proposition 4.3.18 (ii). For example, in type B_l, we have

$$\delta = \alpha_0 + \alpha_1 + 2(\alpha_2 + \alpha_3 + \cdots + \alpha_l)$$

and the longest root is

$$\theta_l = \alpha_1 + 2(\alpha_2 + \alpha_3 + \cdots + \alpha_l).$$

Since only α_1 occurs with coefficient (less than or equal to) 1, this proves that ω_1 is the only minuscule weight in the *dual* root system, C_l. The other cases are dealt with similarly. □

Exercise 5.1.6 Let $\mathfrak{g}$ be an arbitrary simple Lie algebra over $\mathbb{C}$ with Cartan matrix A. Let Γ and $\overline{\Gamma}$ be the Dynkin diagrams of finite and untwisted affine type associated with $\mathfrak{g}$, and let G and $\overline{G}$ be the automorphism groups of Γ and $\overline{\Gamma}$ considered as directed graphs. Show that the following numbers are equal:

(i) $\det(A)$;
(ii) $N+1$, where N is the number of minuscule representations of $\mathfrak{g}$;
(iii) $|\overline{G}|/|G|$.

5.2 Weights and heaps

Definition 5.2.1 Let $\varepsilon : E \to \Gamma$ be a full heap, and let I be a proper ideal of E. We define the *weight of* I, $\mathrm{wt}(I)$, to be the unique element λ of $V_\Gamma{}^*$ with the property that $H_p(v_I) = \lambda(\alpha_p^\vee)v_I$ for all $p \in \Gamma$.

Example 5.2.2 Consider the full heap of Exercise 3.1.19, and the proper ideal $I = \langle 0(0)\rangle$. Notice that we have $H_0(I) = I$, $H_1(I) = -I$ and $H_p(v_I) = 0$ for $p \notin \{0, 1\}$. It follows that, in this case, we have $\mathrm{wt}(I) = \omega_0 - \omega_1$.

Lemma 5.2.3 *Let $\varepsilon : E \to \Gamma$ be a full heap over a Dynkin diagram with generalized Cartan matrix $A = (a_{i,j})$. If $I, I' \in \mathcal{B}(E)$ satisfy $I \prec_q I'$ for some $q \in \Gamma$, then we have*

$$\mathrm{wt}(I') - \mathrm{wt}(I) = \sum_{p\in\Gamma} a_{p,q}\omega_p.$$

Proof Recall from Equations (4.7) and (4.8) of Lemma 4.1.4 that we have

$$H_p \circ X_q - X_q \circ H_p = a_{p,q}X_q.$$

Each operator $H_p \circ X_q$, $X_q \circ H_p$ and X_q sends v_I to a scalar multiple of $v_{I'}$. Inspecting these scalars, we find that

$$\mathrm{wt}(I')(\alpha_p^\vee) - \mathrm{wt}(I)(\alpha_p^\vee) = a_{p,q},$$

from which the assertion follows. □

The above lemma motivates the following definition.

Definition 5.2.4 We define the weight of the fundamental root α_q to be

$$\mathrm{wt}(\alpha_q) := \sum_{p\in\Gamma} a_{p,q}\omega_p.$$

By linear extension, this defines the weight, $\mathrm{wt}(\alpha)$, of an arbitrary element $\alpha \in V_\Gamma$.

Remark 5.2.5 Definition 5.2.4 is compatible with the identification between roots and weights induced by the bilinear form (,). More precisely, we have

$$\begin{aligned}\mathrm{wt}(\alpha_j)(\alpha_i^\vee) &= \left(\sum_{p\in\Gamma} a_{p,j}\omega_p\right)(\alpha_i^\vee)\\ &= a_{i,j}\omega_i(\alpha_i^\vee)\\ &= a_{i,j},\end{aligned}$$

on the one hand, and $(\alpha_j, \alpha_i^\vee) = a_{i,j}$ on the other hand.

Using Definition 5.2.4, Lemma 5.2.3 can be restated to say that $\mathrm{wt}(I') - \mathrm{wt}(I) = \mathrm{wt}(\alpha_q)$. This can be generalized as follows.

Proposition 5.2.6 *Let $\varepsilon : E \to \Gamma$ be a full heap over a Dynkin diagram with finitely many vertices and generalized Cartan matrix $A = (a_{i,j})$. Let $I, I' \in \mathcal{B}(E)$. Then we have*

$$\mathrm{wt}(I') - \mathrm{wt}(I) = \mathrm{wt}(\chi(I, I')).$$

Proof By Proposition 3.2.12 (ii), there exists a finite sequence

$$I = I_0, I_1, \ldots, I_r = I'$$

such that, for each $0 \leq i < r$, we have either $I_i \prec_q I_{i+1}$ or $I_i \succ_q I_{i+1}$ for some $q \in \Gamma$ depending on i. The proof proceeds by induction on i; the case $i = 0$ is trivial and the case $i = 1$ is Lemma 5.2.3. The inductive step follows by Lemma 3.2.17 and linearity of the function wt. □

We will often regard the correspondence between roots and weights given by the function wt as an identification. For example, if $\lambda, \lambda' \in V_\Gamma{}^*$ satisfy $\lambda - \lambda' = \mathrm{wt}(\alpha)$ for some root α, then we may abbreviate this as $\lambda - \lambda' = \alpha$.

This identification between V_Γ and $V_\Gamma{}^*$ allows us to convert the original partial order on V_Γ to a partial order on the weights, as follows.

Definition 5.2.7 Let Γ be a Dynkin diagram. We define a partial order $\leq$ on the weights $V_\Gamma{}^*$ by stipulating that $\lambda \leq \lambda'$ if and only if

$$\lambda' - \lambda = \sum_{p \in \Gamma} c_p \alpha_p,$$

where all the c_p are nonnegative integers. In the context of weights, the terms "high" and "low" refer to this partial order.

Exercise 5.2.8

(i) Confirm that the relation $\leq$ of Definition 5.2.7 is indeed a partial order.
(ii) Give an example of two ideals in a full heap whose weights are not comparable in this partial order.

We are primarily interested in using weights to understand the Lie algebra representations corresponding to parabolic subheaps, as in the situation of Theorem 4.1.6. More precisely, let $\varepsilon : E \to \Gamma$ be a full heap over a Dynkin diagram, let $\mathfrak{g}'(A)$ be the associated derived Kac–Moody algebra. Let Γ' be a proper subgraph of Γ and let $\mathfrak{g} = \mathfrak{g}'(A'))$ be the corresponding Lie algebra. As discussed in Remark 4.1.7, the modules $V_{\Gamma',[I]} = V_{F_{S'}}$ have bases $\{v_J : J \in J(F_{S'})\}$ indexed by the set of all ideals of a certain parabolic subheap $F_{S'}$.

Lemma 5.2.9 *Let $\varepsilon : E \to \Gamma$ be a full heap, let F be a finite parabolic subheap of E, and let V_F be the corresponding Lie algebra representation of the Lie algebra $\mathfrak{g}$.*

(i) *For any $I \in J(F)$, there exists a sequence*

$$\emptyset = I_0 \prec_{p_1} I_1 \prec_{p_2} I_2 \prec_{p_3} \cdots \prec_{p_k} I_k = F$$

of elements of $J(F)$ such that $I = I_l$ for some $0 \leq l \leq k$.

(ii) *Any* $I \in J(F)$ *satisfies* $\mathrm{wt}(\emptyset) \leq \mathrm{wt}(I) \leq \mathrm{wt}(F)$. *In particular, the ideal* F *has the highest weight of all the ideals of* F, *and the ideal* $\emptyset$ *has the lowest weight of all the ideals of* F.

(iii) *If* $I \in J(F)$ *has the property that* $\mathrm{wt}(I)$ *is a nonnegative (respectively, nonpositive) linear combination of fundamental weights* ω_p, *then we must have* $I = F$ *(respectively,* $I = \emptyset$*).*

(iv) *Any two ideals of* F *have distinct weights.*

Proof Let I be an ideal of F. We can refine the partial order on F to a total order in which the elements of I appear as an initial subsequence, in some order. This gives rise to a sequence satisfying the requirements of (i). Taking the weights of each term and applying Proposition 5.2.6 proves assertion (ii).

If $I \neq \emptyset$, then we must have $l > 0$, which means that $I' \prec_p I$ for some I and $H_p(v_I) = v_I$. It follows that $\mathrm{wt}(I)$ contains ω_p with coefficient 1, so $\mathrm{wt}(I)$ fails to be a nonpositive combination of the ω_i. A similar argument shows that if $I \neq F$, then $\mathrm{wt}(I)$ cannot be a nonnegative combination of the ω_i, and this proves (iii).

To prove (iv), suppose that the weight $\mathrm{wt}(I)$ of $I \in J(F)$ is known. By (i), we have

$$\mathrm{wt}(I) = \mathrm{wt}(F) - \sum_{p \in \Gamma} c_p \alpha_p,$$

where the c_p are nonnegative integers. By Proposition 5.2.6, we then have $\chi(I, F) = \sum_{p \in \Gamma} c_p \alpha_p$. It now follows that c_p is the number of elements in $I \cap \varepsilon^{-1}(p)$ for each p, and this information determines I. □

Exercise 5.2.10 Let F be a finite parabolic subheap of a full heap $\varepsilon : E \to \Gamma$, and let $I, I' \in J(F)$. Show that $\mathrm{wt}(I) \leq \mathrm{wt}(I')$ if and only if $I \subseteq I'$.

Exercise 5.2.11 Suppose that we are in the situation of Lemma 5.2.9 but with the extra condition that F is a self-dual parabolic subheap. Show that the set of possible weights of ideals in $J(F)$ is closed under the operation of negation.

Proposition 5.2.12 *Let* F *be a finite parabolic subheap of a full heap* $\varepsilon : E \to \Gamma$. *Then the representation of the Lie algebra* $\mathfrak{g}$ *afforded by the module* V_F *is irreducible.*

Proof Let $\{v_I : I \in J(F)\}$ be the usual basis of V_F, and let $I \in J(F)$. By Lemma 5.2.9 (i), we have

$$v_I = Y_{p_{l+1}} Y_{p_{l+2}} \cdots Y_{p_k} v_F.$$

This shows that the smallest submodule of V_F that contains v_F (in other words, the submodule of V_F generated by v_F) contains all the v_I, and is thus equal to the whole of V_F.

Now let M be a nonzero submodule of V_F. We will be done if we can show that M contains v_F. Let $0 \neq m \in M$, and write

$$m = \sum_{K \in J(F)} c_K v_K.$$

Let $C = \{K \in J(F) : c_K \neq 0\}$; note that C is nonempty by hypothesis. Let I be a minimal element of C with respect to inclusion; recall that by Exercise 5.2.10, this is equivalent to requiring that $\mathrm{wt}(I)$ be minimal in the set $\{\mathrm{wt}(K) : K \in C\}$. By Lemma 5.2.9 (i), we have

$$v_F = X_{p_k} X_{p_{k+1}} \cdots X_{p_{l+1}} v_I,$$

which implies in particular that $|F| = |I| + (l + 2 - k)$.

Suppose that $K \in C$ satisfies

$$X_{p_k} X_{p_{k+1}} \cdots X_{p_{l+1}} v_K \neq 0.$$

This can only happen if

$$X_{p_k} X_{p_{k+1}} \cdots X_{p_{l+1}} v_K = v_{K'}$$

for some $K' \in J(F)$ such that $|K'| = |K| + (l + 2 - k)$. Now, since $K' \subseteq F$, it follows by induction on $l + 2 - k$ that $K \subseteq I$, and then the minimality of I shows that $K = I$.

The previous paragraph implies that

$$X_{p_k} X_{p_{k+1}} \cdots X_{p_{l+1}} m = c_I v_F;$$

in particular, M contains a nonzero multiple of v_F, and also contains v_F itself. The first paragraph of the proof now shows that $M = V_F$. □

Definition 5.2.13 Let $\varepsilon : E \to \Gamma$ be a full heap over an affine Dynkin diagram, let A be the generalized Cartan matrix of Γ, let A^0 be the corresponding finite type matrix (as in Proposition 4.3.18), and let Γ^0 be the Dynkin diagram of A^0. Let F be any parabolic subheap of E associated with the subgraph Γ^0 and let $\mathfrak{g}$ be the simple Lie algebra associated with A^0. In this situation, we call F a *parabolic subheap corresponding to* $\mathfrak{g}$.

Theorem 5.2.14 *Let F be a parabolic subheap corresponding to the simple Lie algebra $\mathfrak{g}$ over $\mathbb{C}$.*

(i) *Over the field $k = \mathbb{C}$, the module V_F affords a finite dimensional irreducible representation of the simple Lie algebra $\mathfrak{g}$.*
(ii) *Under the identification of the usual basis of V_F with a subset of $\mathcal{B}(E)$, a basis element $I \in B(E)$ of V_F is a weight vector of weight* $\mathrm{wt}(I)$.
(iii) *If F' is another parabolic subheap corresponding to $\mathfrak{g}$, then F and F' are isomorphic in* **Heap**(Γ) *(where Γ is the Dynkin diagram of $\mathfrak{g}$) if and only if V_F and $V_{F'}$ are isomorphic as $\mathfrak{g}$-modules.*

Proof The module V_F is finite-dimensional by Lemma 3.3.3 (ii), and is irreducible by Proposition 5.2.12. The other assertions of (i) follow from Theorem 5.1.1. Part (ii) follows from the definitions, because the Lie algebra element h_i acts on V_F via the restriction of the linear operator H_i.

It is clear from construction of the modules V_F and $V_{F'}$ that if F and F' are isomorphic in **Heap**(Γ), then V_F and $V_{F'}$ will be isomorphic as $\mathfrak{g}$-modules, so let us assume that F and F' are not isomorphic in **Heap**(Γ). Without loss of generality, we may assume that $|F| \geq |F'|$. Let $[x]$ and $[x']$ be traces corresponding to F and F', respectively. Since $[x] \neq [x']$ and $[x]$ is at least as long as $[x']$, we can choose a trace

$[y]$ such that (a) there exists $[z]$ with $[x] = [y][z]$ and (b) there does not exist $[z']$ such that $[x'] = [y][z']$. Let $y = s_1 s_2 \cdots s_r$ be a word in S^* representing $[y]$, and let I be the ideal of F corresponding to $[y]$. Let v and v' be the lowest weight vectors of V_F and $V_{F'}$, respectively; these correspond to the empty ideals of F and F' by Lemma 5.2.9 (ii). It follows that

$$X_{p_r} X_{p_{r-1}} \cdots X_{p_1}.v = v_I$$

in V_F, but that

$$X_{p_r} X_{p_{r-1}} \cdots X_{p_1}.v' = 0,$$

from which it follows that V_F and $V_{F'}$ are not isomorphic as $\mathfrak{g}$-modules, as required. □

Example 5.2.15 Let Γ be a Dynkin diagram of type $D_n^{(1)}$, let A be the generalized Cartan matrix associated with Γ, let A^0 be the associated generalized Cartan matrix of finite type, and let Γ^0 be the Dynkin diagram of A. (We abbreviate this by saying that Γ^0 is a Dynkin diagram of type D_n.) Consider the full heap $\varepsilon : E \to \Gamma$ of Figure 3.7, and let $F_{S'}$ be the parabolic subheap of E shown in Figure 3.11.

Recall that Exercise 4.1.10 gives an explicit construction for the module $V_{F_{S'}}$ for the Lie algebra $\mathfrak{g}'(A)$, which is simple by Theorem 5.1.1. By Theorem 5.2.14, $V_{F_{S'}}$ is a simple module for $\mathfrak{g}'(A)$. Furthermore, Lemma 5.2.9 shows that the highest weight of $V_{F_{S'}}$ is ω_1, and the lowest weight is $-\omega_1$.

Exercise 5.2.16 Show that the representation of the simple Lie algebra of type C_n associated with the parabolic subheap shown in Figure 3.10 is irreducible with highest weight ω_1 and lowest weight $-\omega_1$.

Exercise 5.2.17 Show that the natural representation of $\mathfrak{sl}_{l+1}(\mathbb{C})$, which was constructed explicitly in Example 4.1.8 using a parabolic subheap $F_{S'}$, is an irreducible representation with highest weight ω_1 and lowest weight $-\omega_l$. Show that the dual heap of $F_{S'}$ gives rise to another representation of $\mathfrak{sl}_{l+1}(\mathbb{C})$, this time with highest weight ω_l and lowest weight $-\omega_1$.

Exercise 5.2.18 Show that the hypothesis in Exercise 5.2.11 that F be self-dual cannot be dropped. In other words, give an example of a representation associated with a parabolic subheap whose set of weights is not closed under negation.

It turns out that the representations of Example 5.2.15 and Exercises 5.2.16 and 5.2.17 are all minuscule in the sense of Definition 5.1.3. We are not yet in a position to give a conceptual proof of this fact, although it could be proved at this stage by an exhaustive check.

5.3 Periodicity and trivialization

Given two weights λ and μ of a minuscule representation, it turns out that there always exists a set of mutually orthogonal fundamental roots $\{\alpha_1, \alpha_2, \dots, \alpha_k\}$ and an element w of the Weyl group such that the product x of the reflections $s_{w(\alpha_i)}$ associated with the roots $w(\alpha_i)$ satisfies $x(\lambda) = \mu$ and $x(\mu) = \lambda$. This will turn out to be a useful property of minuscule representations for later purposes. One of the main results of this section is the "Trivialization Theorem", Theorem 5.3.16, which can be regarded as an affine version of this property of minuscule representations. In order to understand it, we first need to understand the periodicity properties of full heaps, which we consider next.

Lemma 5.3.1 *Let $\varepsilon : E \to \Gamma$ be a full heap whose elements are labelled $E(i, z)$ in the usual way, and let ϕ be an element of the automorphism group $\mathrm{Aut}_{\mathbf{Heap}(\Gamma)}(E)$. Then there is an element $\alpha = \sum_{i=1}^{n} b_i \alpha_i \in V_\Gamma$ with integer coefficients b_i such that for every element $E(i, z) \in E$ we have*

$$\phi(E(i, z)) = E(i, z + b_i).$$

Furthermore, α is unique up to sign, and the coefficients b_i do not depend on the labelling chosen for E.

Proof By definition of $\mathbf{Heap}(\Gamma)$, the map ϕ induces an isomorphism of partially ordered sets $E \to E$, and this restricts to an automorphism of the vertex chains $\varepsilon^{-1}(p)$. These vertex chains are isomorphic as partially ordered sets to $\mathbb{Z}$, by axiom (F1), so any automorphism of a vertex chain $\varepsilon^{-1}(p)$ can be identified with addition by a fixed number b_p, independently of the choice of labelling of elements of E. Conversely, the automorphism of E may be reconstructed from a knowledge of the automorphisms of the individual vertex chains. The reason that α is unique up to sign is that the automorphism group has precisely two generators, which are inverses of each other. □

Definition 5.3.2 If $\varepsilon : E \to \Gamma$ is a periodic heap, we define the *period* of E to be the period of one of the generators of the automorphism group $\mathrm{Aut}_{\mathbf{Heap}(\Gamma)}(E) \cong \mathbb{Z}$. (The period is well-defined up to sign.)

Example 5.3.3 The heap shown in Figure 2.7 is periodic with period

$$\alpha_0 + \alpha_1 + 2\alpha_2 + 2\alpha_3,$$

which is the content of the convex subheap shown in the dashed box.

Exercise 5.3.4 Show that the heaps shown in Figures 3.2, 3.3, 3.5 and 3.7 are periodic, and find their periods.

Lemma 5.3.5 *Let $\varepsilon : E \to \Gamma$ be a full heap over a finite connected graph, and let I be a proper ideal of E. Let x_0 be a maximal element of I, and let $\{x_i\}_{i \geq 0}$ be an infinite sequence of elements of E with the property that for $i > 0$, x_i is maximal in the set $I \backslash \{x_j : j < i\}$. Then I consists precisely of the set $X = \{x_i\}_{i \geq 0}$.*

Proof Let $x \in I$ and suppose for a contradiction that for all i, we have $x \neq x_i$. The maximality of the x_i shows that we cannot have $x > x_i$ for any i. Since Γ is finite and X is infinite, there must exist a vertex p of Γ for which $\varepsilon^{-1}(p) \cap X$ is infinite. Since the chain $\varepsilon^{-1}(p) \cap X$ is contained in I, the chain is bounded above. Since $\varepsilon^{-1}(p)$ is isomorphic to $\mathbb{Z}$, it follows that given any integer M, there exists an element $E(p, t) \in X$ with $t < M$. On the other hand, Lemma 2.3.2 (ii) shows that we have $x > E(p, t)$ for sufficiently small integers t, which is the required contradiction. □

Lemma 5.3.6 *Let $\varepsilon : E \to \Gamma$ be a full heap over a finite connected graph, and let $\{I_n\}_{n\in\mathbb{Z}}$ be an infinite set of proper ideals for which $I_n \prec_{p_n} I_{n+1}$ for some $p_n \in \Gamma$. Then for any vertex $x \in E$, there exists n for which $\{x\} = I_{n+1}\backslash I_n$.*

Proof For each $n \in \mathbb{Z}$, let x_n be the unique element of $I_{n+1}\backslash I_n$. The ideal I_0 together with the sequence $\{x_{-i}\}_{i\geq 0}$ satisfy the hypotheses of Lemma 5.3.5, and this proves the assertion in the case where $x \in I_0$. If $x \notin I_0$, we consider dual heap E^*, which is also full (see Exercise 2.2.11). Lemma 5.3.5 now applies to the ideal $E\backslash I_0$ and the sequence $\{x_{i+1}\}_{i\geq 0}$, and this proves the assertion for in the case where $x \notin I_0$. □

The next result shows that the partial order on a full heap over a finite connected graph can be refined to a total locally finite partial order in which the elements of some particular finite convex subheap appear consecutively.

Proposition 5.3.7 *Let $\varepsilon : E \to \Gamma$ be a full heap over a finite connected graph, and let F be a finite convex subheap of E.*

(i) *There exists a bijection $f : E \to \mathbb{Z}$ such that*
   (a) *whenever $x, y \in E$ satisfy $x < y$, then $f(x) < f(y)$;*
   (b) *$f(F) = \{1, 2, \dots, n\}$, where n is the size of F.*
(ii) *The full heap E can be reconstructed up to isomorphism from the sequence $\{p_n\}_{n\in\mathbb{Z}}$, where $p_n = \varepsilon(f^{-1}(n))$.*
(iii) *The trace associated with F by Theorem 2.1.20 is $[p_1 p_2 \cdots p_n]$, where the p_i are as above.*

Proof Suppose that F has size n, and let $\{x_1, x_2, \dots, x_n\}$ be a list of elements of F such that the ordering $x_1 <' x_2 <' \cdots <' x_n$ is a refinement of the partial order on F. Let $L, L' \in \mathcal{B}(E)$ be proper ideals of E such that $L \subset L'$ and $F = L'\backslash L$; these exist by Lemma 3.2.4 (vii). For $1 \leq i \leq n$, let $r_i = \varepsilon(x_i)$ and let F_i be the proper ideal $L \cup \{x_1, x_2, \dots, x_i\}$.

Since E has no maximal or minimal elements by Lemma 2.3.2 (v), we can construct infinite sequences of proper ideals

$$L = I_0 \succ_{p_1} I_1 \succ_{p_2} I_2 \cdots$$

and

$$L' = J_0 \prec_{q_1} J_1 \prec_{q_2} J_2 \cdots.$$

Then the sequence $\{K_i\}_{i\in\mathbb{Z}}$ defined by

$$\cdots I_2 \prec_{p_2} I_1 \prec_{p_1} I_0 \prec_{r_1} F_1 \prec_{r_2} \cdots \prec_{r_n} F_n \prec_{q_0} J_0 \prec_{q_1} J_1 \prec_{q_2} J_2 \cdots,$$

indexed so that $F_1 = K_1$, is an infinite sequence of proper ideals in E. We can then define $f : E \to \mathbb{Z}$ by stipulating that $f(x) = n$, where n is the (unique) integer for which $K_n\backslash K_{n-1} = \{x\}$; such an n exists by Lemma 5.3.6. Properties (a) and (b) now hold by construction of the sequence; indeed, we have $f(F) = \{1, 2, \dots, n\}$. This proves the assertions of (i).

The partial order on E can be reconstructed from the refined sequence $\{\varepsilon(f^{-1}(n))\}_{n\in\mathbb{Z}}$ by using Lemma 2.1.5, proving (ii). Part (iii) follows by construction and Theorem 2.1.20. □

Definition 5.3.8 Let $\varepsilon : E \to \Gamma$ be a full heap, let Γ' be a subgraph of Γ and let $I, I' \in \mathcal{B}(E)$. We say that (I, I') is *trivializable in* Γ if the following conditions hold:

(a) the support of $\chi(I, I')$ is contained in Γ';
(b) there exists a sequence of operators S_{k_m} with $k_m \in \Gamma'$ such that the pair

$$(J, J') = (S_{k_1} S_{k_2} \cdots S_{k_l}(I), S_{k_1} S_{k_2} \cdots S_{k_l}(I'))$$

is relatively trivial; and
(c) $\chi(J, J') = \sum_{p \in \Gamma''} \alpha_p$, where $\Gamma'' \subset \Gamma'$ is a set of mutually nonadjacent vertices.

If F' is the trivial subheap $J' \backslash J$ of E, we will also say that (I, I') may be *trivialized to* F' by the sequence $S_{k_1}, \ldots, S_{k_l}$.

Definition 5.3.9 Suppose that (I, I') can be trivialized by the sequence $S_{k_1}, \ldots, S_{k_l}$. For each $0 \le i \le l$, define

$$\gamma_i := \chi(S_{k_{i+1}} \cdots S_{k_l}(I), S_{k_{i+1}} \cdots S_{k_l}(I')).$$

We say that the trivializable pair (I, I') is *monotonically trivializable* with respect to the sequence $S_{k_1}, \ldots, S_{k_l}$ if the following conditions hold:

(a) $I \subseteq I'$;
(b) for each $1 \le i \le l$, the operators S_i that trivialize (I, I') satisfy $\gamma_i \le \gamma_{i+1}$ in the usual order on V_Γ.

Lemma 5.3.10 *Let $\varepsilon : E \to \Gamma$ be a full heap, let Γ' be a subgraph of Γ and let $I, I' \in \mathcal{B}(E)$. If (I, I') is trivializable in Γ', then (I', I) is also trivializable in Γ'.*

Proof Suppose that (I, I') is trivializable in Γ', and maintain the notation of Definition 5.3.8. Since Γ'' consists of mutually nonadjacent vertices, the operators S_p with $p \in \Gamma''$ all commute with each other, and the product $s_{\Gamma''} = \prod_{j \in \Gamma''} s_j$ is then well-defined. Note that $s_{\Gamma''}(\alpha_p) = -\alpha_p$ for all $p \in \Gamma'$. It follows from this that the pair

$$(J', J) = \left(S_{\Gamma''} S_{k_1} \cdots S_{k_l}(I'), S_{\Gamma''} S_{k_1} \cdots S_{k_l}(I)\right)$$

has the property that $\chi(J', J) = \sum_{p \in \Gamma''} \alpha_p$, and it follows that (I', I) is trivializable in Γ', as required. □

Proposition 5.3.11 *Let $\varepsilon : E \to \Gamma$ be a full heap over a finite connected graph. Let I, I' be proper ideals of E such that $I \subset I'$, let $F = I' \backslash I$, and let Γ' be the support of F. Then we have either*

(i) *(I, I') is monotonically trivializable in Γ', or*
(ii) *$\Gamma' = \Gamma$, $\chi(S_p(I), S_p(I')) = \chi(I, I')$ for all $p \in \Gamma$, and there is an automorphism of E of period $\chi(F) - \chi(I, I')$.*

Proof The proof is by induction on the size of F; if $|F| = 1$, then (I, I') is relatively trivial because singleton heaps are trivial, and (i) holds because $\chi(I, I') = \alpha_p$ for some $p \in \Gamma'$. We may assume from now on that $|F| = k > 1$. By Lemma 3.2.28, F is a finite convex subheap of E.

Suppose first that Γ' is not a connected subgraph of Γ. Let $\Gamma' = \Gamma_1 \cup \Gamma_2$ be a decomposition of Γ' into nonempty subgraphs, and define $F_i = F \cap \varepsilon^{-1}(\Gamma_i)$ for

$i = 1, 2$. By Lemma 2.1.5 (ii), no element of F_1 can be comparable in F with any element of F_2, and since F is convex, the same holds for comparability in E. It follows that $I'' = I \cup F_1$ is an ideal of E, which must be proper by Lemma 3.2.4 (iv). Since $|F_1| < |F|$, we may apply the inductive hypothesis to the pair (I, I''); since $\Gamma_1 \neq \Gamma$, possibility (ii) cannot occur. Similarly, we can apply the inductive hypothesis to the pair (I'', I'), where we have $F_2 = I' \backslash I''$, and again possibility (ii) cannot occur. The conclusion of (i) now shows that the pair

$$(J_1, J_2) = (S_{i_1} S_{i_2} \cdots S_{i_k}(I), S_{i_1} S_{i_2} \cdots S_{i_k}(I''))$$

is relatively trivial, where all the i_m come from Γ_1 and $\chi(J_1, J_2)$ is a sum of fundamental roots α_p where $p \in \Gamma_1$. Similarly, the pair

$$(J_3, J_4) = (S_{j_1} S_{j_2} \cdots S_{j_l}(I''), S_{j_1} S_{j_2} \cdots S_{j_l}(I'))$$

is relatively trivial, where all the $j_{m'}$ come from Γ_2 and $\chi(J_3, J_4)$ is a sum of fundamental roots α_q where $q \in \Gamma_2$. Furthermore, both sequences of S_i may be chosen to witness the fact that both (I, I'') and (I'', I') are monotonically trivializable.

Now if $p \in \Gamma_1$ and $q \in \Gamma_2$, then $s_p(\alpha_q) = \alpha_q$, $s_q(\alpha_p) = \alpha_p$, and the operators S_p and S_q commute with each other. It follows that

$$\chi(J_1, J_2) = \chi(S_{j_1} S_{j_2} \cdots S_{j_l}(J_1), S_{j_1} S_{j_2} \cdots S_{j_l}(J_2)),$$

and thus that the pair

$$(J_5, J_6) = (S_{i_1} S_{i_2} \cdots S_{i_k} S_{j_1} S_{j_2} \cdots S_{j_l}(I), S_{i_1} S_{i_2} \cdots S_{i_k} S_{j_1} S_{j_2} \cdots S_{j_l}(I'')),$$

is relatively trivial, and $\chi(J_5, J_6)$ is a sum of elements α_p for $p \in \Gamma_1$. Similarly, the pair

$$(J_6, J_7) = (S_{i_1} S_{i_2} \cdots S_{i_k} S_{j_1} S_{j_2} \cdots S_{j_l}(I''), S_{i_1} S_{i_2} \cdots S_{i_k} S_{j_1} S_{j_2} \cdots S_{j_l}(I'))$$

is relatively trivial, and $\chi(J_5, J_6)$ is a sum of elements α_q for $q \in \Gamma_2$. Since no element of Γ_1 is adjacent to any element of Γ_2, the pair (J_5, J_7) is also relatively trivial. The concatenation of the two sequences preserves monotonicity, from which (i) follows. This completes the inductive step in this case.

From now on, we may assume that Γ' is connected, which means in particular that the maximal and minimal elements of F are disjoint (recall that we have assumed $|F| > 1$). Let $f : E \to \mathbb{Z}$ be a bijection with the properties listed in Proposition 5.3.7, so that $[p_1 p_2 \cdots p_n]$ is the trace corresponding to F; in particular, all the p_i lie in Γ'.

Since F has a minimal element β' labelled p_1, it follows that $S_{p_1}(I) \succ_{p_1} I$.

Suppose first that $S_{p_1}(I') = I'$. In this case, we still have $S_{p_1}(I) \subseteq S_{p_1}(I')$, because F is nonempty, and we may define $F' = S_{p_1}(I') \backslash S_{p_1}(I)$. If p_1 does not lie in the support of F', then the support of F' is not the whole of Γ. The inductive hypothesis then shows that the situation in (i) must apply to F', and by extension also to F, completing the proof in this case. On the other hand, if p_1 does lie in the support of F', then the situation in (ii) still cannot apply to F', because the second condition of (ii) fails. In this case too, situation (i) applies to F', and by extension to F'.

Next we suppose that $S_{p_1}(I') \prec_{p_1} I'$. Since F has no minimal elements that are also maximal, we still have $S_{p_1}(I) \subseteq S_{p_1}(I')$. We can then apply the argument of the previous paragraph to the subheap $F'' = S_{p_1}(I') \backslash S_{p_1}(I)$.

The third and final possibility is that $S_{p_1}(I') \succ_{p_1} I'$. If this is the case, we may repeat the above argument involving p_1 but apply it to the maximal element α' of F labelled p_n. This will either complete the proof by induction, or show that we must have both $S_{p_n}(I) \prec_{p_n} I$ and $S_{p_n}(I') \prec_{p_n} I'$. In the latter case, we still have $S_{p_n}(I) \subseteq S_{p_n}(I')$, and furthermore, I has a maximal element α with $\varepsilon(\alpha) = p_n$ and $E \backslash I'$ has a minimal element β with $\varepsilon(\beta) = p_1$. This means that E has a convex subheap F^+ with trace

$$[p_n p_1 p_2 \cdots p_{n-1} p_n p_1].$$

We can then repeat the whole argument either by (a) shifting our attention one step to the left of F and considering the convex subheap $F_{\mathrm{L}} = (F \backslash \{\beta'\}) \cup \{\alpha\}$, or by (b) shifting our attention one step to the right of F and considering the convex subheap $F_{\mathrm{R}} = (F \backslash \{\alpha'\}) \cup \{\beta\}$; note that we have $|F_{\mathrm{L}}| = F = |F_{\mathrm{R}}|$. We will only reach a total impasse if this repeating sequence can be extended indefinitely far in each direction, and in this case, Lemma 5.3.6 shows that every element of E must correspond to an entry in the repeating sequence; in particular, we have $\chi(F) = \Gamma$. We can reconstruct E from this sequence by Proposition 5.3.7 (ii). Since the operations of shifting n places to the left or right are automorphisms of the sequence, it follows that they induce automorphisms of E of periods $\pm\chi(F)$, and the assertions follow. □

Corollary 5.3.12 *Let $\varepsilon : E \to \Gamma$ be a full heap and let α be a positive real root. There exists a generator s_i of the Weyl group $W(\Gamma)$ such that $s_i(\alpha) < \alpha$. Furthermore, if α is not a fundamental root then we have $s_i(\alpha) > 0$.*

Proof If α is a fundamental root, then we can take s_i to be the reflection corresponding to α, so we may assume that α is not a fundamental root. By Theorem 3.2.30, α is representable in E, so we can choose $I, I' \in \mathcal{B}(E)$ such that $\chi(I, I') = \alpha$. By Proposition 5.3.11, (I, I') is monotonically trivializable to F by some sequence $S_{k_1}, \ldots, S_{k_r}$. With γ_i as in Definition 5.3.9, we have

$$\gamma_0 \leq \gamma_1 \leq \gamma_2 \leq \cdots \leq \gamma_l = \alpha.$$

Since α is a root, F is a singleton and γ_0 is a fundamental root. It follows that there is a maximal integer i with $0 \leq i < l$ such that $\gamma_i < \gamma_{i+1} = \alpha$. Since $\gamma_{i+1} = s_{i+1}(\gamma_i)$, we have $s_{i+1}(\alpha) < \alpha$, as required. □

Theorem 5.3.13 *Let $\varepsilon : E \to \Gamma$ be a full heap over an affine Dynkin diagram.*

(i) *The Dynkin diagram Γ cannot be of type $A_{2l}^{(2)}$ for any $l \geq 1$. It follows that the coefficient of α_0 in δ is 1, and that we have $\delta = \theta + \alpha_0$, where θ is the highest root of the corresponding simple Lie algebra.*

(ii) *The element $\delta > 0$ of Theorem 4.2.3 (v), interpreted as a linear combination of fundamental roots, is representable in E.*

(iii) *If $\phi \in \mathrm{Aut}_{\mathbf{Heap}(\Gamma)}(E)$ then the period of ϕ must be an integral multiple of δ.*

(iv) *The heap E is periodic with period δ.*

Proof We know that the highest root θ is representable in E from Theorem 3.2.30, so let $I, I' \in \mathcal{B}(E)$ be such that $\chi(I, I') = \theta$. Proposition 4.3.18 (iii) shows that $s_0(\theta) = \theta + 2\lambda_0\alpha_0$, which means that $\chi(s_0(I), s_0(I')) = \theta + 2\lambda_0\alpha_0$ by Theorem 3.2.21. This can only happen if $s_0(I) \prec_0 I$, $I' \prec s_0(I')$, and $\lambda_0 = 1$; in particular, Γ cannot be of type $A_{2l}^{(2)}$ for $l \geq 1$, which proves the first assertion of (i). The other assertions of (i) now follow from Proposition 4.3.18 (ii). Furthermore, Proposition 4.3.18 (iii) implies that

$$\chi(s_0(I), I') = \chi(I, s_0(I')) = \theta + \alpha_0 = \delta,$$

which means that δ is representable in E, proving (ii).

To prove (iii), let α be the period of ϕ. Let $I \in \mathcal{B}(E)$ and let $I' = \phi(I)$, so that $\chi(I, I') = \alpha$. Let $p \in \Gamma$. Since ϕ is an automorphism, we have $s_p(I) \prec_p I$ if and only if $s_p(I') \prec_p I'$, and similar identities hold for the relation $\succ_p$ and the equality relation. It follows that $\chi(s_p(I), s_p(I')) = \alpha$. By Theorem 3.2.21, we have

$$s_p(\alpha) = s_p(\chi(I, I')) = \chi(s_p(I), s_p(I')) = \alpha.$$

Since p was arbitrary, we have $s_p(\alpha) = \alpha$ for all p. It follows from Theorem 4.2.3 that we have $(\alpha, \alpha_p) = 0$ for all p, and thus that α lies in the radical of the bilinear form $(,)$. By part (v) of that result, α must be a multiple of δ. However, periods must by definition be integer linear combinations of the α_p, and since α_0 occurs in δ with coefficient 1 by the above paragraph, this can only happen if α is an integer multiple of δ, which proves (iii).

To prove (iv), let $J, J' \in \mathcal{B}(E)$ be such that $\chi(J, J') = \delta$. By Theorem 4.2.3, we have $s_i(\delta) = \delta$ for all $i \in \Gamma$. It follows from Theorem 3.2.21 that

$$\chi(w(J), w(J')) = w(\delta) = \delta$$

for all $w \in W$, and this is incompatible with Proposition 5.3.11 (i) because every fundamental root occurs in δ with nonzero coefficient. Proposition 5.3.11 (ii) now shows that there exists an automorphism $\phi \in \mathrm{Aut}_{\mathbf{Heap}(\Gamma)}(E)$ with period δ. It follows from (iii) that any other automorphism of $\mathrm{Aut}_{\mathbf{Heap}(\Gamma)}(E)$ is an integral power of ϕ, and thus that E is periodic, proving (iv). □

Exercise 5.3.14 Show that if $\varepsilon : E \to \Gamma$ is a full heap over an affine Dynkin diagram, then there will always exist distinct proper ideals with the same weight.

Exercise 5.3.15 Show that the statement of Corollary 5.3.12 is always false if we replace the word "real" by "imaginary".

Theorem 5.3.16 *(Trivialization Theorem) Let $\varepsilon : E \to \Gamma$ be a full heap over an affine Dynkin diagram, and let $I, I' \in \mathcal{B}(E)$ be distinct proper ideals. Then precisely one of the following two situations occurs:*

(i) *$\chi(I, I') = k\delta$ for some integer k, or*
(ii) *the pair (I, I') is trivializable in Γ', where Γ' is the support of $\chi(I, I')$.*

Proof Observe first that cases (i) and (ii) are mutually exclusive by Theorem 3.2.21 and Theorem 4.2.3 (v).

We next note that if the conditions of the theorem hold for the ordered pair (I, I'), then they also hold for (I', I). This is clear for condition (i) by replacing k with $-k$, and it holds for condition (ii) by Lemma 5.3.10.

Suppose first that $I \subset I'$. By Theorem 5.3.13 (iii), the only way E can have an automorphism ϕ of period $\alpha = \chi(I, I')$ is for $\alpha = k\delta$ for some integer k, which is covered by condition (i). Otherwise, Proposition 5.3.11 (i) applies, and (I, I') is trivializable in Γ'. If $I' \subset I$, then we can argue by symmetry as in the previous paragraph.

We may assume from now on that the proper ideal $I \cap I'$ (see Lemma 3.2.4 (ii)) is not equal either to I or to I'. By Lemma 3.2.4 (viii), no vertex in the support Γ_1 of $\chi(I, I \cap I')$ can be adjacent to any vertex in the support Γ_2 of $\chi(I \cap I', I)$. By assumption, both Γ_1 and Γ_2 are nonempty; note also that their union is the support of $\chi(I, I')$.

Proposition 5.3.11 (i) then shows both that $(I, I \cap I')$ is trivializable in Γ_1 and that $(I \cap I', I')$ is trivializable in Γ_2. More precisely, there is a sequence of operators S_{a_r} with $a_r \in \Gamma_1$ such that the pair

$$(J_1, J_2) = (S_{a_1} S_{a_2} \cdots S_{a_l}(I), S_{a_1} S_{a_2} \cdots S_{a_l}(I \cap I'))$$

is relatively trivial and satisfies $\chi(J_1, J_2) = \sum_{p \in \Gamma'_1} \alpha_p$, and also a sequence of operators S_{b_r} with $b_r \in \Gamma_2$ such that the pair

$$(J_3, J_4) = (S_{b_1} S_{b_2} \cdots S_{b_m}(I), S_{b_1} S_{b_2} \cdots S_{b_m}(I \cap I'))$$

is relatively trivial and satisfies $\chi(J_3, J_4) = \sum_{q \in \Gamma'_1} \alpha_q$. The nonadjacency condition between Γ_1 and Γ_2 implies that each operator S_{a_i} commutes with each operator S_{b_j}, and furthermore that $S_{a_i}(\alpha_q) = \alpha_q$ and $S_{b_j}(\alpha_p) = \alpha_p$ for each α_p and α_q. It follows that the pair

$$(J_5, J_6) = (S_{a_1} S_{a_2} \cdots S_{a_l} S_{b_1} S_{b_2} \cdots S_{b_m}(I), S_{a_1} S_{a_2} \cdots S_{a_l} S_{b_1} S_{b_2} \cdots S_{b_m}(I'))$$

has the property that

$$\chi(J_5, J_6) = \sum_{p \in \Gamma'_1} \alpha_p + \sum_{q \in \Gamma'_2} \alpha_q,$$

and that the latter is a sum of mutually nonadjacent distinct fundamental roots. It follows that (J_5, J_6) is relatively trivial, which completes the proof that (I, I') is trivializable in Γ'. □

Exercise 5.3.17 Let $\varepsilon : E \to \Gamma$ be a full heap over an affine Dynkin diagram, and let $W = W(\Gamma)$ be the associated Weyl group. Let $I, I' \in \mathcal{B}(E)$.

(i) Show that the following are equivalent:
 (a) $\chi(I, I')$ is a (real or imaginary) root associated with Γ;
 (b) for each $w \in W$, either $w(I) \subseteq w(I')$ or $w(I) \supseteq w(I')$.

(ii) Show that the following are equivalent:
 (a) $\chi(I, I')$ is not an imaginary root associated with Γ;
 (b) there is an element $w \in W$ exchanging I with I', and furthermore, that w may be taken to be a product of commuting reflections.

(iii) Show that not every element in a Weyl group is a product of commuting reflections. (Consider type A_2.)

Example 5.3.18 Consider the full heap $\varepsilon : E \to A_l^{(1)}$ of Figure 3.2, where $S(\Gamma) = \{s_0, s_1, \ldots, s_l\}$. Since the generalized Cartan matrix A is symmetric, it follows from Theorem 4.2.3 (iii) that (α_i, α_i) is constant for all fundamental roots, so there is only one possible root length for real roots. Let F be a nonempty finite convex subheap of E, and let $I, I' \in \mathcal{B}(E)$ be such that $F = I' \backslash I$; recall that I and I' exist by Lemma 3.2.28. By the Trivialization Theorem (Theorem 5.3.16), either $\chi(F)$ is an imaginary root (as in Definition 4.2.14) or the pair (I, I') is trivializable to F', where F' is a trivial subheap. However, since E is totally ordered, F' must be a singleton, and $\chi(F') = \alpha_p$, a fundamental root. Since $\alpha_p = \chi(F') = \chi(w(I), w(I'))$ for some $w \in W$, Theorem 3.2.21 shows that $\chi(I, I') = w^{-1}(\alpha_p)$; in other words, $\chi(I, I')$ is a real root. It follows that the positive real and imaginary roots of type affine A are precisely the contents of the various nonempty finite convex subheaps of E.

Example 5.3.19 Consider the full heap $\varepsilon : E \to A_l^{(1)}$ of Figure 3.2, where $S(\Gamma) = \{s_0, s_1, \dots, s_l\}$, and let E_0 be the parabolic subheap of E shown in Figure 3.9; recall that this corresponds to $S' = S(\Gamma)\backslash\{s_0\}$. Let F be a nonempty convex subheap of E_0. As in the proof of Lemma 3.2.4 (vii), there exist ideals I_0 and I_0' of E_0 with $F = I_0' \backslash I_0$. If I and I' are the elements of $\mathcal{B}(E)$ corresponding by Lemma 3.3.3 (v) to I_0 and I_0' respectively, then we have $F = I_0' \backslash I_0$. Since $F \cap \varepsilon^{-1}(0)$ is empty, part (i) of Theorem 5.3.16 cannot apply, and we conclude that the pair (I, I') is trivializable in Γ' to a trivial subheap F' of E. Since this trivialization is performed only with operators S_i with $i \neq 0$, the subheap $I' \backslash I$ must be a subheap of E_0 by Lemma 3.3.3. Since E_0 is totally ordered, F' must be a singleton. The argument of Example 5.3.18 now shows that the positive roots of the parabolic subgroup $W_{S'}$ are precisely the contents of the nonempty convex subheaps of E_0. Let $I, I' \in \mathcal{B}(E)$ satisfy $E_0 = I' \backslash I$. Using the definition of the action of S_i, we see that $\chi(S_i(I), S_i(I')) \leq \chi(I, I')$. It then follows from Theorem 4.3.10 (vii) that the highest root, θ, is the content of E_0 itself, namely

$$\theta = \alpha_1 + \alpha_2 + \cdots + \alpha_l.$$

Exercise 5.3.20

(i) Show how to adapt the argument of Example 5.3.19 to show that the positive roots of the root system of A_l are precisely the elements of the form

$$\{\alpha_i + \alpha_{i+1} + \cdots + \alpha_j \text{ for } 1 \leq i \leq j \leq l\}.$$

(ii) Let $\varepsilon_1, \varepsilon_2, \dots, \varepsilon_{l+1}$ be linearly independent vectors in $\mathbb{R}^{l+1}$, and identify α_i with $\varepsilon_i - \varepsilon_{i+1}$. Show that the positive and negative roots of A_l are precisely those of the form $\varepsilon_i - \varepsilon_j$, where $1 \leq i, j \leq l+1$ and $i \neq j$.

Exercise 5.3.21

(i) Use Exercise 5.3.20 to show that a root system of type A_l has $l(l+1)/2$ positive roots.
(ii) Use Proposition 4.2.6 to show that the dimension of a simple Lie algebra of type A_l is the same as $\dim(\mathfrak{sl}_{l+1}(\mathbb{C}))$, that is, $l(l+2)$.
(iii) Deduce that the operators E_i, F_i and H_i of Example 1.3.3 generate the whole of $\mathfrak{sl}_{l+1}(\mathbb{C})$.

Exercise 5.3.22 Consider the full heap $\varepsilon : E \to C_l^{(1)}$ of Figure 3.3, where $S(\Gamma) = \{s_0, s_1, \dots, s_l\}$. Let $I, I' \in \mathcal{B}(E)$, and suppose that $I \subset I'$ with $F = I' \backslash I$.

(i) Show that the fundamental roots α_0 and α_l are long, and that the other fundamental roots are short. Deduce that there are two possible lengths for real roots associated to Γ.
(ii) Use the Trivialization Theorem to show that either $\chi(F)$ is an imaginary root or (I, I') is trivializable to a singleton subheap, $F' = \{\alpha\}$.
(iii) Show that if $\varepsilon(\alpha) = 0$ or $\varepsilon(\alpha) = l$, then (a) $\chi(F)$ is a long root, (b) the subheap F is self-dual, and (c) the middle element of F is α. (The meaning of (c) is that if $* : F \cong F^*$ is the duality isomorphism, then α is the unique fixed point of $*$.)
(iv) Show that if $\varepsilon(\alpha) \notin \{0, l\}$, then (a) $\chi(F)$ is a short root and (b) the subheap F is not self-dual, unless F is a singleton.
(v) Show that no element of W can send a fundamental root in the set $\{\alpha_0, \alpha_l\}$ to a fundamental root in the set $\{\alpha_1, \alpha_2, \ldots, \alpha_{l-1}\}$, or vice versa.
(vi) Show that no element of W can send the root α_0 to the root α_l, or vice versa, even though these two roots have the same length.

Exercise 5.3.23 Consider the full heap $\varepsilon : E \to C_l^{(1)}$ of Figure 3.3, where $S(\Gamma) = \{s_0, s_1, \ldots, s_l\}$, and let E_0 be the parabolic subheap of E shown in Figure 3.10; recall that this corresponds to $S' = S(\Gamma)\backslash\{s_0\}$.

(i) Show that the roots of the parabolic subgroup $W_{S'}$ are precisely the contents of the nonempty convex subheaps of E_0.
(ii) Show that a root is long if and only if it is the content of a self-dual convex subheap F of E_0 whose middle element, α, satisfies $\varepsilon(\alpha) = l$.
(iii) Show that the content of E_0 itself, namely

$$2(\alpha_1 + \alpha_2 + \cdots + \alpha_{l-1}) + \alpha_l,$$

is the highest root θ_l, and that θ_l is long.
(iv) Show that there are two convex subheaps of E_0 whose content is equal to the highest short root, θ_s, and that we have

$$\theta_s = \alpha_1 + 2(\alpha_2 + \cdots + \alpha_{l-2} + \alpha_{l-1}) + \alpha_l.$$

Exercise 5.3.24 Maintain the notation of Exercise 5.3.23. Let $\varepsilon_1, \varepsilon_2, \ldots, \varepsilon_l$ be a linearly independent set of vectors in $\mathbb{R}^l$. If $i < l$, we identify α_i with $\varepsilon_i - \varepsilon_{i+1}$, and we identify α_l with $2\varepsilon_l$.

(i) Show that the long positive roots in type C_l are those of the form

$$\beta_i := 2(\alpha_i + \alpha_{i+1} + \cdots + \alpha_{l-1}) + \alpha_l$$

for $1 \leq i \leq l$, where we interpret $\beta_l = \alpha_l$.
(ii) Show that the short positive roots in type C_l are those of the form

$$\alpha_i + \alpha_{i+1} + \cdots + \alpha_j$$

for $1 \leq i \leq j \leq l$ together with those of the form

$$\alpha_i + \alpha_{i+1} + \cdots + \alpha_l + \alpha_{l-1} + \alpha_{l-2} + \cdots + \alpha_j$$

for $1 \leq i < j \leq l$.

(iii) Show that, under the identifications above, the long roots in type C_l are those of the form $\{\pm 2\varepsilon_i : 1 \leq i \leq l\}$, and the short roots in type C_l are those of the form $\{\pm\varepsilon_i \pm \varepsilon_j : 1 \leq i < j \leq l\}$, where the two sign choices are independent.

Exercise 5.3.25

(i) Use Exercise 5.3.24 to show that a root system of type C_l has l^2 positive roots.
(ii) Use Proposition 4.2.6 and Example 1.2.6 to show that the dimension of a simple Lie algebra of type C_l is the same as $\dim(\mathfrak{sp}_{2l}(\mathbb{C}))$, that is, $l(2l+1)$.
(iii) Deduce that the operators E_i, F_i and H_i of Example 1.3.4 generate the whole of $\mathfrak{sp}_{2l}(\mathbb{C})$.

Exercise 5.3.26 Consider the full heap $\varepsilon : E \to D_l^{(1)}$ of Figure 3.7, where $S(\Gamma) = \{s_0, s_1, \ldots, s_l\}$. Let $I, I' \in \mathcal{B}(E)$, and suppose that $I \subset I'$ with $F = I' \backslash I$.

(i) Show that all the fundamental roots have the same length.
(ii) Use the Trivialization Theorem to show that either $\chi(F)$ is an imaginary root or that (I, I') is trivializable to a trivial subheap F' of size 1 or 2.
(iii) Show that if F' has size 2 then (a) $\chi(F)$ is the sum of two orthogonal roots, but is not itself a root, (b) the subheap F is self-dual, and (c) the middle two elements of F are the elements of F'.
(iv) Show that if F' is a singleton then $\chi(F)$ is a root.
(v) Show that no element of W can send $\alpha_0 + \alpha_1$ to $\alpha_{l-1} + \alpha_l$ or vice versa, even though these two elements have the same length.

Exercise 5.3.27 Maintain the notation of Exercise 5.3.26, and let $\varepsilon_1, \varepsilon_2, \ldots, \varepsilon_l$ be a linearly independent set of vectors in $\mathbb{R}^l$. If $i < l$, we identify α_i with $\varepsilon_i - \varepsilon_{i+1}$, and we identify $\alpha_l = \varepsilon_{l-1} + \varepsilon_l$. (Notice that these vectors are linearly independent, so this identification makes sense.)

(i) Show that the positive roots in type D_l are those that have the form

$$\{\alpha_i + \alpha_{i+1} + \cdots + \alpha_{l-2} + \alpha_{l-1} \text{ such that } 1 \leq i \leq l-1\},$$

or

$$\{\alpha_i + \alpha_{i+1} + \cdots + \alpha_{l-2} + \alpha_l \text{ such that } 1 \leq i \leq l-1\}$$

(including the element α_l itself), or

$$\{\alpha_i + \alpha_{i+1} + \cdots + \alpha_{l-2} + \alpha_{l-1} + \alpha_l \text{ such that } 1 \leq i \leq l-2\},$$

or

$$\{\alpha_i + \alpha_{i+1} + \cdots + \alpha_l + \alpha_{l-2} + \alpha_{l-3} + \cdots + \alpha_j \text{ such that } 1 \leq i < j \leq l-2\}.$$

(ii) Show that, under the identifications above, the positive roots in type D_l are those of the form $\{\varepsilon_i \pm \varepsilon_j : 1 \leq i < j \leq l\}$. Deduce that there are $2l(l-1)$ roots in type D_l.
(iii) Show that the highest root in type D_l is

$$\theta_l = \alpha_1 + 2(\alpha_2 + \alpha_3 + \ldots + \alpha_{l-2}) + \alpha_{l-1} + \alpha_l,$$

or $\varepsilon_1 + \varepsilon_2$ in coordinates.

Exercise 5.3.28

(i) Use Exercise 5.3.27 to show that a root system of type D_l has $l(l-1)$ positive roots.
(ii) Use Proposition 4.2.6 and Example 1.2.8 to show that the dimension of a simple Lie algebra of type C_l is the same as $\dim(\mathfrak{so}_{2l}(\mathbb{C}))$, that is, $l(2l-1)$.
(iii) Deduce that the operators E_i, F_i and H_i of Example 1.3.5 generate the whole of $\mathfrak{so}_{2l}(\mathbb{C})$.

Exercise 5.3.29 Consider the parabolic subheap F of Figure 3.11, corresponding to the Dynkin diagram of type D_l. Recall from the proof of Proposition 4.4.3 that the pair $\{\alpha_1, \theta_l\}$ is a strongly orthogonal pair of roots in type D_l.

(i) Show that the content $\chi(F)$ of F satisfies $\chi(F) = \alpha_1 + \theta_l$.
(ii) Use the coordinates of Exercise 5.3.27 to show that a pair of positive roots is a strongly orthogonal pair if and only if they are of the form $\{\varepsilon_i - \varepsilon_j, \varepsilon_i + \varepsilon_j\}$ for suitable $1 \leq i \neq j \leq l$.

Exercise 5.3.30 Use Proposition 4.3.18 (iv), Exercise 5.3.27 and induction to show that the order of the Weyl group $W(D_n)$ is $2^{n-1}n!$.

Exercise 5.3.31 Recall the result of Exercise 5.3.24.

(i) Use dual root systems to show that we can identify the long positive roots in type B_l with vectors in $\mathbb{R}^l$ of the form $\{\varepsilon_i \pm \varepsilon_j : 1 \leq i < j \leq l\}$, and we can identify the short positive roots in type B_l with the vectors in $\mathbb{R}^l$ of the form $\{\varepsilon_i : 1 \leq i \leq l\}$.
(ii) Show that the usual scalar product in $\mathbb{R}^l$, when restricted to the span of the roots in B_l, is a nonzero scalar multiple of the bilinear form of Theorem 4.2.3.

Exercise 5.3.32 Consider the full heap $\varepsilon : E \to D_l^{(1)}$ of Figure 3.7, where $S(\Gamma) = \{s_0, s_1, \ldots, s_l\}$, and let E_0 be the parabolic subheap of E shown in Figure 3.11; recall that this corresponds to $S' = S(\Gamma)\backslash\{s_0\}$.

(i) Show that the roots of the parabolic subgroup $W_{S'}$ are precisely the contents of the nonempty convex subheaps of the self-dual heap E_0 that are not invariant under the duality $* : E_0 \to E_0$.
(ii) Show that the content of E_0 itself, namely

$$2(\alpha_1 + \alpha_2 + \cdots + \alpha_{l-2}) + \alpha_{l-1} + \alpha_l,$$

is the unique element of the fundamental chamber $\overline{C}$ conjugate to $\alpha_{l-1} + \alpha_l$.
(iii) Show that there are two convex subheaps of E_0 whose content is equal to the highest root, θ, and that we have

$$\theta = \alpha_1 + 2(\alpha_2 + \alpha_3 + \cdots + \alpha_{l-2}) + \alpha_{l-1} + \alpha_l.$$

5.4 Reflections

Recall from Section 4.2 that a real root α associated with a Weyl group W is an element of the form $\alpha = w(\alpha_i)$, where $w \in W$ and where α_i is a fundamental root.

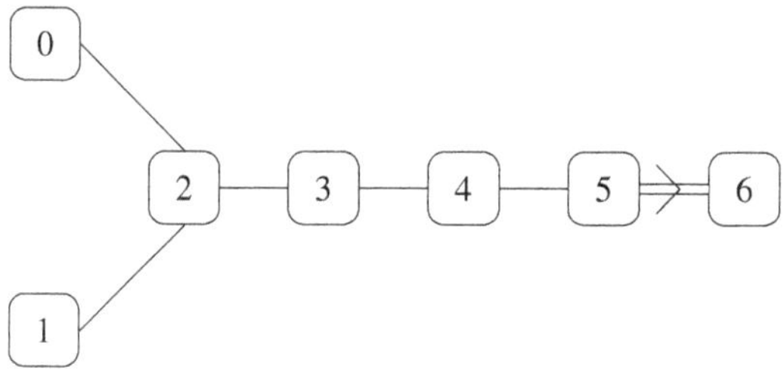

Figure 5.1 Dynkin diagram of type $B_6^{(1)}$

The Weyl group contains an element s_i that sends α_i to its negative and fixes the hyperplane orthogonal to α_i with respect to the symmetric bilinear form $(,)$; such an orthogonal transformation is called a *reflection*. The group W also contains a reflection s_α associated with the real root α, as follows; the proof is left as an exercise.

Lemma 5.4.1 *Let $w \in W$ and let α_i be a fundamental root with corresponding reflection s_i. Then the element $s_\alpha = w s_i w^{-1} \in W$ is a reflection sending α to $-\alpha$ and fixing pointwise the orthogonal complement of α with respect to $(,)$.*

Exercise 5.4.2 Prove Lemma 5.4.1.

Remark 5.4.3 Since s_α is conjugate in W to s_i, it is clear that s_α has order 2 in W.

The action of a reflection s_α on the ideals of a full heap has a simple description. To describe it, it is convenient to extend our previous notation of $\prec$ and $\succ$.

Definition 5.4.4 Let $\varepsilon : E \to \Gamma$ be a full heap, let $I \in \mathcal{B}$ and let L be a finite convex subheap of E with $\chi(L) = \alpha$ (in other words, $L \in \mathcal{L}_\alpha(E)$). Then

(a) we write $L \succ I$ (or $I \succ_\alpha I'$) to mean that both $I' := I \cup L \in \mathcal{B}$ and $I \cap L = \emptyset$, and
(b) we write $L \prec I$ (or $I' \prec_\alpha I$) to mean that both $L \leq I$ and $I' := I \backslash L \in \mathcal{B}$.

Remark 5.4.5 Definition 5.4.4 generalizes the previous definition of $\prec$ and $\succ$ in the sense that $I \prec_p I$ if and only if $I \prec_{\alpha_p} I'$.

Example 5.4.6 Figure 5.1 shows the Dynkin diagram of type $B_6^{(1)}$, and Figure 5.2 shows a full heap $\varepsilon : \mathrm{FH}(B_6^{(1)}(6)) \to B_6^{(1)}$.

Figure 5.3 shows a proper ideal, J, of the full heap $\mathrm{FH}(B_6^{(1)}(6))$ in Figure 5.2. The elements of the ideal are drawn as unshaded.

Figure 5.4 shows another proper ideal, I, of the full heap in Figure 5.2. The elements of the ideal are drawn as unshaded; the two different types of shading emphasise the fact that $I \subset J$.

Note that the set $L = J \backslash I$ contains five elements: one labelled 3, one labelled 4, one labelled 5 and two labelled 6. In this case, we have $\chi(L) = \alpha$, where $\alpha = \alpha_3 + \alpha_4 + \alpha_5 + 2\alpha_6$. We express this by saying that $I \prec_\alpha J$.

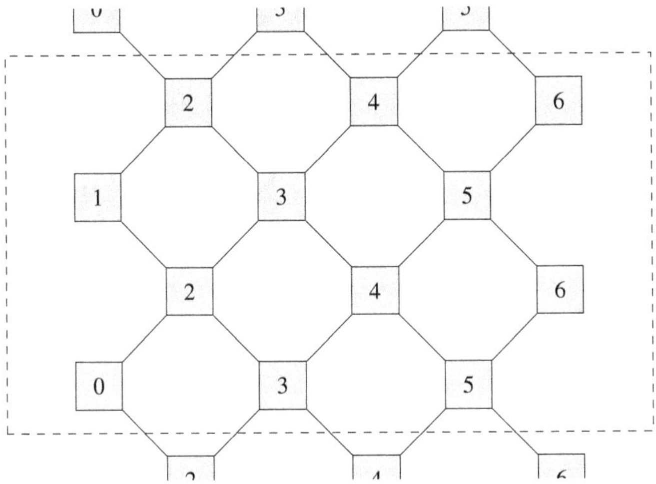

Figure 5.2 A full heap $\varepsilon : \mathrm{FH}(B_6^{(1)}(6)) \to B_6^{(1)}$

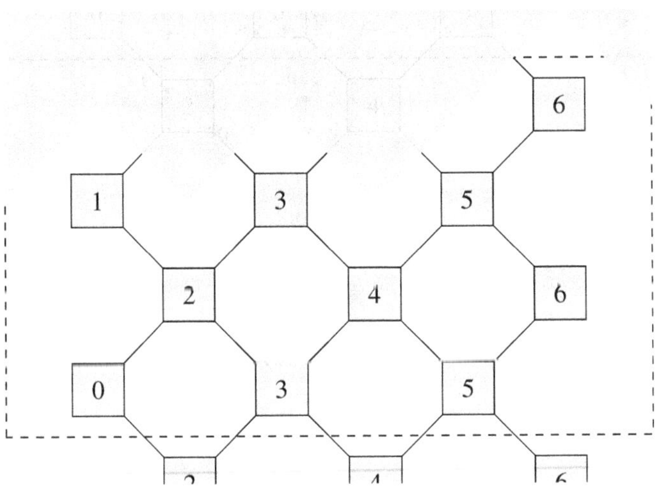

Figure 5.3 A proper ideal, J, of the full heap $\mathrm{FH}(B_6^{(1)}(6))$ in Figure 5.2

Exercise 5.4.7 Suppose that $\varepsilon : E \to \Gamma$ is a full heap, $I \in \mathcal{B}(E)$, and $\alpha \in V_\Gamma$. Show that there is at most one $J \in \mathcal{B}(E)$ such that $I \prec_\alpha J$, and at most one $J' \in \mathcal{B}(E)$ such that $I \succ_\alpha J'$.

Exercise 5.4.8 Let $\varepsilon : E \to \Gamma$ be a full heap and let $W = W(\Gamma)$ be the associated Weyl group. Let α be a positive real root, and suppose that the proper ideals $J, I \in \mathcal{B}(E)$ satisfy $J \prec_\alpha I$. Show that $w(J)$ and $w(I)$ are comparable with respect to inclusion. More precisely, prove that

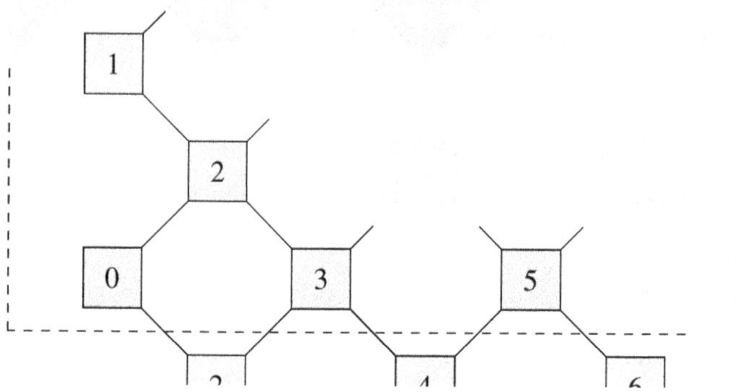

Figure 5.4 Another proper ideal, I, of the full heap in Figure 5.2

(i) if $w(\alpha) > 0$ then $w(J) \subset w(I)$ and $w(J) \prec_{w(\alpha)} w(I)$;
(ii) if $w(\alpha) < 0$ then $w(I) \subset w(J)$ and $w(J) \succ_{-w(\alpha)} w(I)$.

Deduce that if $J \prec_i I$ then $\ell(ws_i) > \ell(w)$ if and only if $w(J) \subset w(I)$.

We can also generalize the raising and lowering operators X_p, Y_p and H_p defined in Definition 3.1.4.

Definition 5.4.9 Let $\varepsilon : E \to \Gamma$ be a full heap. Let $\mathcal{B}$ be the set of proper ideals of E, and let V_E be the k-vector space with $\mathcal{B}$ as a basis. Let L be a finite convex subheap of E. We define linear operators X_L, Y_L and H_L on V_E as follows:

$$X_L(v_I) = \begin{cases} v_{I \cup L} & \text{if } L \succ I, \\ 0 & \text{otherwise,} \end{cases}$$

$$Y_L(v_I) = \begin{cases} v_{I \setminus L} & \text{if } L \prec I, \\ 0 & \text{otherwise,} \end{cases}$$

$$H_L(v_I) = \begin{cases} v_I & \text{if } L \prec I \text{ and } L \not\succ I, \\ -v_I & \text{if } L \succ I \text{ and } L \not\prec I, \\ 0 & \text{otherwise.} \end{cases}$$

Proposition 5.4.10 *Let $\varepsilon : E \to \Gamma$ be a full heap and $I \in \mathcal{B}(E)$ a proper ideal of E. Suppose that there exists $\alpha \in V_\Gamma$ and proper ideals $J, J' \in \mathcal{B}(E)$ such that $J \prec_\alpha I$ and $I \prec_\alpha J'$. Then α is an integer multiple of the lowest positive imaginary root, δ.*

Proof Suppose for a contradiction that α is not an integer multiple of δ. Applying the Trivialization Theorem (Theorem 5.3.16) to the pair (J, J'), we find that there exists a sequence of operators S_{k_m} such that the pair

$$(J, J') = (S_{k_1} S_{k_2} \cdots S_{k_l}(I), S_{k_1} S_{k_2} \cdots S_{k_l}(I'))$$

is relatively trivial. It follows that

$$\gamma = s_{k_1} s_{k_2} \cdots s_{k_l}(\chi(J, J'))$$

is a sum of distinct orthogonal fundamental roots. However, we also have $\chi(J, J') = 2\alpha$, which means that the coefficients of the fundamental roots in γ are all even integers. This is a contradiction. □

We can now generalize Definition 3.1.1 and Lemma 3.1.2, as follows. (The aforementioned results deal with the case where α is a fundamental root.)

Definition 5.4.11 Let $\varepsilon : E \to \Gamma$ be a full heap and let α be a real root associated with Γ. We define subsets $J_\alpha^+(E)$, $J_\alpha^-(E)$ and $J_\alpha^0(E)$ of $J(E)$ as follows:

$$\begin{aligned} J_\alpha^+(E) &= \{I \in J(E) : I \succ_\alpha I' \text{ for some } I' \in J(E)\};\\ J_\alpha^-(E) &= \{I \in J(E) : I \prec_\alpha I' \text{ for some } I' \in J(E)\};\\ J_\alpha^0(E) &= J(E) \backslash (J_\alpha^+(E) \cup J_\alpha^-(E)). \end{aligned}$$

Lemma 5.4.12 *Let $\varepsilon : E \to \Gamma$ be a full heap and let α be a real root associated with Γ. The sets $J_\alpha^+(E)$, $J_\alpha^-(E)$ and $J_\alpha^0(E)$ of $J(E)$ appearing in Definition 5.4.11 partition $J(E)$. Furthermore, each $I \in J_\alpha^-(E)$ satisfies $I \prec_\alpha I'$ for a unique I' in $J_\alpha^+(E)$, and conversely each $I \in J_\alpha^+(E)$ satisfies $I \succ_\alpha I'$ for a unique I' in $J_\alpha^-(E)$.*

Proof The only nontrivial thing to check is that $J_\alpha^+(E)$ is disjoint from $J_\alpha^-(E)$, and this follows from Proposition 5.4.10. □

Theorem 5.4.13 *Let $\varepsilon : E \to \Gamma$ be a full heap, let $I \in \mathcal{B}(E)$ a proper ideal of E, and let α be a positive real root. We have*

$$s_\alpha(v_I) = \begin{cases} v_{J'} & \text{if } J' \succ_\alpha I, \\ v_J & \text{if } J \prec_\alpha I, \text{ and} \\ v_I & \text{otherwise.} \end{cases}$$

Proof By Proposition 5.4.10, the cases in the statement of the theorem are mutually exclusive. Recall from Remark 5.4.3 that s_α has order 2.

It remains to identify which pairs of elements of $\mathcal{B}(E)$ are exchanged by the action of s_α. Choose $w \in W$ and $s_i \in S$ such that $s_\alpha = w s_i w^{-1}$. Suppose that $J, J' \in \mathcal{B}(E)$ are exchanged by the action of S_α, let $\beta = \chi(J, J')$, let $\gamma = w^{-1}(\beta)$, and define $I = w^{-1}(J)$ and $I' = w^{-1}(J')$. Since s_α interchanges J and J', Theorem 3.2.21 shows that

$$s_\alpha(\beta) = s_\alpha(\chi(J, J')) = \chi(s_\alpha(J), s_\alpha(J')) = \chi(J', J) = -\beta.$$

It follows that $s_i(\gamma) = -\gamma$, and the formula for the action of s_i on the roots (Theorem 4.2.3 (i)) then shows that γ is a multiple of α_i. By Theorem 4.2.3 (vi), we have $\gamma = \pm\alpha_i$, from which it follows that $\chi(J, J') = \pm\alpha$.

A similar, but easier, argument shows that if $\chi(J, J') = \pm\alpha$, then s_α interchanges J and J'. Since s_α interchanges J and J' if and only if $\chi(J, J') = \pm\alpha$, the statement follows. □

Proposition 5.4.14 *Let $\varepsilon : E \to \Gamma$ be a full heap over an affine Dynkin diagram, let $I \in \mathcal{B}(E)$ a proper ideal of E. Let θ be the highest root of the corresponding finite type Dynkin diagram Γ^0 and let δ be the lowest positive imaginary root.*

(i) *We have $s_0(I) = I$ if and only if $s_\theta(I) = I$.*
(ii) *We have $s_0(I) \succ_{\alpha_0} I$ if and only if $s_\theta(I) \prec_\theta (I)$, and we have $s_0(I) \prec_{\alpha_0} I$ if and only if $s_\theta(I) \succ_\theta (I)$.*
(iii) *We have $s_0([I]) = s_\theta([I])$.*

Proof Let $\phi : E \to E$ be the automorphism of period δ of Theorem 5.3.13. By Theorem 5.3.13 (i), we have $\delta = \theta + \alpha_0$. It follows that if $J, J' \in \mathcal{B}(E)$ satisfy $J \prec_{\alpha_0} J'$, then we have $J' \prec_\theta \phi(J)$, and the same is true with the roles of α_0 and θ reversed. The conclusions of (i) and (ii) now follow from Theorem 5.4.13, and the conclusion of (iii) follows from the above argument by quotienting by the action of the group generated by ϕ. □

5.5 Minuscule representations from heaps

The results of Section 5.3 allow us to obtain a more precise picture of the situation studied in Definition 5.2.13.

Lemma 5.5.1 *Let $\varepsilon : E \to \Gamma$ be a full heap over an affine Dynkin diagram, let A be the generalized Cartan matrix of Γ, let δ be the lowest positive imaginary root and let p be a vertex of Γ such that*

(a) *$\Gamma \backslash \{p\}$ is (path-)connected and*
(b) *the coefficient of α_p in δ is 1.*

Then for any parabolic subheap $F = F_{S\backslash\{p\}}$, we have

(i) *for any vertex $q \neq p$ of Γ, the set $F \cap \varepsilon^{-1}(q)$ is nonempty,*
(ii) *F has a unique maximal element, α_1, and a unique minimal element, α_2, and*
(iii) *if $\varepsilon(\alpha_1) = p_1$ and $\varepsilon(\alpha_2) = p_2$, then the highest weight of V_F is ω_{p_1} and the lowest weight of V_F is $-\omega_{p_2}$.*

Proof By Lemma 3.3.3 (vii), there are ideals $I, I' \in \mathcal{B}(E)$ and an integer z such that $I \subset I'$, $F \cong I' \backslash I$,

$$I = \langle E(p, z) \rangle$$

and

$$E \backslash I' = \{x \in E : x \geq E(p, z+1)\}.$$

By hypothesis (b) and Theorem 5.3.13 (iv), there exists $\phi \in \mathrm{Aut}_{\mathbf{Heap}(\Gamma)}(E)$ such that $\phi(E(p, z)) = E(p, z+1)$; it follows that $\phi(x) > x$ for all $x \in E$. For any $q \neq p$, Lemma 2.3.2 (ii) shows that there is a unique maximal element α in

$$\langle E(p, z) \rangle \cap \varepsilon^{-1}(q).$$

It follows that $\phi(\alpha) \leq E(p, z+1)$ and $\phi(\alpha) \not\leq \alpha$. In turn, this implies that $\phi(\alpha) \in I' \backslash I$ and $F \cap \varepsilon^{-1}(q)$ is nonempty, proving (i).

By Lemma 3.3.3 (ii), F is finite, so certainly F has a maximal element. Suppose that α and α' are distinct maximal elements of F. Since α and α' are incomparable, we must have $\varepsilon(\alpha) \neq \varepsilon(\alpha')$. Since $S\backslash\{p\}$ is connected, there is a path

$$\varepsilon(\alpha) = p_0, p_1, \ldots, p_k = \varepsilon(\alpha')$$

of distinct vertices in Γ such that $k > 1$ and none of the p_i is equal to p. For each i with $0 \leq i \leq k$, part (i) guarantees the existence of a maximal element α_i of F such that $\varepsilon(\alpha_i) = p_i$. Since α and α' are maximal, there must exist a j with $0 < j < k$ such that $\alpha_{j-1} > \alpha_j < \alpha_{j+1}$. Applying Lemma 2.3.13 with $\beta = \alpha_j$ shows that there is a minimal element γ of $E\backslash I'$ with $\varepsilon(\gamma) = p_j$, which contradicts Lemma 3.3.3 (vi).

It follows that F has a unique maximal element, and a dual argument shows that F has a unique minimal element. This completes the proof of (ii). Part (iii) follows from (ii) and the construction of V_F. □

Lemma 5.5.2 *Let $\varepsilon : E \to \Gamma$ be a full heap over an affine Dynkin diagram, let A be the generalized Cartan matrix of Γ, let A^0 be the corresponding finite type matrix (as in Proposition 4.3.18), and let Γ^0 be the Dynkin diagram of A^0.*

(i) *All parabolic subheaps $F_{S\backslash\{s_0\}}$ arising from E are isomorphic in* **Heap**(Γ).
(ii) *The isomorphism type in* **Heap**(Γ) *of $\varepsilon : E \to \Gamma$ is determined by the isomorphism type in* **Heap**(Γ) *of $\varepsilon' : F_{S\backslash\{s_0\}} \to \Gamma^0$.*

Proof Let F be a parabolic subheap of E as described in the statement of (i).

By Lemma 3.3.3 (vii), there are ideals $I, I' \in \mathcal{B}(E)$ and an integer z such that $I \subset I'$, $F \cong I'\backslash I$,

$$I = \langle E(0, z) \rangle$$

and

$$E\backslash I' = \{x \in E : x > E(0, z + 1)\}.$$

By Theorem 5.3.13 (iv), we know E is periodic with period δ, where δ is the lowest positive imaginary root. By Theorem 5.3.13 (i), the coefficient of α_0 in δ is 1. It follows from this that, given any integer z', there is an automorphism (in **Heap**(Γ)) of E taking $E(0, z)$ and $E(0, z + 1)$ to $E(0, t)$ and $E(0, t + 1)$, respectively. Since the structure of F is determined by z, this means that the structure of F depends only on E, completing the proof of (i).

We now turn to the problem of reconstructing E from F, regarding F as a heap over Γ. Let $I, I' \in \mathcal{B}(E)$ be as in the proof of (i), so that $I'\backslash I \cong_{\mathbf{Heap}(\Gamma)} F$. The argument of the previous paragraph shows that I has a maximal element, $E(0, z + 1)$ labelled 0, and that $E\backslash I'$ has a minimal element, $E(0, z)$. Let $\phi \in \mathrm{Aut}_{\mathbf{Heap}(\Gamma)}(E)$ be an automorphism of period δ such that $\phi(\alpha) \geq \alpha$ for all $\alpha \in E$. By Theorem 5.3.13, ϕ exists (and is unique); the argument in the previous paragraph then shows that $\phi(E(0, z)) = E(0, z + 1)$.

By Lemma 2.3.2 (ii), every element of E has an upper and a lower bound in the chain $\varepsilon^{-1}(0)$. It follows that for any $\alpha \in E$, there exists an integer t such that $\phi^t(\alpha)$ lies in the half-open interval $[E(0, z), E(0, z + 1))$. Since ϕ sends $E(0, z)$ to $E(0, z + 1)$, the integer t is unique. This proves that the content of the convex subheap $F' = [E(0, z), E(0, z+1))$ is δ; let $[p_1 p_2 \cdots p_h]$ be a trace representing F'. We now argue as in the last paragraph (the "third and final possibility") of the proof of Proposition 5.3.11 to reconstruct E using Proposition 5.3.7 (ii): note that for each natural number k, there is a finite convex subheap of E with trace $[p_1 p_2 \cdots p_h]^k$. □

Definition 5.5.3 If $\varepsilon : E \to \Gamma$ is a full heap over an affine Dynkin diagram, we call the heap $\varepsilon' : F \to \Gamma^0$ associated with E as in Lemma 5.5.2 the *principal subheap* associated with E. It is well-defined up to isomorphism in $\mathbf{Heap}(\Gamma)$.

Example 5.5.4 Up to isomorphism in $\mathbf{Heap}(\Gamma)$ (ignoring unused labels), the principal subheaps of the full heaps $\mathrm{FH}(A_l^{(1)}(l))$ (Figure 3.2), $\mathrm{FH}(C_l^{(1)}(1))$ (Figure 3.3) and $\mathrm{FH}(D_l^{(1)}(1))$ (Figure 3.7) are shown in Figures 3.9, 3.10 and 3.11, respectively. The principal subheap of the full heap $\mathrm{FH}(A_{2l-1}^{(2)}(l))$ (Figure 3.5) is also isomorphic to the heap of Figure 3.10.

Proposition 5.5.5 *Let $\varepsilon : E \to \Gamma$ be a full heap over an affine Dynkin diagram, let $\mathfrak{g}$ be the corresponding simple Lie algebra, let $F = \psi(E)$ be the principal subheap of E corresponding to $\mathfrak{g}$, and let $V_F = \rho(F)$ be the $\mathfrak{g}$-module corresponding to F. Then the module V_F is minuscule.*

Proof From the classification of affine Dynkin diagrams given in Appendix B, we see that either the Dynkin diagram is a circuit, or the Dynkin diagram is a tree with 0 as an extremal vertex. In either case, $\Gamma \backslash \{0\}$ is connected. By Theorem 5.3.13 (i) and Proposition 4.3.18 (ii), α_0 occurs with coefficient 1 in δ. It follows from Lemma 5.5.1 (ii) that F has a unique maximal element β; let $p = \varepsilon(\beta)$.

Let v_0 be the highest weight vector of V_F. It follows from Lemma 5.2.9 (ii) that we have $H_i(v_0) = \delta_{i,p} v_0$, and this shows that the highest weight of V_F is ω_p.

Suppose now that α is a positive root of $\mathfrak{g}$ and that μ is a weight of V_F; recall from Lemma 5.2.9 (ii) that μ corresponds to an ideal of F. The formula for a reflection now shows that

$$s_\alpha(\mu) = \mu - (\mu, \alpha^\vee)\alpha.$$

By Remark 5.2.5 and Theorem 5.4.13, we have

$$\mathrm{wt}(s_\alpha(\mu)) = \mathrm{wt}(\mu) + c\,\mathrm{wt}(\alpha),$$

where $c \in \{-1, 0, 1\}$. It follows that for any root α and any weight μ of V_F, we have $(\mu, \alpha^\vee) \leq 1$. Specializing μ to the highest weight of V_F, we see that V_F is a minuscule representation. This completes the proof. □

Theorem 5.5.6 *Let $\varepsilon : E \to \Gamma$ be a full heap over an affine Dynkin diagram, let $\mathfrak{g}$ be the corresponding simple Lie algebra, let $F = \psi(E)$ be the principal subheap of E corresponding to $\mathfrak{g}$, and let $V_F = \rho(F)$ be the $\mathfrak{g}$-module corresponding to F. If $\varepsilon' : E' \to \Gamma$ is also a full heap over Γ, then the following are equivalent:*

(i) *the heaps E and E' are isomorphic in* $\mathbf{Heap}(\Gamma)$*;*
(ii) *the heaps F and $\psi(E')$ are isomorphic in* $\mathbf{Heap}(\Gamma)$*;*
(iii) *the minuscule $\mathfrak{g}$-modules V_F and $V_{\rho(\psi((E')))}$ are isomorphic.*

Proof The implications (i) $\Rightarrow$ (ii) and (ii) $\Rightarrow$ (iii) follow by construction.

Let us therefore assume that the $\mathfrak{g}$-modules V_F and $V_{\rho(\psi(E'))}$, which are minuscule by Proposition 5.5.5, are isomorphic. By Theorem 5.2.14 (iii), $F = \psi(E)$ and $\psi(E')$ are isomorphic in **Heap**(Γ), and by Lemma 5.5.2 (ii), this implies that E and E' are isomorphic in **Heap**(Γ). The implication (iii) $\Rightarrow$ (i) now follows. □

Remark 5.5.7 It should be noted that Theorem 5.5.6 does not say that each minuscule representation corresponds to a unique full heap. It may be the case that a Dynkin diagram Γ of finite type can be extended to nonisomorphic Dynkin diagrams Γ_1 and Γ_2 of affine type in such a way that there are full heaps over each of Γ_1 and Γ_2 giving rise to the same minuscule representation. As we shall see in Chapter 6, this can happen when one of Γ_1 and Γ_2 is untwisted and the other is twisted.

5.6 Invariant bilinear forms

An interesting feature of many minuscule representations is that they support invariant linear forms of the corresponding Lie algebras. In this section, we will investigate some invariant bilinear forms of this type; recall that these were defined in Section 1.2.

Definition 5.6.1 Let F be a self-dual principal subheap corresponding to a minuscule representation M. Let λ be a weight of M, and let I be the corresponding ideal of F. We define the *sign*, $\sigma(\lambda)$ of λ to be $+1$ (respectively, -1) if $|I|$ is even (respectively, odd).

Lemma 5.6.2 *Let F be a self-dual principal subheap corresponding to a minuscule representation M, and let λ be a weight of M. Then $-\lambda$ is also a weight of M. Furthermore, the signs of λ and $-\lambda$ are equal if $|F|$ is even, and opposite if $|F|$ is odd.*

Proof The weights of M are closed under negation by Exercise 5.2.11. If λ corresponds to the ideal I of F, then $-\lambda$ corresponds to $(F \backslash I)^*$, where $*$ is the duality of F. The second assertion now follows from the fact that $|I|$ and $|F \backslash I|$ have equal parities if and only if $|F|$ is even. □

Definition 5.6.3 Let M be a minuscule representation associated with a self-dual principal subheap, F. Define a bilinear form $p : M \otimes M \to \mathbb{C}$ by its effect on basis elements as follows:

$$p(m, m') = \begin{cases} \sigma(m) & \text{if the weights of } m \text{ and } m' \text{ are mutual negatives;} \\ 0 & \text{otherwise.} \end{cases}$$

Theorem 5.6.4 *Let F be a self-dual principal subheap corresponding to a minuscule representation M of the simple Lie algebra $\mathfrak{g}$ over $\mathbb{C}$, and let λ be a weight of M.*

(i) *The bilinear form p of Definition 5.6.3 is nondegenerate and $\mathfrak{g}$-invariant.*
(ii) *The bilinear form p is symplectic if $|F|$ is odd, and orthogonal if $|F|$ is even.*

Proof Part (ii) follows from Lemma 5.6.2, and the nondegeneracy assertion of (i) follows from Exercise 1.2.3 and the simplicity of minuscule representations.

It remains to establish the assertion about $\mathfrak{g}$-invariance. It is enough, by Exercise 1.2.2, to check the invariance conditions for the generators h_i, x_{α_i} and $x_{-\alpha_i}$. Let λ and λ' be the weights of m and m', respectively.

Consider first the case of h_i. If m and m' have opposite weights, then $h_i.m = \lambda m$ and $h_i.m' = \lambda' m'$, where $\lambda = -\lambda'$, and then the claim follows by bilinearity. On the other hand, if $\lambda \neq -\lambda'$, then $p(m, m')$, $p(h_i.m, m')$ and $p(m, h_i.m')$ are all zero, completing the proof of the claim.

The cases of x_{α_i} and $x_{-\alpha_i}$ follow a similar case analysis; we consider the situation involving x_{α_i}, because the other is similar. Suppose the first term in the sum

$$p(x_{\alpha_i}.m, m') + p(m, x_{\alpha_i}.m')$$

is nonzero. This means that $\lambda' = -\lambda - \alpha_i$, which in turn implies that $x_{\alpha_i}.m$ is the basis vector of weight $-\lambda$, and also that $x_{\alpha_i}.m'$ is the basis vector of weight $-\lambda$. Since the weights of $x_{\alpha_i}.m$ and m have opposite signs (because α_i is a fundamental root) it follows that the two terms in the sum cancel, as required. The other calculations are similar or easier than this. □

Example 5.6.5 Consider the principal subheap F of Figure 3.10, which is associated with the simple Lie algebra $\mathfrak{g}$ of type C_n. This corresponds to the $2n$-dimensional minuscule representation $L(C_n, \omega_1)$. Since F has an odd number of vertices, Theorem 5.6.4 shows that $L(C_n, \omega_1)$ is endowed with (a nondegenerate) symplectic $\mathfrak{g}$-invariant bilinear form. As we saw in Example 1.2.6, the usual definition of the simple Lie algebra of type C_n is in terms of this symplectic structure.

Exercise 5.6.6 Consider the $2n$-dimensional representation $L(D_n, \omega_1)$ associated with the principal subheap F shown in Figure 3.11. Use Theorem 5.6.4 to show that $L(D_n, \omega_1)$ is endowed with a (nondegenerate) orthogonal $\mathfrak{g}$-invariant bilinear form. As we saw in Example 1.2.8, the usual definition of the simple Lie algebra of type D_n is in terms of this orthogonal structure.

We will be able to give more applications of Theorem 5.6.4 in Chapter 6, when we will have more examples of principal subheaps.

5.7 Notes and references

1 The classification of minuscule weights in Theorem 5.1.5 is well-known. There are other equivalent ways to define minuscule weights. The most commonly found alternative definition is to call a dominant integral weight λ minuscule if the weights of $L(\lambda)$ form a single orbit under the action of the Weyl group.

De Concini [18, chapter 2] gives the following characterization: if λ is a dominant integral weight then it is minuscule if whenever $\mu \leq \lambda$ is a dominant integral weight of $L(\lambda)$, then we necessarily have $\mu = \lambda$. This criterion may be proved from the observations that (a) any weight is in the same W-orbit as a dominant weight (Theorem 4.3.10 (iii)), and (b) every weight is less than or equal to the highest weight in the usual order on weights.

Another equivalent criterion, which appears in Bourbaki [8, VIII section 7.3, proposition 6(ii)], is for all the weights to have the same norm. (Bourbaki gives six equivalent definitions of minuscule weights.)

2 A weight λ is said to be *quasi-minuscule* if it has the property that

$$(\lambda, \alpha^\vee) = 2\frac{(\lambda, \alpha)}{(\alpha, \alpha)} \leq 2$$

for all (real) positive roots α. Although minuscule weights do not exist for all simple Lie algebras over $\mathbb{C}$, quasi-minuscule weights do (see Lassalle [43, chapter 2]).

3 The number associated with $\mathfrak{g}$ appearing in Exercise 5.1.6 appears in a fourth context, as the index of the root lattice of $\mathfrak{g}$ in the weight lattice of $\mathfrak{g}$. (The *root lattice* is the $\mathbb{Z}$-span of the roots, and the *weight lattice* is the $\mathbb{Z}$-span of the weights.) The reason for this is nontrivial; see [8, VIII section 7.3].

4 Most of Theorem 5.3.13 is proved in Green [28, chapter 7], but the proof given here is less complicated.

5 Theorem 5.3.16 and Theorem 5.5.6 are new.

6 The idea of Exercise 5.4.8 can be used to give explicit descriptions of the Bruhat order in Weyl groups of various types. In the classical types, such explicit descriptions are well known; see for example Björner and Brenti [4, chapter 8]. The ideas here also work for types E_6 and E_7, although we do not do this here because the results are hard to state concisely.

7 Bourbaki [8, VIII chapter 7, proposition 12] proves a more general result than Theorem 5.6.4 concerning the vector space of invariant bilinear forms on a finite dimensional simple $\mathfrak{g}$-module M. Bourbaki proves that there are no nonzero invariant bilinear forms on M unless w_0 negates the highest weight of M. If w_0 does negate the highest weight of M, then there is a one-dimensional space of invariant bilinear forms on M, and all the nonzero bilinear forms are nondegenerate. Furthermore, each nonzero bilinear form is either orthogonal or symplectic, and Bourbaki gives a criterion to distinguish the two cases based on the parity of a certain integer associated with the module. This integer is given by (λ, γ), where λ is the highest weight and γ is the sum of the positive coroots; in other words, the sum of the positive roots of the dual system. As we will later prove in Theorem 8.3.10, this always agrees with the number of elements in the principal subheap; see also Gross [31, chapter 1].

6

Full heaps over affine Dynkin diagrams

It will turn out that the correspondence of Theorem 5.5.6 extends to a bijection between the isomorphism types of full heaps over untwisted affine Dynkin diagrams and the isomorphism classes of minuscule modules; the latter were classified in Theorem 5.1.5.

In order to do this we need to construct a corresponding full heap for each minuscule representation of a simple Lie algebra. Most of Chapter 6 is devoted to constructing the necessary heaps. We first deal with the difficult (nonranked) case of type A in Section 6.1, and classify the relevant full heaps in Theorem 6.1.18. The related problem of classifying the proper ideals in type A is considered in Section 6.2. The full heaps corresponding to the spin representations in type D are constructed in Section 6.3, and in turn this enables the constructions in types B and twisted D in Section 6.4. The three exceptional full heaps are constructed in Section 6.5.

The main result of Chapter 6 is Theorem 6.6.2, which classifies all full heaps over affine Dynkin diagrams. (There are other examples of full heaps, some of which are described in Section 6.1.) In addition to giving a construction of all minuscule representations, this theorem has a number of other interesting corollaries, such as a description of all the lowest weights of minuscule representations in Proposition 6.6.8. Many of the constructions in this chapter will be of key importance when studying particular examples later on.

6.1 Full heaps in type $A_l^{(1)}$

Definition 6.1.1 Let $\Gamma(\mathbb{Z})$ be the graph whose vertices are indexed by $\mathbb{Z}$, and where vertices i and j are adjacent if and only if $|i - j| = 1$. Let $E(\mathbb{Z})$ be the set

$$\{(x, y) : x, y \in \mathbb{Z} \text{ and } x - y \text{ is even}\}.$$

We equip $E(\mathbb{Z})$ with the partial order $\leq$ defined by $(a, b) \leq (c, d)$ if and only if both $d > b$ and $|c - a| \leq |d - b|$.

Exercise 6.1.2 Prove that the relation $\leq$ of Definition 6.1.1 is a locally finite partial order. Prove also that if (c, d) covers (a, b), then we must have $d = b + 1$ and $c = a \pm 1$.

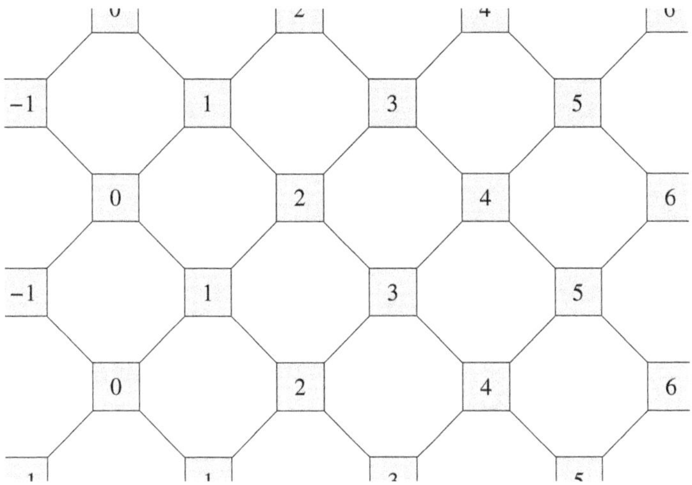

Figure 6.1 Part of the heap $\varepsilon : E(\mathbb{Z}) \to \Gamma(\mathbb{Z})$

Exercise 6.1.3 Suppose that $(a, b) \leq (c, d)$ in $(E(\mathbb{Z}), \leq)$.

(i) Show that if $a \neq c$, then there is a finite chain of covering relations

$$(a, b) = (x_0, y_0) < (x_1, y_1) < \cdots < (x_r, y_r) = (c, d)$$

such that all x_i satisfy $a \leq x_i \leq c$ if $a < c$, and $c \leq x_i \leq a$ if $c < a$.

(ii) Show that if $a = c$, then for any e with $|e - a| = 1$, there exists a finite chain of covering relations

$$(a, b) = (x_0, y_0) < (x_1, y_1) < \cdots < (x_r, y_r) = (c, d)$$

such that $x_i \in \{a, e\}$ for all i.

Part of the heap $\varepsilon : E(\mathbb{Z}) \to \Gamma(\mathbb{Z})$ is shown in Figure 6.1.

Lemma 6.1.4 *The map $\varepsilon : E(\mathbb{Z}) \to \Gamma(\mathbb{Z})$ given by $\varepsilon((x, y)) = x$ is a locally finite full heap.*

Proof Suppose that (a, b) and (c, d) are elements of $E(\mathbb{Z})$ satisfying $(a, b) \leq (c, d)$. It follows from the definition of $\leq$ that any element (x, y) in the open interval $((a, b), (c, d))$ satisfies $a < x < c$, from which it follows that every interval is finite, and that $(E(\mathbb{Z}), \leq)$ is a locally finite poset.

We next prove that ε is a heap. Axiom (H1) follows from the fact that if (x, y_1) and (x, y_2) are elements of $E(\mathbb{Z})$ satisfying $y_1 < y_2$, then we have $(x, y_1) \leq (x, y_2)$. Since $E(\mathbb{Z})$ is locally finite, to prove Axiom (H2), it is enough by Lemma 2.1.5 (iv) to show that every covering relation is contained in an edge chain. This follows by Exercise 6.1.2, which shows that if (c, d) covers (a, b), then $c = a \pm 1$.

Finally, we prove that ε is a full heap. The argument in the previous paragraph shows that each vertex chain of $E(\mathbb{Z})$ consists of either the even integers or the odd integers under the usual partial order; it follows that each vertex chain is isomorphic to $\mathbb{Z}$, which

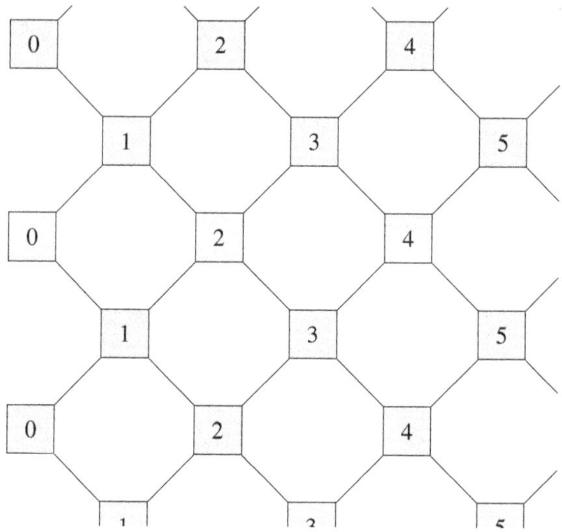

Figure 6.2 Part of the full heap $\varepsilon : E^+ \to B_\infty$ of Exercise 6.1.5

proves Axiom (F1). Axiom (F2) follows from the observation that each element (a, b) in $E(\mathbb{Z})$ covers $(a+1, b-1)$ and $(a-1, b-1)$, and is covered by $(a+1, b+1)$ and $(a-1, b+1)$. Axiom (F3) follows from the fact that $a_{p,p-1} = a_{p,p+1} = -1$ together with the observation that a p-interval in $E(\mathbb{Z})$ consists of four elements of the form $\{(p, c+1), (p-1, c), (p+1, c), (p, c-1)\}$. □

Exercise 6.1.5 Let E^+ be the underlying poset of the subheap of $\varepsilon : E(\mathbb{Z}) \to \Gamma(\mathbb{Z})$ given by $\varepsilon^{-1}(\mathbb{Z}^{\geq 0}) \cap E(\mathbb{Z})$. (See Figure 6.2.) Let $\varepsilon : E^+ \to \mathbb{Z}$ be the function $\varepsilon((x, y)) = x$.

Let B_∞ be the graph whose vertices are indexed by nonnegative integers, for which there is an edge between i and j if and only if $|i - j| = 1$. The edge is undecorated unless $\{i, j\} = \{0, 1\}$, in which case the edge is doubled and points towards 0.

Show that $\varepsilon : E^+ \to B_\infty$ is a full heap.

Exercise 6.1.6 Let D_∞ be a graph whose vertices are indexed by the set $\mathbb{Z}^{\geq 0} \cup \{\overline{0}\}$, whose edges (all undecorated) are defined by the following rules. If $\overline{0} \notin \{i, j\}$, then there is an edge between i and j if $|i - j| = 1$, and no edge otherwise. The vertex $\overline{0}$ is connected only to vertex 1.

Let E^+ be the poset of Exercise 6.1.5. Let $\varepsilon : E^+ \to D_\infty$ (see Figure 6.3) be the function

$$\varepsilon((x, y)) = \begin{cases} \overline{0} & \text{if } x = 0 \text{ and } (x - y)/2 \text{ is odd,} \\ x & \text{otherwise.} \end{cases}$$

Show that $\varepsilon : E^+ \to D_\infty$ is a full heap.

A good way to construct the full heaps over $A_l^{(1)}$ is as quotients of the heap $\varepsilon : E(\mathbb{Z}) \to \Gamma(\mathbb{Z})$.

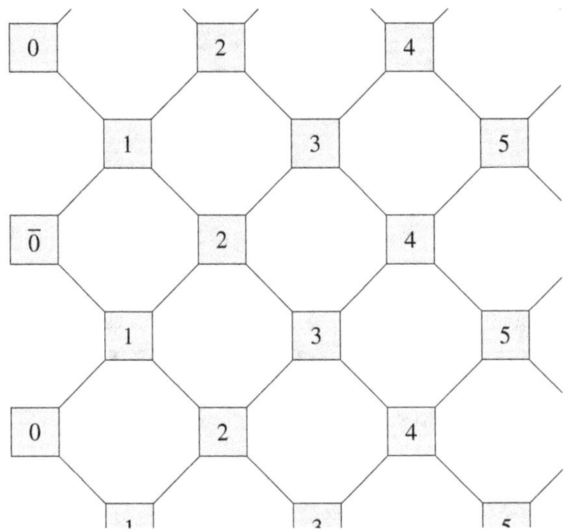

Figure 6.3 Part of the full heap $\varepsilon : E^+ \to D_\infty$ of Exercise 6.1.6

Definition 6.1.7 Let $\varepsilon : E(\mathbb{Z}) \to \Gamma(\mathbb{Z})$ be the heap defined above. Consider the Dynkin diagram $A_l^{(1)}$ for $l \geq 1$, let $n = l + 1$, and choose an integer k with $1 \leq k \leq l$. Let G_k be the additive group of integers, and let $g_k \in G_k$ be a generator of G_k. Let G_k act on $E(\mathbb{Z})$ by

$$g_k.(x, y) = (x + n, y - n + 2k).$$

Lemma 6.1.8 *Maintain the notation of Definition 6.1.7. Let $p \in E(\mathbb{Z})$ and $g \in G_k$.*

(i) *The group G_k acts by poset automorphisms on $E(\mathbb{Z})$.*
(ii) *If p is comparable to $g(p)$, then g is the identity.*
(iii) *We have $g(\varepsilon^{-1}(\varepsilon(p))) = \varepsilon^{-1}(\varepsilon(g(p)))$.*
(iv) *The quotient heap*

$$\overline{\varepsilon} : G_k \backslash E(\mathbb{Z}) \to G_k \backslash \Gamma(\mathbb{Z})$$

exists, in the sense of Theorem 2.4.6.
(v) *The quotient graph $G_k \backslash \Gamma(\mathbb{Z})$ is isomorphic to the underlying graph of the Dynkin diagram $A_l^{(1)}$.*

Proof Part (i) holds because any $g \in G_k$ acts on $E(\mathbb{Z})$ by adding a constant vector to each point.

Suppose for $p = (x, y) \in E(\mathbb{Z})$ and $g \in G_k$, p is comparable to $g(p)$. Define m so that $g = g_k^m$. We then have

$$g.(x, y) = (x + mn, y - mn + 2mk).$$

If $m \neq 0$ then

$$y - mn < y - mn + 2mk < y + mn,$$

which means that $|mn| > |mn - 2mk|$. The definition of the partial order on $E(\mathbb{Z})$ then means that p and $g(p)$ are not comparable. We conclude that $m = 0$ and g is the identity, proving (ii).

Part (iii) follows from the fact that if $g.(x, y) = (x', y')$, then $x \equiv x' \bmod n$ and $y \equiv y' \bmod n$.

It follows from (i) and (ii) that the action of G_k on $E(\mathbb{Z})$ satisfies the hypotheses of Proposition A.2.4. Part (iv) then follows by combining (iii) with Theorem 2.4.6.

Part (v) follows from the construction of the quotient graph $G_k \backslash \Gamma(\mathbb{Z})$ combined with the fact that the generator g_k satisfies $g_k.(x, y) = (x + n, y')$. □

Definition 6.1.9 Recall from Lemma 5.5.2 and Proposition 5.5.5 that if Γ is an affine Dynkin diagram and $\Gamma_0 = \Gamma \backslash \{0\}$, then each minuscule representation of the simple Lie algebra corresponding to Γ_0 determines up to isomorphism at most one full heap over Γ. If this minuscule representation has highest weight ω_p, then we will denote the corresponding full heap by $\varepsilon : \mathrm{FH}(\Gamma(p)) \to \Gamma$.

We denote the quotient heap defined by Lemma 6.1.8 (iv) by

$$\varepsilon_k : \mathrm{FH}(A_l^{(1)}(k)) \to A_l^{(1)};$$

this terminology will be justified by Theorem 6.1.18 and Proposition 6.2.6 below.

Exercise 6.1.10 Show that if we have $k < 1$ or $k > l$, the construction of $\mathrm{FH}(A_l^{(1)}(k))$ fails.

Lemma 6.1.11 *Let $\varepsilon_k : \mathrm{FH}(A_l^{(1)}(k)) \to A_l^{(1)}$ be the heap of Definition 6.1.9, where $l \geq 1$ and $1 \leq k \leq l$. If $[x, y]$ is a p-interval in $\mathrm{FH}(A_l^{(1)}(k))$, and q is a vertex of $A_l^{(1)}$ adjacent to p, then there is a single element in $\varepsilon_k^{-1}(q) \cap [x, y]$.*

Proof Let $n = l + 1$, and let $\alpha_0 = (x_0, y_0)$ and $\alpha_1 = (x_1, y_1)$ be elements of $E(\mathbb{Z})$ that project under $\pi_{E(\mathbb{Z})}$ to x and y respectively. By applying an element of G_k if necessary, we may assume that $x_1 = x_0$. Since (x_0, y_0) and (x_0, y_1) lie in the same vertex chain of $E(\mathbb{Z})$, it must be the case that $y_1 = y_0 + 2$, otherwise $(x_0, y_0 + 2)$ would project to an element of (x, y) with label $p = x_0 \bmod n$.

Since $l > 1$, there is a unique integer x' such that $|x' - x_0| = 1$ and $x' = q \bmod n$. If we have $(x', y') \in E(\mathbb{Z})$ and $y' > y_0 + 2$ then (x', y') is greater than both (x_0, y_0) and $(x_0, y_0 + 2)$. Dually, if we have $(x', y') \in E(\mathbb{Z})$ and $y' < y_0$ then (x', y') is less than both (x_0, y_0) and $(x_0, y_0 + 2)$. It follows that the only way (x', y') can project to an element in $[x, y]$ is if $y' = y_0 + 1$. Furthermore, $(x', y_0 + 1)$ actually does project to an element of $[x, y]$, and this is the unique element required by the statement. □

Lemma 6.1.12 *Let $\varepsilon_k : \mathrm{FH}(A_l^{(1)}(k)) \to A_l^{(1)}$ be the heap of Definition 6.1.9, where $l \geq 1$ and $1 \leq k \leq l$. If $\alpha_0 < \alpha_1$ is a covering pair in $\mathrm{FH}(A_l^{(1)}(k))$, and β_0 is any element of $E(\mathbb{Z})$ for which $\pi_{E(\mathbb{Z})}(\beta_0) = \alpha_0$, then there exists a covering pair $\beta_0 < \beta_1$ in $E(\mathbb{Z})$ such that $\pi_{E(\mathbb{Z})}(\beta_1) = \alpha_1$.*

Proof Since $\pi_{E(\mathbb{Z})}$ is surjective, we can choose an element β_1 such that $\pi_{E(\mathbb{Z})}(\beta_1) = \alpha_1$. Since $\mathrm{FH}(A_l^{(1)}(k))$ is a locally finite heap, $\varepsilon_k(\alpha_0)$ and $\varepsilon_k(\alpha_1)$ must be adjacent vertices of $A_l^{(1)}$. By Lemma 6.1.8 (v), we may replace β_1 by $g(\beta_1)$ for some element $g \in G_k$ such that $\varepsilon(\beta_1) = \varepsilon(\beta_0) \pm 1$.

We may now assume that β_0 and β_1 are in the same edge chain of $E(\mathbb{Z})$, and we have $\beta_0 < \beta_1$ because $\alpha_0 < \alpha_1$. If there exists $\beta' \in E(\mathbb{Z})$ with $\beta_0 < \beta' < \beta_1$, then we must have $\alpha_0 < \pi_{E(\mathbb{Z})}(\beta') < \alpha_1$, a contradiction. It follows that $\beta_0 < \beta_1$ is a covering pair with the required properties. □

Exercise 6.1.13 Show that, under the assumptions of Lemma 6.1.12, it is possible for $\beta_0 < \beta_1$ to be a covering relation in $E(\mathbb{Z})$ and yet for $\pi_{E(\mathbb{Z})}(\beta_0) < \pi_{E(\mathbb{Z})}(\beta_1)$ not to be a covering relation in $\mathrm{FH}(A_l^{(1)}(k))$.

Lemma 6.1.14 *Let $\varepsilon_k : \mathrm{FH}(A_l^{(1)}(k)) \to A_l^{(1)}$ be the heap of Definition 6.1.9, and assume that $1 \leq k \leq l$. If $[x, y]$ is a p-interval and*

$$x = \alpha_0 < \alpha_1 < \cdots < \alpha_m = y$$

is a chain of covering relations in $\mathrm{FH}(A_l^{(1)}(k))$ with $m > 2$, then we must have $k = 1$ or $k = l$.

Proof Let $E = \mathrm{FH}(A_l^{(1)}(k))$, and let r be an integer for which $x = E(p, r)$ and $y = E(p, r + 1)$.

Since $l \geq 2$, there are two distinct vertices of $A_l^{(1)}$ adjacent to p; let us call these q_1 and q_2. By Lemma 6.1.11, the interval $[x, y]$ contains a unique element z_1 with label q_1, a unique element z_2 with label q_2, and no other elements with labels adjacent to p.

We deal first with the case in which α_1 has label $q_1 = p + 1 \bmod n$ and α_{m-1} has label $q_2 = p - 1 \bmod n$. Since the sequence

$$\varepsilon_k(\alpha_0), \varepsilon_k(\alpha_1), \ldots, \varepsilon_k(\alpha_n)$$

is a path in $A_l^{(1)}$, but none of the elements $\alpha_1, \ldots, \alpha_{m-1}$ have label p, we must have $m \geq n$.

By repeatedly applying Lemma 6.1.12, there is a chain

$$\beta_0 < \beta_1 < \cdots < \beta_m$$

of covering relations in $E(\mathbb{Z})$ such that $\pi_{E(\mathbb{Z})}(\beta_i) = \alpha_i$. Let $\beta_0 = E(\mathbb{Z})(p', r_0)$ for some integer r_0. Since $m \geq n$, Exercise 6.1.2 shows that we have $\beta_m = E(\mathbb{Z})(p'', r_m)$ for some $r_m \geq r_0 + n$. Since the elements $\alpha_1, \ldots, \alpha_{m-1}$ do not have label p, we must have $p'' = p' + n$, which implies that $\beta_m \geq E(\mathbb{Z})(p' + n, r_0 + n)$.

Now the element $E(\mathbb{Z})(p' + n, r_0 + n)$ is identified under $\pi_{E(\mathbb{Z})}$ with $E(\mathbb{Z})(p', r_0 + 2n - 2k)$. Arguing as in the proof of Lemma 6.1.11, we must have $r_0 + 2n - 2k = r_0 + 2$. It follows that $k = l$, as required.

The other possibility is that α_1 has label $q_1 = p - 1 \bmod n$ and α_{m-1} has label $q_2 = p + 1 \bmod n$. The argument of the preceding three paragraphs may then be altered to prove that $k = 1$ in this case. □

Lemma 6.1.15 *Let $\varepsilon_k : \mathrm{FH}(A_l^{(1)}(k)) \to A_l^{(1)}$ be the heap of Definition 6.1.9, and assume that $1 < k < l$ (so in particular $l > 2$). If $[x, y]$ is a p-interval, then $[x, y]$ consists of four elements, $\{x, y, z_1, z_2\}$, where $\varepsilon(z_1) = p - 1 \bmod n$, $\varepsilon(z_2) = p + 1 \bmod n$, and the chains $x < z_1 < y$ and $x < z_2 < y$ consist entirely of covering relations.*

Proof The existence and uniqueness of z_1 and z_2 is guaranteed by Lemma 6.1.11, and the assertion about covering relations follows from Lemma 6.1.14. □

Lemma 6.1.16 *Let $l \geq 1$ and let $\Gamma = A_l^{(1)}$. Then the heaps*

$$\varepsilon_k : \mathrm{FH}(A_l^{(1)}(k)) \to A_l^{(1)},$$

as k ranges over the set $\{1, \ldots, l\}$, are pairwise nonisomorphic in **Heap**(Γ).

Proof If $l = 1$ there is only one heap in the set, so the statement holds vacuously.
Suppose that $l > 1$ and let $n = l + 1$. Choose a sequence of elements

$$\alpha_0 < \alpha_1 < \cdots < \alpha_{n-1} < \alpha_n$$

in $\mathrm{FH}(A_l^{(1)}(k))$ such that $\varepsilon(\alpha_i) = i \bmod n$ and such that for each i with $0 \leq i < n, \alpha_{i+1}$ is the smallest element of its vertex chain that is greater than α_i. (The existence of α_{i+1} follows from the definition of edge chains.) Let $\beta_0 = E(\mathbb{Z})(0, r)$ be the element of $E(\mathbb{Z})$ for which $\pi_{E(\mathbb{Z})}(\beta_0) = \alpha_0$. It follows by induction on i that we have

$$\pi_{E(\mathbb{Z})}(E(\mathbb{Z})(i, r + i)) = \alpha_i$$

for all $0 \leq i \leq n$.

The element $E(\mathbb{Z})(n, r + n)$ is identified by $\pi_{E(\mathbb{Z})}$ with $E(\mathbb{Z})(0, r + 2n - 2k)$. It follows that if we have $\alpha_0 = \mathrm{FH}(A_l^{(1)}(k))(0, r')$ for $r' \in \mathbb{Z}$, then we also have $\alpha_n = \mathrm{FH}(A_l^{(1)}(k))(0, r' + 2n - 2k)$. This procedure shows how to determine k using structural properties of the heap $\mathrm{FH}(A_l^{(1)}(k))$, and completes the proof. □

Exercise 6.1.17 Show that the chain

$$\alpha_0 < \alpha_1 < \cdots < \alpha_{n-1} < \alpha_n$$

in the proof of Lemma 6.1.16 may or may not consist of covering relations.

Theorem 6.1.18 (McGregor-Dorsey) *Let $l \geq 1$ be an integer and let $\Gamma = A_l^{(1)}$. The set of heaps $\varepsilon_k : \mathrm{FH}(A_l^{(1)}(k)) \to \Gamma$ for $k \in \{1, \ldots, l\}$ is a complete and irredundantly described set of* **Heap**(Γ)*-isomorphism classes of full heaps over Γ.*

Proof If $l = 1$, then Γ has two vertices and Theorem 2.3.15 (iv) shows that there is a unique full heap over Γ. The repeating motif of this heap corresponds to the trace [01], and by inspection, this heap is isomorphic in **Heap**(Γ) to $\varepsilon_1 : \mathrm{FH}(A_1^{(1)}(1)) \to \Gamma$. This completes the proof in the case $l = 1$, so we assume that $l > 1$ from now on.

The heap $\varepsilon_l : \mathrm{FH}(A_l^{(1)}(l)) \to \Gamma$ is isomorphic in **Heap**(Γ) to the full heap E shown in Figure 3.2, and $\varepsilon_1 : \mathrm{FH}(A_l^{(1)}(1)) \to \Gamma$ is isomorphic to the dual of E, which is also a full heap.

Suppose now that $1 < k < l$. Because no two elements of a the vertex chain of $E(\mathbb{Z})$ are identified with each other by the quotient map $\pi_{E(\mathbb{Z})}$, it follows that the vertex chains of $\varepsilon_k : \mathrm{FH}(A_l^{(1)}(k)) \to \Gamma$ are each isomorphic to $\mathbb{Z}$, satisfying Axiom (F1). By Lemma 6.1.15, every element of $\varepsilon_k^{-1}(p) \cap \mathrm{FH}(A_l^{(1)}(k))$ both covers and is covered by an element labelled with label $p - 1$ and an element with label $p + 1$ (where indices are taken modulo $n = l + 1$). It follows that Axiom (F2) holds. Lemma 6.1.15 also shows that Axiom (F3) holds, which shows that the heaps $\varepsilon_k : \mathrm{FH}(A_l^{(1)}(k)) \to \Gamma$ are full heaps.

The heaps in the statement are nonisomorphic by Lemma 6.1.16. Theorems 5.1.5 and 5.5.6, together with Lemma 5.5.2 (i), show that there are at most l **Heap**(Γ)-isomorphism classes of full heaps over $A_l^{(1)}$, so it follows that the set in the statement is complete. □

Exercise 6.1.19 Show that the poset $\mathrm{FH}(A_l^{(1)}(k))$ is ranked if $k = 1$ or $k = l$ or $k = (l + 1)/2$ (if l is odd), but is not ranked otherwise.

6.2 Proper ideals in type $A_l^{(1)}$

Definition 6.2.1 Let $l \geq 1$ and $n = l + 1$. Let $1 \leq k \leq l$, let

$$\varepsilon_k : \mathrm{FH}(A_l^{(1)}(k)) \to A_l^{(1)}$$

be the full heap of Definition 6.1.9, and let I be a proper ideal of $\mathrm{FH}(A_l^{(1)}(k))$. For each $1 \leq i \leq n$, let α_i be the largest element of $\varepsilon^{-1}(i) \cap I$, with indices taken modulo n. We define a word $\sigma(I) = u_1 u_2 \cdots u_n$ in the alphabet $\{+, -\}$ by the condition that $u_i = +$ (respectively, $u_i = -$) if $\alpha_i < \alpha_{i+1}$ (respectively, $\alpha_i > \alpha_{i+1}$).

Proposition 6.2.2 *Maintain the notation of Definition 6.2.1. There is a bijection between the proper ideals of $E = \mathrm{FH}(A_l^{(1)}(k))$ and pairs (t, σ), where $t \in \mathbb{Z}$ and σ is a word of length n in the alphabet $\{+, -\}$ containing precisely k occurrences of $+$. The pair (t, σ) corresponds to the ideal I such that $\sigma = \sigma(I)$ and $E(0, t)$ is the maximal element of $\varepsilon^{-1}(0) \cap I$ with respect to some fixed labelling of the vertex chains of E.*

Proof Let I be a proper ideal of E. Let α_0 be the largest element of $\varepsilon^{-1}(0) \cap I$, and let $\beta_0 = E(0, r)$ be an element of $E(\mathbb{Z})$ for which $\pi_{E(\mathbb{Z})}(\beta_0) = \alpha_0$. If $\alpha_1 > \alpha_0$ (respectively, $\alpha_1 < \alpha_0$), then it follows by Lemma 6.1.11 that there is an element β_1 of $E(\mathbb{Z})$ such that $\pi_{E(\mathbb{Z})}(\beta_1) = \alpha_1$ and $\beta_1 > \beta_0$ (respectively, $\beta_1 < \beta_0$) is a covering relation. We can iterate this procedure to produce a sequence

$$\beta_0, \beta_1, \ldots, \beta_n,$$

and it will follow that $\pi_{E(\mathbb{Z})}(\beta_0) = \pi_{E(\mathbb{Z})}(\beta_n)$. This means that $\beta_n = E(n, r - 2n + 2k)$, and hence that there are k occurrences of $+$ and $n - k$ occurrences of $-$ in $\sigma(I)$. Conversely, any string of k occurrences of $+$ and $n - k$ occurrences of $-$ will define a proper ideal of E, once the maximal element of $\varepsilon^{-1}(p) \cap I$ is specified for some vertex p. This establishes the claimed bijection. □

Although the heaps $\mathrm{FH}(A_l^{(1)}(k))$ are hard to visualize directly in general, their principal subheaps are easy to draw.

Proposition 6.2.3 *Let $\varepsilon_k : \mathrm{FH}(A_l^{(1)}(k)) \to A_l^{(1)}$ be the full heap of Definition 6.1.9, using the notational conventions of Definition 6.2.1, and let $\pi : E \to \mathrm{FH}(A_l^{(1)}(k))$ be the usual projection map of heaps. Let F be the subheap of $E(\mathbb{Z})$ consisting of those* *elements within the rectangular region bounded by the lines*

$$\begin{aligned} y &= x + 1, \\ y &= x + 1 - 2n + 2k, \\ x + y &= 1, \\ x + y &= 1 + 2k, \end{aligned}$$

whose corners are the points

$$(0, 1), (d, 1 - d), (k, k + 1), \text{ and } (n, 1 + 2k - n).$$

Then $\pi(F)$ is the principal subheap of $\mathrm{FH}(A_l^{(1)}(k))$, *and* $\pi(F) \cong_{\mathbf{Heap}} F$.

Proof The region below one or both of the lines $x + y = 1$ and $y = x + 1 - 2n + 2k$ picks out the vertices α of E satisfying the condition

$$\pi_{E(\mathbb{Z})}(\alpha) \leq \pi_{E(\mathbb{Z})}(0, 0).$$

Similarly, the region above one or both of the lines $x + y = 1 + 2k$ and $y = x$ picks out the vertices α of E satisfying the condition

$$\pi_{E(\mathbb{Z})}(\alpha) \geq \pi_{E(\mathbb{Z})}(0, 2).$$

The complement of the union of the aforementioned two regions is a disjoint union of a countably infinite number of rectangles, namely the one described in the statement together with all of its translates by integer multiples of the vector $(n, 2k - n)$. (Each rectangle shares one point with its two adjacent translates, and is disjoint from the others.)

Applying π to this picture has the effect of identifying all the translates of the above rectangles. If we parametrize $E = \mathrm{FH}(A_l^{(1)}(k))$ in such a way that $E(0, 0) = \pi_{E(\mathbb{Z})}(0, 0)$, then the union of rectangles projects to the subheap of $\mathrm{FH}(A_l^{(1)}(k))$ consisting of the elements

$$\{\alpha \in E : \alpha \not\leq E(0, 0) \text{ and } \alpha \not\geq E(0, 1)\},$$

which by Lemma 3.3.3 (vii) is precisely the principal subheap of E. Furthermore, since the labels of the rectangular subheap F described in the statement all lie in the range $1 \leq p < n$, it follows that π is injective on vertices and labels when restricted to F. □

Proposition 6.2.4 *Maintain the notation of Definition 6.2.1. There is a bijection between the ideals of the principal subheap F of* $\mathrm{FH}(A_l^{(1)}(k))$ *and words σ of length n in the alphabet $\{+, -\}$ containing precisely k occurrences of $+$.*

Proof By Lemma 3.3.3 (v) and Proposition 6.2.2, the ideals of F are in correspondence with the proper ideals (t, σ) of $\mathrm{FH}(A_l^{(1)}(k))$ for a fixed value of t, from which the statement follows. □

Example 6.2.5 Figure 6.4 shows a rectangular subheap F of $E(\mathbb{Z})$ corresponding to the principal subheap of $\mathrm{FH}(A_4^{(1)}(3))$. The dashed region is the ideal of F corresponding to the sequence $+ - + + -$. This may be identified with a walk from the left endpoint

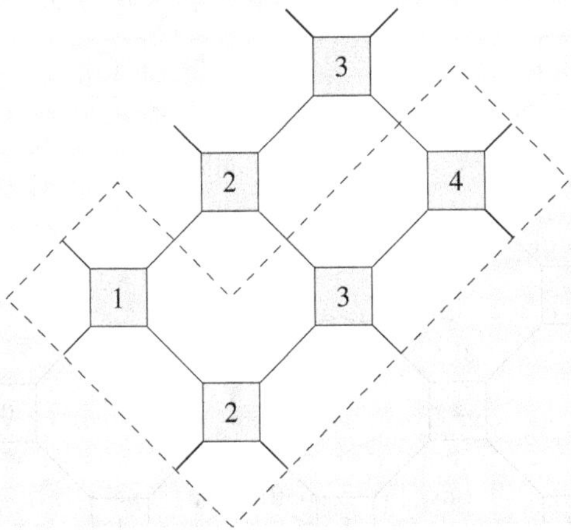

Figure 6.4 A rectangular subheap F of $E(\mathbb{Z})$ corresponding to the principal subheap of $\mathrm{FH}(A_4^{(1)}(3))$

of the rectangle to the right endpoint, using the sequence "up, down, up, up, down". The two extreme ends of the rectangle are identified under the quotient map π.

Proposition 6.2.6 *Let F be the principal subheap of the heap $\varepsilon_k : \mathrm{FH}(A_l^{(1)}(k)) \to A_l^{(1)}$, where $l \geq 1$, $n = l + 1$ and $1 \leq k \leq l$. Then the minuscule module V_F has highest weight ω_k and lowest weight $-\omega_{n-k}$.*

Proof Proposition 6.2.3 describes the bounding lines of the rectangular heap corresponding to F, and it follows from this that the maximal (respectively, minimal) element of F is labelled by k (respectively, $n - k$). The result now follows from Lemma 5.5.1 (iii). □

Exercise 6.2.7 Show that the minuscule module of highest weight ω_k of the simple Lie algebra of type A_l has dimension $\binom{l+1}{k}$.

Exercise 6.2.8 Let $\mathfrak{g}$ be a Lie algebra and let V be a $\mathfrak{g}$-module. Show that the k-th exterior power $\bigwedge^k(V)$ has a unique $\mathfrak{g}$-module structure such that for every $x \in \mathfrak{g}$ and $v_1, v_2 \ldots, v_k \in V$, we have

$$x.(v_1 \wedge v_2 \wedge \cdots \wedge v_k) = \sum_{i=1}^{k} v_1 \wedge \cdots \wedge v_{i-1} \wedge (x.v_i) \wedge v_{i+1} \wedge \cdots \wedge v_k.$$

Exercise 6.2.9 Now let $\mathfrak{g}$ be the simple Lie algebra over $\mathbb{C}$ of type A_l, let V be the $\mathfrak{g}$-module of highest weight ω_1, corresponding to $\mathrm{FH}(A_l^{(1)}(1))$, and suppose that $1 \leq k \leq l$. Arguing as in Example 4.1.8, we can equip V with a basis $\{v_1, v_2, \ldots, v_n\}$ corresponding to a nested sequence of proper ideals, such that v_1 is the highest weight vector and v_n is the lowest. This means that the generators e_i and f_i are identified with the matrix units $E_{i,i+1}$ and $E_{i+1,i}$, respectively. Let V_F be the minuscule module corresponding to the heap $\mathrm{FH}(A_l^{(1)}(k))$. If σ is a word of length $n = l + 1$ containing k occurrences of the symbol $+$ and $n - k$ occurrences of the symbol $-$, then there is a basis element $v_\sigma = v_{\sigma(I)}$ of V_F, and a basis element

$$x_\sigma = v_{i_1} \wedge v_{i_2} \wedge \cdots \wedge v_{i_k}$$

of $\bigwedge^k(V)$, where $i_1 < i_2 < \cdots < i_k$ and the indices i_r are precisely those where σ has an occurrence of $+$. Show that there is an isomorphism of $\mathfrak{g}$-modules between V_F and $\bigwedge^k(V)$ that identifies v_σ with x_σ for all σ.

Exercise 6.2.10 Let $\mathfrak{g}$ be a simple Lie algebra of type A_n, and let F be the principal subheap associated with $L(A_n, \omega_j)$.

(i) Show that F is self-dual if and only if both n is odd and $j = (n + 1)/2$. Show that, if these conditions hold, we have $|F| = j^2$.
(ii) Show that if the conditions of (i) hold, then the bilinear form p of Theorem 5.6.4 is symplectic if j is odd, and orthogonal if j is even.
(iii) Show that if the conditions of (i) hold, then under the identifications of Exercise 6.2.9, the form p is given by

$$\begin{aligned} &p(v_{i_1} \wedge v_{i_2} \wedge \cdots \wedge v_{i_j}, v_{i_{j+1}} \wedge v_{i_{j+2}} \wedge \ldots \wedge v_{i_{n+1}}) \\ &\quad = v_{i_1} \wedge v_{i_2} \wedge \cdots \wedge v_{i_j} \wedge v_{i_{j+1}} \wedge v_{i_{j+2}} \wedge \ldots \wedge v_{i_{n+1}}. \end{aligned}$$

Exercise 6.2.11 Let $\Gamma = A_l^{(1)}$, and consider the faithful action of the affine Weyl group $W(A_l^{(1)})$ on the proper ideals of the full heap $\varepsilon_k : \mathrm{FH}(A_l^{(1)}(k)) \to \Gamma$. Maintain the notation of Proposition 6.2.2, and let σ be a word of length $n = l + 1$ in the alphabet $\{+, -\}$.

(i) Show that if $i \neq 0$, the generator s_i sends (t, σ) to $(t, s_i(\sigma))$, where the action of s_i on words of length n is by place permutation by the transposition $(i, i + 1)$.
(ii) Let τ be the word of length $n - 2$ for which $\sigma = u\tau u'$ and $u, u' \in \{+, -\}$. Show that we have

$$s_0((t, \sigma)) = \begin{cases} (t + 1, u'\tau u) & \text{if } u = + \text{ and } u' = -, \\ (t - 1, u'\tau u) & \text{if } u = - \text{ and } u' = +, \\ (t, \sigma) & \text{if } u = u'. \end{cases}$$

Exercise 6.2.12 (McGregor-Dorsey) Consider the full heap $E = \varepsilon_k : \mathrm{FH}(A_l^{(1)}(k)) \to A_l^{(1)}$.

(i) Show that E has a convex subheap F corresponding to the trace

$$[012 \cdots j][l(l - 1)(l - 2) \cdots (j + 1)]$$

if and only if $j = (l + 1) - k$.
(ii) Assuming that $j = (l + 1) - k$, show that for any $N > 0$, E has a convex subheap corresponding to the trace

$$([012 \cdots j][l(l - 1)(l - 2) \cdots (j + 1)])^N.$$

6.3 Spin representations in type D_l

Definition 6.3.1 Suppose that $a, b \in \mathbb{Z}$ satisfy $a \leq b$. Let $E_{a,b}$ be the subset of $E(\mathbb{Z})$ consisting of the elements

$$\{(x, y) : x, y \in \mathbb{Z}, a \leq x \leq b, \text{ and } x - y \text{ is even.}\}$$

Let $\Gamma = D_l^{(1)}$ be the Dynkin diagram of type affine D_l, numbered as in Appendix B. For each $k \in \{l-1, l\}$, we define the partially ordered set $\mathrm{FH}(D_l^{(1)}(k))$ to be $E_{1,l-1}$, and we define the function $\varepsilon_k : \mathrm{FH}(D_l^{(1)}(k)) \to \Gamma$ by

$$\varepsilon_k((x, y)) = \begin{cases} 1 & \text{if } x = 1 \text{ and } y = 1 \bmod 4, \\ 0 & \text{if } x = 1 \text{ and } y = 3 \bmod 4, \\ k & \text{if } x = l-1 \text{ and } y = x \bmod 4, \\ 2l-1-k & \text{if } x = l-1 \text{ and } y = x+2 \bmod 4, \\ x & \text{otherwise.} \end{cases}.$$

(This notation will be justified by Propositions 6.3.2 and 6.3.3.)

Proposition 6.3.2 *If $l \geq 4$ and $k \in \{l-1, l\}$, then the function $\varepsilon_k : \mathrm{FH}(D_l^{(1)}(k)) \to D_l^{(1)}$ of Definition 6.3.1 is a full heap.*

Proof The assertion in Axiom (H1) about vertex chains follows from the corresponding property for the heap $\varepsilon : E(\mathbb{Z}) \to \Gamma(\mathbb{Z})$ in Lemma 6.1.4. The same is true of the edge chains, except in the case where one of the vertices p involved in the edge chain $\varepsilon_k^{-1}(\{p, q\})$ is an endpoint of $\Gamma = D_l^{(1)}$. In this case, the chain can be identified with a subset of one of the edge chains of $E(\mathbb{Z})$, and is therefore a chain by restriction.

To prove Axiom (H2), we can argue as in the proof of Lemma 6.1.4 that every covering relation is involved in an edge chain. It only remains to be seen that if (c, d) covers (a, b) in $\mathrm{FH}(D_l^{(1)}(k))$, then $d = b+1$ and $c = a \pm 1$; the corresponding property for $E(\mathbb{Z})$ holds by Exercise 6.1.2. Let us therefore assume that (c, d) covers (a, b). Since the partial order on $\mathrm{FH}(D_l^{(1)}(k))$ is obtained from $E(\mathbb{Z})$ by restriction, there is a finite chain of covering relations

$$(a, b) = (x_0, y_0) < (x_1, y_1) < \cdots < (x_r, y_r) = (c, d)$$

in $E(\mathbb{Z})$. However, using Exercise 6.1.3, we can arrange for all the points (x_i, y_i) to lie in the subset $\mathrm{FH}(D_l^{(1)}(k))$. This proves the assertion about covering relations, and Axiom (H2) follows.

Axiom (F1) follows because every vertex chain in $\mathrm{FH}(D_l^{(1)}(k))$ consists of either a complete vertex chain of $E(\mathbb{Z})$, or (in the case where the vertex is an endpoint) every other vertex in a vertex chain of $E(\mathbb{Z})$.

Axiom (F2) follows by a case by case check. The key observation here is that if $l > 4$ and p_1 and p_2 are distinct endpoints of $D_l^{(1)}$ connected to a common vertex q, then every vertex of $\mathrm{FH}(D_l^{(1)}(k)) \cap \varepsilon_k^{-1}(q)$ either (a) covers an element labelled p_1 and is covered by an element labelled p_2, or (b) vice versa. The situation for $l = 4$ can be dealt with by an ad hoc verification.

Axiom (F3) also follows by a case by case check. It is similar to the corresponding proof in Lemma 6.1.4, except where q is a branch point or an end point. In any case,

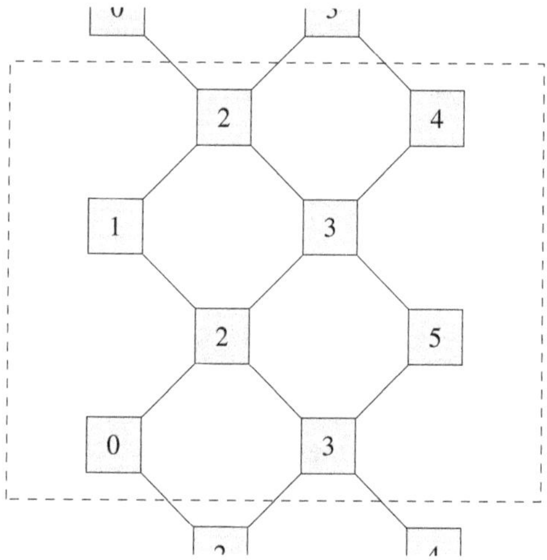

Figure 6.5 The full heap $\varepsilon_5 : \mathrm{FH}(D_5^{(1)}(5)) \to D_5^{(1)}$

we find that each q-interval contains precisely two elements with labels adjacent to q, as required. □

Proposition 6.3.3 *Let $l \geq 4$, let $k \in \{l-1, l\}$, and let k' be the integer for which $\{l-1, l\} = \{k, k'\}$.*

(i) *A principal subheap F of $E = \mathrm{FH}(D_l^{(1)}(k))$ is the triangle-shaped finite subset of E bounded by the lines $y = x + 1$ and $x + y = 1$.*
(ii) *The highest weight of the minuscule module V_F corresponding to F is ω_k.*
(iii) *The lowest weight of V_F is $-\omega_k$ if l is even, and $-\omega_{k'}$ if l is odd.*

Proof We argue as in the proof of Proposition 6.2.3 that the principal subheap of E is given by

$$F = \{\alpha : \alpha \not\leq (0, 0) \text{ and } \alpha \not\geq (0, 2)\},$$

and part (i) follows from this.

The subheap F always has a maximal element labelled k. If l is even, then the minimal element of F is also labelled k; if l is odd, then the label of the minimal element is the other element in the set $\{l, l-1\}$. Parts (ii) and (iii) now follow as in Proposition 6.2.6. □

Example 6.3.4 Figure 6.5 shows the full heap $\varepsilon_5 : \mathrm{FH}(D_5^{(1)}(5)) \to D_5^{(1)}$, and Figure 6.6 shows a triangle-shaped principal subheap of $\mathrm{FH}(D_5^{(1)}(5))$.

Remark 6.3.5 The principal subheap F of the full heap E over $D_l^{(1)}$ in Figure 3.7 is shown in Figure 3.11. Since the maximal element of F has label 1, Lemma 5.5.1 shows that V_F has highest weight ω_1 and that E is nothing other than the full heap $\varepsilon : \mathrm{FH}(D_l^{(1)}(1)) \to D_l^{(1)}$.

Another example of a full heap over $D_l^{(1)}$ is shown in Example 7.1.10 below.

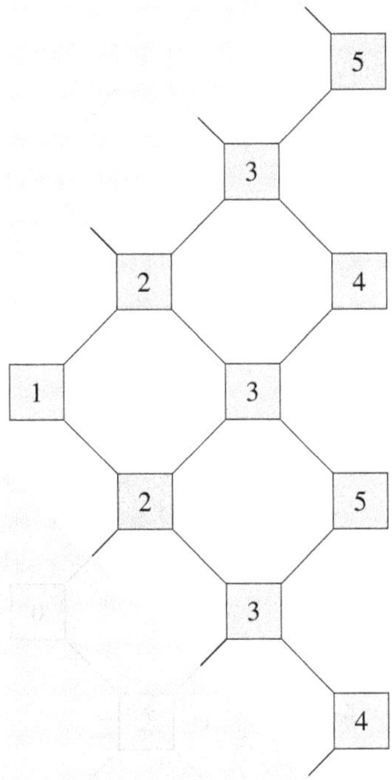

Figure 6.6 A principal subheap of $\varepsilon_5 : \mathrm{FH}(D_5^{(1)}(5)) \to D_5^{(1)}$

The description of the proper ideals of the full heaps of Proposition 6.3.2 is similar to, but easier than, the corresponding problem in type $A_l^{(1)}$.

Definition 6.3.6 For $l \geq 4$ and $k \in \{l-1, l\}$, let $\varepsilon_k : E = \mathrm{FH}(D_l^{(1)}(k)) \to D_l^{(1)}$ be the full heap defined in Definition 6.3.1. Fix labellings $E(i, m)$ of the vertex chains of E, and let I be a proper ideal of E. For each integer i with $1 \leq i < l$, we define r_i to be the largest integer such that $(i, r_i) \in I$. We then join the points $(i, r_i + 1)$ into a path in order of increasing x coordinate to create a path, P'. We extend P' into a path P by connecting the point $(1, r_1 + 1)$ to the point $(0, y) \in \{(0, r_i), (0, r_i + 2)\}$ for which $y = 1 \bmod 4$. We identify P with the pair (t, σ), where $t \in \mathbb{Z}$ is the largest integer such that $E(0, t) \in I$, and $\sigma = \sigma(I) = u_1 u_2 \ldots u_{l-1}$ is the word of length $l - 1$ in the alphabet $\{+, -\}$ for which $u_i = +$ (respectively, $u_i = -$) if $r_i = r_i + 1$ (respectively, $r_i = r_i - 1$).

Example 6.3.7 Figure 6.7 shows the principal subheap F of $\varepsilon_5 : \mathrm{FH}(D_5^{(1)}(5)) \to D_5^{(1)}$. Let I be the ideal of F consisting of all entries except the highest occurrences of 3 and of 5. The dashed line in the figure is the path P corresponding to I; note that I corresponds to the region underneath the path. In this case, we have $\sigma(I) = {+}{+}{-}{+}$;

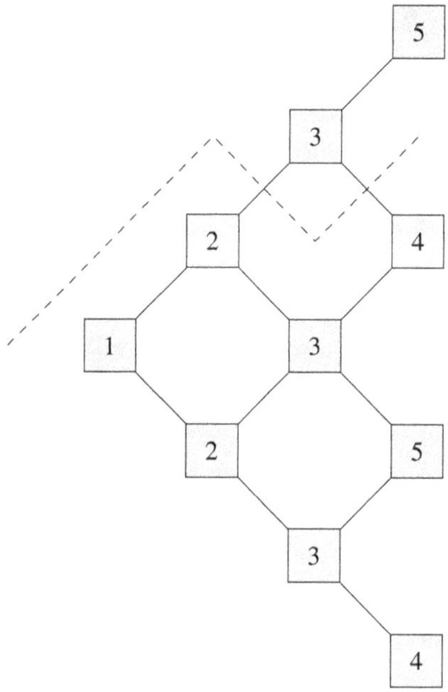

Figure 6.7 An ideal of the principal subheap of $\varepsilon_5 : \mathrm{FH}(D_5^{(1)}(5)) \to D_5^{(1)}$

this describes the up-down movement of the path from left to right (in this case, "up, up, down, up").

Proposition 6.3.8 *Maintain the notation of Definition 6.3.6.*

(i) *There is a bijection between the proper ideals of $E = \mathrm{FH}(D_l^{(1)}(k))$ and pairs (t, σ), where $t \in \mathbb{Z}$ and σ is a word of length $l - 1$ in the alphabet $\{+, -\}$. The pair (t, σ) corresponds to the ideal I such that $\sigma = \sigma(I)$ and $E(0, t)$ is the maximal element of $\varepsilon^{-1}(0) \cap I$ with respect to some fixed labelling of the vertex chains of E.*

(ii) *There is a bijection between the ideals of the principal subheap F of $\mathrm{FH}(D_l^{(1)}(k))$ and words σ of length $l - 1$ in the alphabet $\{+, -\}$.*

(iii) *The dimensions of the minuscule modules of weights ω_{l-1} and ω_l of the simple Lie algebra of type D_l are both equal to 2^{l-1}.*

Proof Part (i) follows from Definition 6.3.6 and an argument similar to, but easier than, the proof of Proposition 6.2.2. Part (ii) follows from (i) by modifying the proof of Proposition 6.2.4. Part (iii) holds because the set in part (ii) has cardinality 2^{l-1}, and this set indexes a basis for V_F by construction. □

Definition 6.3.9 The two minuscule modules $L(D_l, \omega_{l-1})$ and $L(D_l, \omega_l)$ of dimension 2^{l-1} are known as the *spin representations* of the simple Lie algebra of type D_l.

Exercise 6.3.10 Let $l \geq 4$ and $\Gamma = D_l^{(1)}$, and consider the faithful action of the affine Weyl group $W(D_l^{(1)})$ on the proper ideals of the full heap $\varepsilon_l : \mathrm{FH}(D_l^{(1)}(l)) \to \Gamma$. With each word σ of length $l-1$ in the alphabet $\{+,-\}$, we associate a word σ^+ of length l in the same alphabet, by appending one symbol to the right of σ so that the total number of occurrences of $-$ is even. (For example, if $\sigma = ++-+$ then $\sigma^+ = ++-+-$, and if $\sigma = -+-+$ then $\sigma^+ = -+-++$.)

Maintain the notation of Proposition 6.3.8, but re-index the ideals by symbols (t, σ^+) where $t \in \mathbb{Z}$ and σ^+ is a word of length l as above.

(i) Show that if $i \notin \{0, l\}$, the generator s_i sends (t, σ^+) to $(t, s_i(\sigma^+))$, where the action of s_i on words of length l is by place permutation by the transposition $(i, i+1)$.

(ii) Show s_l acts on (t, σ^+) by acting by the transposition $(l-1, l)$ followed by altering each of the $l-1$-st and l-th symbols. (For example, $s_5((t, ++-+-)) = (t, ++-+-)$ and $s_5((t, -+-++)) = (t, -+---)$.)

(iii) Let τ be the word of length $l-2$ for which $\sigma^+ = uu'\tau$ and $u, u' \in \{+,-\}$. Show that we have

$$s_0((t, \sigma^+)) = \begin{cases} (t+1, --\tau) & \text{if } u = u' = +, \\ (t-1, ++\tau) & \text{if } u = u' = -, \\ (t, \sigma^+) & \text{if } u \neq u'. \end{cases}$$

Exercise 6.3.11 Maintain the notation of Exercise 6.3.10. Denote the Chevalley generators of the simple Lie algebra $\mathfrak{g}$ of type D_l by e_i, f_i and h_i for $1 \leq i \leq l$, and consider the action of $\mathfrak{g}$ on the minuscule module V_F of highest weight ω_l. Let us write

$$\sigma^+ = u_1 u_2 \cdots u_l.$$

We write v_{σ^+} for the basis element of V_F indexed by the ideal corresponding to the orbit $\{(t, \sigma^+) : t \in \mathbb{Z}\}$. If $1 \leq i < l$ let us write $\sigma^+ = \tau_i u u' \tau_i'$, where $u, u' \in \{+,-\}$ and τ_i has length $i-1$.

(i) Show that if $1 \leq i < l$, we have

$$e_i(v_{\sigma^+}) = \begin{cases} v_{\tau_i+-\tau_i'} & \text{if } \sigma^+ = \tau_i-+\tau_i', \\ 0 & \text{otherwise} \end{cases}$$

and

$$f_i(v_{\sigma^+}) = \begin{cases} v_{\tau_i-+\tau_i'} & \text{if } \sigma^+ = \tau_i+-\tau_i', \\ 0 & \text{otherwise.} \end{cases}$$

(ii) Show that we have

$$e_l(v_{\sigma^+}) = \begin{cases} v_{\tau_{l-1}++} & \text{if } \sigma^+ = \tau_{l-1}--, \\ 0 & \text{otherwise} \end{cases}$$

and

$$f_l(v_{\sigma^+}) = \begin{cases} v_{\tau_{l-1}--} & \text{if } \sigma^+ = \tau_{l-1}++, \\ 0 & \text{otherwise.} \end{cases}$$

Exercise 6.3.12 Formulate and prove an analogue of Exercise 6.3.10 for the full heap $\varepsilon_{l-1} : \mathrm{FH}(D_l^{(1)}(l-1)) \to D_l^{(1)}$.

Exercise 6.3.13 Formulate and prove an analogue of Exercise 6.3.11 for the full heap $\varepsilon_{l-1} : \mathrm{FH}(D_l^{(1)}(l-1)) \to D_l^{(1)}$.

Exercise 6.3.14 Let $l \geq 4$, and let E be one of the full heaps $\text{FH}(D_l^{(1)}(i))$ for $i \in \{l-1, l\}$.

(i) Show that if l is odd, then E has no unique largest antichain.
(ii) Show that if l is even, then E has a unique largest antichain F whose content, $\chi(F)$, is equal either to

$$\alpha_1 + \alpha_3 + \cdots + \alpha_{l-3} + \alpha_{l-1}$$

or to

$$\alpha_1 + \alpha_3 + \cdots + \alpha_{l-3} + \alpha_l.$$

(iii) Show that if l is even, there is no element of $W(D_l)$ that sends the set of roots

$$\{\alpha_1, \alpha_3, \ldots, \alpha_{l-3}, \alpha_{l-1}\}$$

to

$$\{\alpha_1, \alpha_3, \ldots, \alpha_{l-3}, \alpha_l\}.$$

6.4 Types $B_l^{(1)}$ and $D_{l+1}^{(2)}$

Definition 6.4.1 Let $l \geq 3$ and $k \in \{l, l+1\}$, and let $\varepsilon_k : E = \text{FH}(D_{l+1}^{(1)}(k)) \to D_{l+1}^{(1)}$ be the full heap defined in Definition 6.3.1. Let $\pi : D_{l+1}^{(1)} \to B_l^{(1)}$ be the function on vertices such that

$$\pi(p) = \begin{cases} p & \text{if } p \notin \{l, l+1\}, \\ l & \text{otherwise.} \end{cases}$$

We define the function $\varepsilon_B : \text{FH}(B_l^{(1)})(l) \to B_l^{(1)}$ to be $\pi \circ \varepsilon_k$.

Proposition 6.4.2 *If $l \geq 3$ then the function $\varepsilon_B : \text{FH}(B_l^{(1)}(l)) \to B_l^{(1)}$ of Definition 6.4.1 is a full heap.*

Proof The proof is almost identical to the proof of Proposition 6.3.2. The main difference concerns the verification of Axiom (F3) for a p-interval with $p = l$. In this case, the Cartan matrix satisfies $a_{l,l-1} = -2$, and the axiom holds because each open p-interval in $\text{FH}(B_l^{(1)}(l))$ is an occurrence of $l-1$. □

Proposition 6.4.3 *Let $l \geq 3$.*

(i) *A principal subheap F of $E = \text{FH}(B_l^{(1)}(l))$ is the triangle-shaped finite subset of E bounded by the lines $y = x + 1$ and $x + y = 1$.*
(ii) *The highest weight of the minuscule module V_F corresponding to F is ω_l, and the lowest weight is $-\omega_l$.*

Proof The proof follows the same lines as that of Proposition 6.3.3, but is simpler because the minimal element of F is always labelled l. □

Example 6.4.4 Figure 6.8 shows the full heap $\varepsilon_B : \text{FH}(B_4^{(1)}(4)) \to B_4^{(1)}$, and Figure 6.9 shows a triangle-shaped principal subheap of $\text{FH}(B_4^{(1)}(4))$.

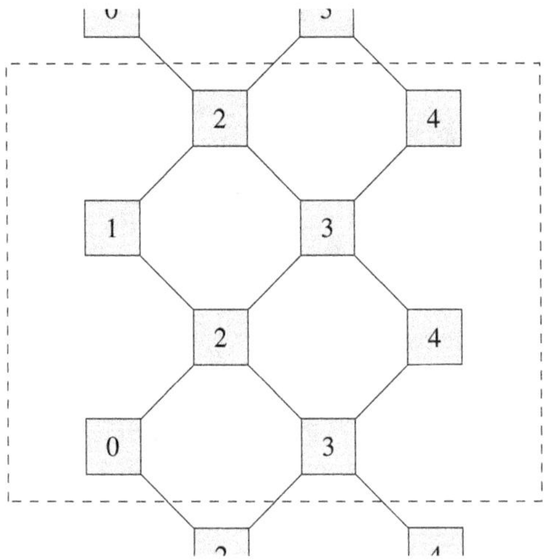

Figure 6.8 The full heap $\varepsilon_B : \mathrm{FH}(B_4^{(1)}(4)) \to B_4^{(1)}$

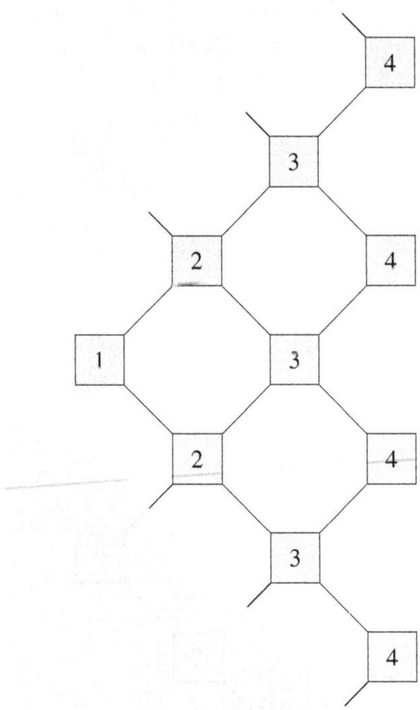

Figure 6.9 A principal subheap of $\varepsilon_4 : \mathrm{FH}(B_4^{(1)}(4)) \to B_4^{(1)}$

Proposition 6.4.5

(i) *There is a bijection between the proper ideals of* $E = \mathrm{FH}(B_l^{(1)}(l))$ *and pairs* (t, σ)*, where* $t \in \mathbb{Z}$ *and* σ *is a word of length* l *in the alphabet* $\{+, -\}$*. The pair* (t, σ) *corresponds to the ideal* I *such that* $\sigma = \sigma(I)$ *(as in Definition 6.3.6), and* $E(0, t)$ *is the maximal element of* $\varepsilon^{-1}(0) \cap I$ *with respect to some fixed labelling of the vertex chains of* E.

(ii) *There is a bijection between the ideals of the principal subheap* F *of* $\mathrm{FH}(B_l^{(1)}(l))$ *and words* σ *of length* l *in the alphabet* $\{+, -\}$.

(iii) *The dimension of the minuscule module of weight* ω_l *of the simple Lie algebra of type* B_l *is* 2^l.

Proof Since the heaps $\mathrm{FH}(B_l^{(1)}(l))$, $\mathrm{FH}(D_{l+1}^{(1)}(l))$ and $\mathrm{FH}(D_{l+1}^{(1)}(l+1))$ are isomorphic as abstract posets, the assertions follow from Proposition 6.3.8. □

Definition 6.4.6 The minuscule module $L(B_l, \omega_l)$ of dimension 2^l is known as the *spin representation* of the simple Lie algebra of type B_l.

Exercise 6.4.7 Let $l \geq 3$ and $\Gamma = B_l^{(1)}$, and consider the faithful action of the affine Weyl group $W(B_l^{(1)})$ on the proper ideals of the full heap $\varepsilon_B : \mathrm{FH}(B_l^{(1)}(l)) \to \Gamma$. Maintain the notation of Proposition 6.3.8.

(i) Show that if $i \notin \{0, l\}$, the generator s_i sends (t, σ) to $(t, s_i(\sigma))$, where the action of s_i on words of length l is by place permutation by the transposition $(i, i+1)$.

(ii) Show s_l acts on (t, σ^+) by altering the rightmost symbol.

(iii) Let τ be the word of length $l-2$ for which $\sigma^+ = uu'\tau$ and $u, u' \in \{+, -\}$. Show that we have

$$s_0((t, \sigma^+)) = \begin{cases} (t+1, --\tau) & \text{if } u = u' = +, \\ (t-1, ++\tau) & \text{if } u = u' = -, \\ (t, \sigma^+) & \text{if } u \neq u'. \end{cases}$$

Exercise 6.4.8 Maintain the notation of Exercise 6.4.7. Denote the Chevalley generators of the simple Lie algebra $\mathfrak{g}$ of type B_l by e_i, f_i and h_i for $1 \leq i \leq l$, and consider the action of $\mathfrak{g}$ on the minuscule module V_F of highest weight ω_l. Let us write

$$\sigma = u_1 u_2 \cdots u_l.$$

We write v_σ for the basis element of V_F indexed by the ideal corresponding to the orbit $\{(t, \sigma) : t \in \mathbb{Z}\}$. If $1 \leq i < l$ let us write $\sigma = \tau_i u u' \tau_i'$, where $u, u' \in \{+, -\}$ and τ_i has length $i - 1$.

(i) Show that if $1 \leq i < l$, we have

$$e_i(v_\sigma) = \begin{cases} v_{\tau_i+-\tau_i'} & \text{if } \sigma = \tau_i-+\tau_i', \\ 0 & \text{otherwise} \end{cases}$$

and

$$f_i(v_\sigma) = \begin{cases} v_{\tau_i-+\tau_i'} & \text{if } \sigma = \tau_i+-\tau_i', \\ 0 & \text{otherwise.} \end{cases}$$

(ii) Show that we have

$$e_l(v_\sigma) = \begin{cases} v_{\tau_l+} & \text{if } \sigma = \tau_l-, \\ 0 & \text{otherwise} \end{cases}$$

and

$$f_l(v_\sigma) = \begin{cases} v_{\tau_l-} & \text{if } \sigma = \tau_l+, \\ 0 & \text{otherwise.} \end{cases}$$

Exercise 6.4.9 Let F be the principal subheap of the full heap $\varepsilon_B : \mathrm{FH}(B_l^{(1)})(l) \to B_l^{(1)}$, and let I be the ideal of F consisting of the lowest occurrences of each of the elements with labels $1, 2, \ldots, l$ (so that $|I| = l$).

(i) Show that $\chi(I)$ is W-conjugate to α_l, and is therefore a short root.
(ii) Using the equation $s_i(\emptyset, I) = (s_i(\emptyset), s_i(I))$, or otherwise, show that $s_i(\chi(I)) \leq \chi(I)$ for all $1 \leq i \leq l$.
(iii) Show that the highest short root in type B_l is $\alpha_1 + \alpha_2 + \cdots + \alpha_l$.

Exercise 6.4.10 Show that $\binom{n+1}{2}$ is odd if $n = 1$ or $2 \bmod 4$, and is even if $n = 0$ or $3 \bmod 4$. By applying Theorem 5.6.4 to the principal subheap F associated with the representation $L(B_n, \omega_n)$, show that the bilinear form p associated with $L(B_n, \omega_n)$ is symplectic if $n = 1$ or $2 \bmod 4$, and orthogonal if $n = 0$ or $3 \bmod 4$.

Exercise 6.4.11 Adapt Exercise 6.4.10 to show that if n is even, then the bilinear form p associated with either of the spin representations $L(D_n, \omega_{n-1})$ or $L(D_n, \omega_n)$ is symplectic if $n = 2 \bmod 4$, and orthogonal if $n = 0 \bmod 4$.

Definition 6.4.12 Let $l \geq 4$ and let $\varepsilon_k : E = \mathrm{FH}(D_l^{(1)}(l)) \to D_l^{(1)}$ be the full heap defined in Definition 6.3.1. Let $\pi' : D_l^{(1)} \to D_{l-1}^{(2)}$ be the function on vertices such that

$$\pi'(p) = \begin{cases} 0 & \text{if } p \in \{0, 1\}, \\ l-2 & \text{if } p = l, \\ p-1 & \text{otherwise.} \end{cases}$$

We define the function $\varepsilon_D : \mathrm{FH}(D_{l-1}^{(2)}(l-2)) \to D_{l-1}^{(2)}$ to be $\pi' \circ \varepsilon_k$.

Remark 6.4.13 Notice that we would obtain the same function in Definition 6.4.12 if we started with the heap $\mathrm{FH}(D_l^{(1)}(l-1))$.

Proposition 6.4.14 *If $l \geq 3$ then the function $\varepsilon_D : \mathrm{FH}(D_{l-1}^{(2)}(l-1)) \to D_{l-1}^{(2)}$ of Definition 6.4.12 is a full heap.*

Proof The proof is almost identical to the proof of Proposition 6.4.2. □

Example 6.4.15 Figure 6.10 shows the full heap $\varepsilon_D : \mathrm{FH}(D_4^{(2)}(3)) \to D_4^{(2)}$. Note that the repeating motif contains half as many vertices as the corresponding heaps in types $B_4^{(1)}$ and $D_5^{(1)}$.

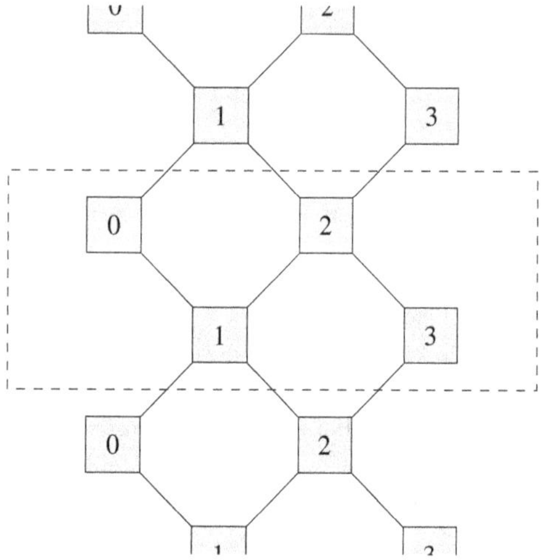

Figure 6.10 The full heap $\varepsilon_D : \mathrm{FH}(D_4^{(2)}(3)) \to D_4^{(2)}$

Exercise 6.4.16 Show that the principal subheap of the full heap $\varepsilon_D : \mathrm{FH}(D_l^{(2)}(l - 1)) \to D_l^{(2)}$ affords the spin representation of the simple Lie algebra of type B_{l-1}.

Proposition 6.4.17

(i) *There is a bijection between the proper ideals of* $E = \mathrm{FH}(D_l^{(2)}(l-1))$ *and pairs* (t, σ), *where* $t \in \mathbb{Z}$ *and* σ *is a word of length* l *in the alphabet* $\{+, -\}$. *The pair* (t, σ) *corresponds to the ideal* I *such that* $\sigma = \sigma(I)$ *(as in Definition 6.3.6), and* $E(0, t)$ *is the maximal element of* $\varepsilon^{-1}(0) \cap I$ *with respect to some fixed labelling of the vertex chains of* E.

(ii) *There is a bijection between the ideals of the principal subheap* F *of* $\mathrm{FH}(D_l^{(2)}(l - 1))$ *and words* σ *of length* l *in the alphabet* $\{+, -\}$.

Proof This follows from Proposition 6.3.8; see also the proof of Proposition 6.4.5. □

Exercise 6.4.18 Let $l \geq 3$ and $\Gamma = D_l^{(2)}$, and consider the faithful action of the affine Weyl group $W(D_l^{(2)})$ on the proper ideals of the full heap $\varepsilon_D : \mathrm{FH}(D_l^{(2)}(l-1)) \to \Gamma$. Maintain the notation of Proposition 6.4.17.

(i) Show that if $i \not\in \{0, l\}$, the generator s_i sends (t, σ) to $(t, s_i(\sigma))$, where the action of s_i on words of length l is by place permutation by the transposition $(i, i+1)$.

(ii) Let τ be the word of length $l - 2$ for which $\sigma^+ = uu'\tau$ and $u, u' \in \{+, -\}$. Show that we have

$$s_0((t, \sigma^+)) = \begin{cases} (t+1, --\tau) & \text{if } u = u' = +, \\ (t-1, ++\tau) & \text{if } u = u' = -, \\ (t, \sigma^+) & \text{if } u \neq u'. \end{cases}$$

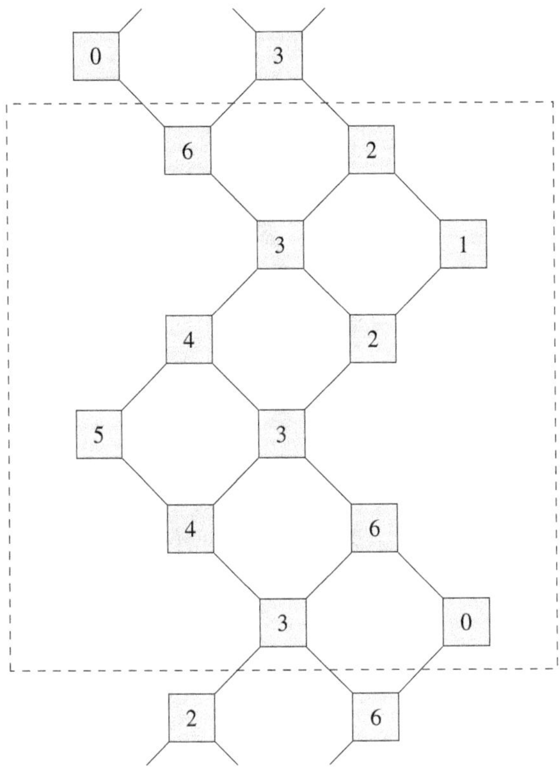

Figure 6.11 Part of the full heap $\varepsilon : E_6^{(1)}(5) \to E_6^{(1)}$

(iii) Let τ be the word of length $l - 2$ for which $\sigma^+ = \tau u u'$ and $u, u' \in \{+, -\}$. Show that we have

$$s_l((t, \sigma^+)) = \begin{cases} (t+1, \tau--) & \text{if } u = u' = +, \\ (t-1, \tau++) & \text{if } u = u' = -, \\ (t, \sigma^+) & \text{if } u \neq u'. \end{cases}$$

6.5 Full heaps in type $E_6^{(1)}$ and $E_7^{(1)}$

Definition 6.5.1 The full heap $\varepsilon : \mathrm{FH}(E_6^{(1)}(5)) \to E_6^{(1)}$ is shown in Figure 6.11, and its dual is the full heap $\varepsilon : \mathrm{FH}(E_6^{(1)}(1)) \to E_6^{(1)}$.

The principal subheap of $E_6^{(1)}(5))$ is shown in Figure 6.12, and its dual is the principal subheap of $E_6^{(1)}(1)$.

Exercise 6.5.2 Prove that the full heaps of Definition 6.5.1 are indeed full, and that their principal subheaps are as described. Prove also that the full heaps correspond to the highest weights ω_5 and ω_1, as the notation suggests.

Remark 6.5.3 In order to represent the partial order well on the page, the heap in Figure 6.11 has been drawn so that the top edge of the dashed rectangle is identified with the bottom edge via a twist.

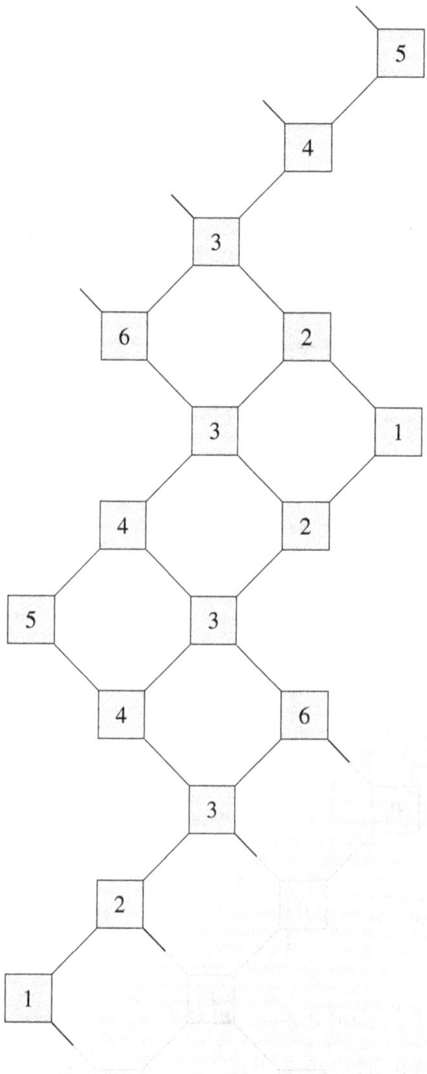

Figure 6.12 The principal subheap of $\varepsilon : E_6^{(1)}(5) \to E_6^{(1)}$

We defer discussion of the proper ideals of $E_6^{(1)}(5)$ because such a description at this point would appear to be ad hoc.

Definition 6.5.4 The full heap $\varepsilon : \text{FH}(E_7^{(1)}(6)) \to E_7^{(1)}$ is shown in Figure 6.13, and its principal subheap is shown in Figure 6.14.

Exercise 6.5.5 Prove that the full heap of Definition 6.5.4 is indeed full, and that its principal subheap is as described. Prove also that the full heap corresponds to the highest weight ω_6.

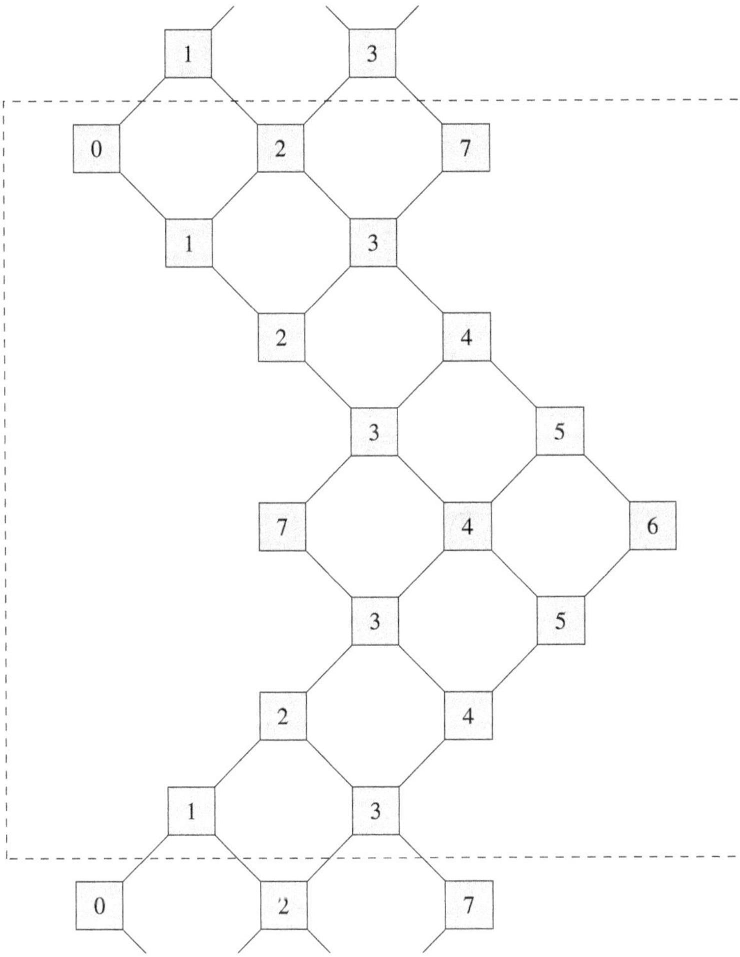

Figure 6.13 Part of the full heap $\varepsilon : E_7^{(1)}(6) \to E_7^{(1)}$

Remark 6.5.6 In order to represent the partial order well on the page, the heap in Figure 6.13 has been drawn so that the nodes labelled 7 are not vertically aligned.

There is a more convenient way to think of the orbits of the finite Weyl group $W(E_7)$ on the proper ideals of the full heap $E = E_7^{(1)}(6)$. Indexing the elements of E in the usual way by symbols $E(p, z)$, we can fix an integer t and define a finite subheap F' of E (shown in Figure 6.15) by

$$F' = \{x \in E : x \not\geq E(7, t+1) \text{ and } x \not\leq E(7, t-1)\}.$$

Since the period, δ, of E contains α_7 with coefficient 2, it follows that the ideals of F' are in bijection with orbits of ideals of E, by adapting the argument of Lemma 3.3.3 (v).

The advantage of this approach is that the ideals of F' are much easier to classify. There are two types of ideal: those that contain the unique occurrence of 7, and those

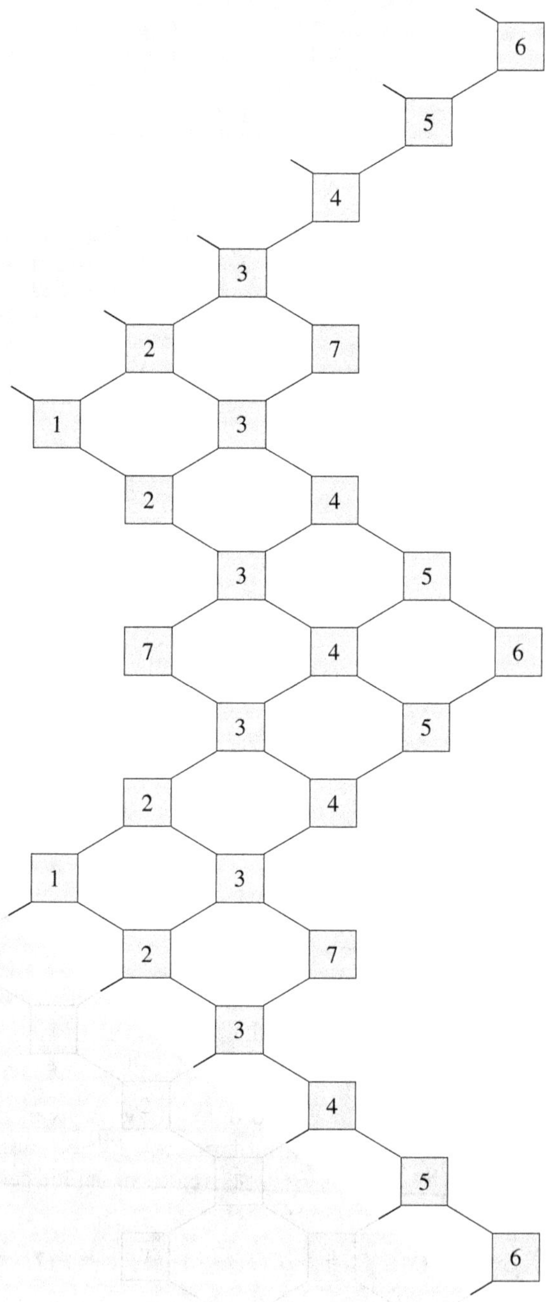

Figure 6.14 The principal subheap of $\varepsilon : E_7^{(1)}(6) \to E_7^{(1)}$

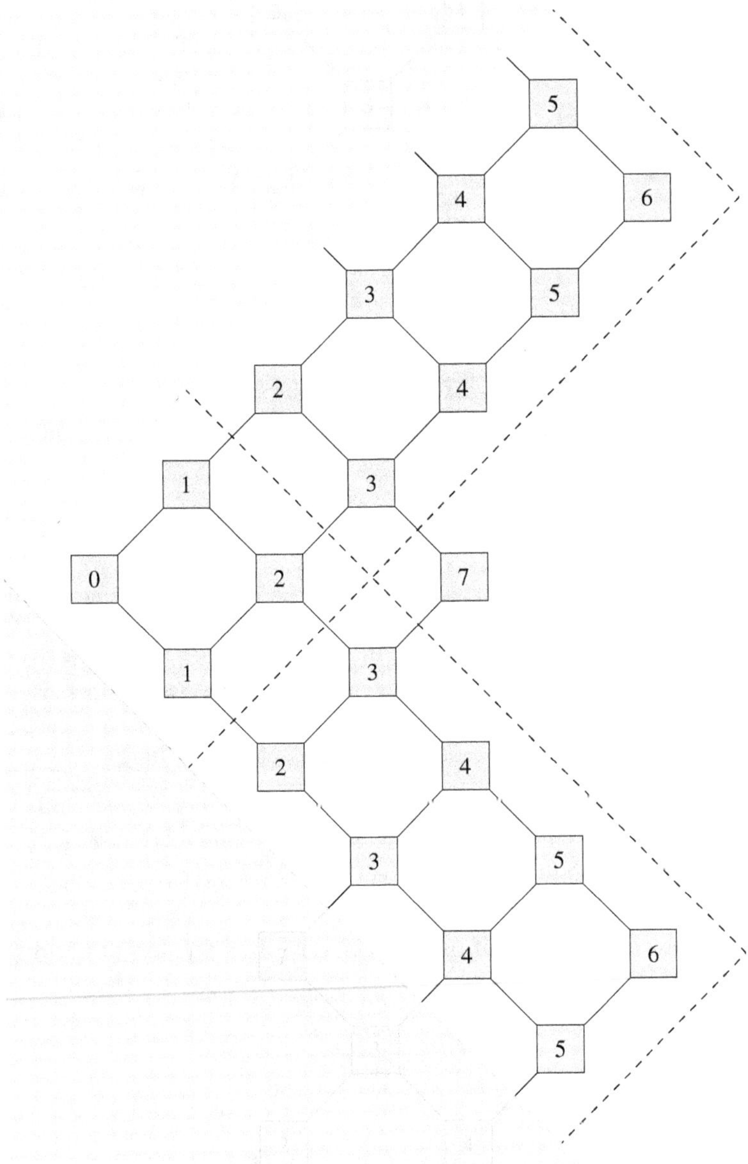

Figure 6.15 The finite subheap F' of $E_7^{(1)}(6)$

that do not. The ideals that contain (respectively, do not contain) an occurrence of 7 correspond to the ideals of the subheap in the upper (respectively, lower) dashed box in Figure 6.15.

Consider the heap F_1 corresponding to upper dashed box in the subheap F'. The 12 elements of F_1 consist of two diagonal rows of six elements; the elements $u_0, \ldots, u_5$ of the upper row are labelled $0, \ldots, 5$ respectively, and the elements $l_1, \ldots, l_6$ in the lower row are labelled $1, \ldots, 6$ respectively. If I is an ideal of F_1, we can label I by a pair of integers $(i, j)^+$, where $0 \leq i < j \leq 7$. Here, i is the smallest integer such that $u_i \notin I$ (or 6 if no such element exists) and j is the smallest integer such that $l_i \notin I$ (or 7 if no such integer exists).

The heap F_2 corresponding to the lower dashed box in the subheap F' is dual to F_1. If J is an ideal of F_2, then we can consider $F' \backslash J^*$ as an ideal of F_1. If $F' \backslash J^*$ is labelled by the pair $(i, j)^+$, then we label J by the pair $(i, j)^-$.

Example 6.5.7 The ideal of F' generated by the highest occurrence of 3 is $(4, 5)^+$, and the ideal of F' generated by the unique occurrence of 0 is $(0, 7)^-$.

Exercise 6.5.8 Use the above construction to show that the minuscule module in type E_7 has dimension 56.

Exercise 6.5.9 Identify the finite Weyl group $W(E_6)$ with the subgroup of $W(E_7)$ generated by all the s_i except s_6.

(i) Show that there are four orbits of $W(E_6)$ on the ideals of the principal subheap of $E_7^{(1)}(6)$, and that they have sizes 1, 27, 27 and 1.
(ii) Show that the action of $W(E_6)$ on each of the two orbits of size 27 is equivalent to the action of $W(E_6)$ on each of the principal subheaps corresponding to the two minuscule representations in type E_6.
(iii) Deduce that the minuscule modules in type E_6 each have dimension 27.

It is possible to use the above constructions to label directly the ideals of the principal subheaps corresponding to minuscule representations in types E_6 and E_7 by symbols $(i, j)^{\pm}$, but the results are rather unenlightening.

6.6 The classification of full heaps over affine Dynkin diagrams

Lemma 6.6.1 *The full heaps appearing in Figures 3.3, 3.5 and 3.7 are*

$$\varepsilon : \mathrm{FH}(C_l^{(1)}(1)) \to C_l^{(1)},$$
$$\varepsilon : \mathrm{FH}(A_{2l-1}^{(2)}(1)) \to A_{2l-1}^{(2)}$$

and

$$\varepsilon : \mathrm{FH}(D_l^{(1)}(1)) \to D_l^{(1)},$$

respectively. The principal subheaps of the first two full heaps above are isomorphic and shown in Figure 3.10; the principal subheap of the third full heap is shown in Figure 3.11.

Proof This follows from Lemma 5.5.1. □

Theorem 6.6.2 *Any full heap over an affine Dynkin diagram* Γ *is isomorphic, in* **Heap**(Γ), *to one of the following:*

$$\begin{aligned}
\varepsilon : \mathrm{FH}(A_l^{(1)}(k)) &\to A_l^{(1)}, \text{ where } l \geq 1 \text{ and } 1 \leq k \leq l;\\
\varepsilon : \mathrm{FH}(B_l^{(1)}(l)) &\to B_l^{(1)}, \text{ where } l \geq 3;\\
\varepsilon : \mathrm{FH}(C_l^{(1)}(1)) &\to C_l^{(1)}, \text{ where } l \geq 2;\\
\varepsilon : \mathrm{FH}(D_l^{(1)}(1)) &\to D_l^{(1)}, \text{ where } l \geq 4;\\
\varepsilon : \mathrm{FH}(D_l^{(1)}(k)) &\to D_l^{(1)}, \text{ where } l \geq 4 \text{ and } k \in \{l-1, l\};\\
\varepsilon : \mathrm{FH}(E_6^{(1)}(k)) &\to E_6^{(1)}, \text{ where } k \in \{1, 5\};\\
\varepsilon : \mathrm{FH}(E_7^{(1)}(6)) &\to E_7^{(1)};\\
\varepsilon : \mathrm{FH}(A_{2l-1}^{(2)}(1)) &\to A_{2l-1}^{(2)}, \text{ where } l \geq 3;\\
\varepsilon : \mathrm{FH}(D_l^{(2)}(l-1)) &\to D_l^{(2)}, \text{ where } l \geq 4.
\end{aligned}$$

Proof A complete list of affine Dynkin diagrams is given in Appendix B. Let Γ be such a diagram, and let Γ^0 be the finite-type subdiagram obtained by removing the vertex labelled 0. Note that if Γ is of the form $X_l^{(1)}$ for some X and l then Γ^0 is of type X_l, if Γ is of type $A_{2l-1}^{(2)}$ then Γ^0 is of type C_l, and if Γ is of type $D_{l+1}^{(2)}$ then Γ^0 is of type B_l.

By Theorem 5.5.6, each isomorphism class of full heaps over Γ corresponds to a minuscule representation of the simple Lie algebra corresponding to Γ^0. By Theorem 5.1.5, this means that the number of isomorphism classes of full heaps over Γ is at most l (respectively, 1, 1, 3, 2, 1, 1, 1) if Γ is of type $A_l^{(1)}$ (respectively, $B_l^{(1)}$, $C_l^{(1)}$, $D_l^{(1)}$, $E_6^{(1)}$, $E_7^{(1)}$, $A_{2l-1}^{(2)}$, $D_l^{(2)}$). (If Γ is of a type not listed, there are no minuscule representations.) The earlier results of Chapter 6 show that all the heaps in the statement are indeed full and, counting them, we find that the upper bound for the number of isomorphism classes is achieved in each case. It follows that the list is complete. □

Corollary 6.6.3 *Every minuscule representation of a simple Lie algebra over* $\mathbb{C}$ *may be constructed as a module* V_F *for some principal subheap* F *of a full heap over an affine Dynkin diagram.*

Proof By Theorem 6.6.2 and Theorem 5.1.5, the full heaps over affine Dynkin diagrams of the form $X_l^{(1)}$ correspond irredundantly to the minuscule modules for simple Lie algebras. □

The following result is very well-known.

Corollary 6.6.4 *The Weyl group acts transitively on the weights of any minuscule representation.*

Proof This follows by combining Corollary 6.6.3 with Proposition 3.3.8. □

Corollary 6.6.5 *The weight spaces of any minuscule representation are one-dimensional.*

Proof This follows by combining Corollary 6.6.3 with Lemma 5.2.9 (iv). (Alternatively, this follows from Corollary 6.6.4 together with the general fact that the weight space of the highest weight of an irreducible module for a simple Lie algebra over $\mathbb{C}$ is one-dimensional.) □

The following is a sharper version of the original definition of minuscule (Definition 5.1.3.)

Corollary 6.6.6 *If λ is a minuscule weight, then we have*

$$(\lambda, \alpha^\vee) = 2\frac{(\lambda, \alpha)}{(\alpha, \alpha)} \in \{-1, 0, 1\}$$

for all roots α.

Proof It follows by Corollary 6.6.3 and Theorem 5.4.13 that the reflection s_α for any root α satisfies

$$s_\alpha(\lambda) \in \{\lambda - \alpha, \lambda, \lambda + \alpha\}.$$

The result now follows from the formula for a reflection combined with Definition 5.1.3. □

Corollary 6.6.7 *Let F be a principal subheap corresponding to a minuscule representation, and let α (respectively, β) be the maximal (respectively, minimal) elements of F. If $\rho : F \to \mathbb{Z}$ is a rank function for F, then we have $\rho(\alpha) - \rho(\beta) = h - 2$, where h is the Coxeter number.*

Proof This follows by inspection using a case by case check; recall that the values of h are listed in Appendix B. □

Proposition 6.6.8 *The lowest weights of the minuscule representations of the simple Lie algebras are as follows.*

Type	p : highest weight is ω_p	q : lowest weight is $-\omega_q$
A_l	k	$l+1-k$
B_l	l	l
C_l	1	1
D_l	1	1
D_l, l even	$k \in \{l-1, l\}$	k
D_l, l odd	$k \in \{l-1, l\}$	$2l-1-k$
E_6	$k \in \{1, 5\}$	$6-k$
E_7	6	6

Proof For each line of the table, let F be the principal subheap giving rise to the corresponding minuscule representation; the existence of F is guaranteed by Corollary 6.6.3. The second column gives the label of the maximal element of F, and the third gives the label of the minimal element of F. The proof follows by a case by case check, using the constructions of the various heaps listed in Theorem 6.6.2. □

Exercise 6.6.9

(i) Show that the minuscule representation corresponding to the full heap $\varepsilon : \mathrm{FH}(A_{2l-1}^{(2)}(1)) \to A_{2l-1}^{(2)}$ ($l \geq 3$) is the natural representation in type C_l.

(ii) Show that the minuscule representation corresponding to the full heap ε : $\text{FH}(D_l^{(2)}(l-1)) \to D_l^{(2)}$ $(l \geq 4)$ is the spin representation in type B_{l-1}.
(iii) Deduce that it is possible for two full heaps over distinct Dynkin diagrams to give rise to the same minuscule representation using the construction of Theorem 5.5.6.
(iv) Show that the situation in (iii) cannot occur unless one of the Dynkin diagrams is of twisted type.

6.7 Notes and references

1 Full heaps over affine Dynkin diagrams are described in Green [28], except in type A, where they are described in McGregor-Dorsey [57]. Theorem 6.1.18 is the main result of [57], although the proof here is new. McGregor-Dorsey's original proof uses only combinatorics. Theorem 6.6.2 is new.
2 Exercise 6.2.9 can be rephrased in terms of Young tableaux with one column. The tableaux can be put into canonical bijection with the ideals of the principal subheap; this idea is also discussed in Wildberger [90].
3 Many texts on Lie algebras (for example, Carter [11]) describe the construction of the spin representations in types B and D using Clifford algebras. The constructions in this chapter show that this is unnecessary.
4 The representations $L(D_l, \omega_{l-1})$ and $L(D_l, \omega_l)$ are also called *half-spin representations*, especially in the physics literature. Each of them has half the dimension of the spin representation $L(B_l, \omega_l)$.
5 The Dynkin diagrams B_∞ and D_∞ are discussed briefly in Kac's book [37, exercise 4.14]. Kac also discusses the notion of the integers as a Dynkin diagram, denoted by A_∞.
6 The constructions of all minuscule representations in Corollary 6.6.3 equip each minuscule module with a distinguished basis. It turns out that this is the canonical basis in the sense of Lusztig [50], and it naturally gives rise to a crystal basis in the sense of Kashiwara [39]. This basis is far easier to calculate in the case of minuscule representations, because the weight spaces are all one-dimensional, as mentioned in Corollary 6.6.5. The canonical bases for minuscule modules in types A, B and D were constructed by Marsh in [55, theorem 2.16], [56, theorem 3.16] and [56, theorem 2.18] respectively, in terms of distinguished coset representatives of parabolic subgroups. For details on the relationship with crystals, see [28, chapter 8].
7 Although the weight spaces of minuscule representations are one-dimensional, there are also non-minuscule representations of simple Lie algebras with this property. One of these is the $2n+1$-dimensional natural representation in type B_n, namely $L(B_n, \omega_1)$. This module is equipped with a nondegenerate orthogonal form and, in fact, this is how the simple Lie algebra of type B_n is usually defined. More examples of nonminuscule representations with one-dimensional weight spaces are given in Ginzburg [27, corollary 1.4.1].

7

Chevalley bases

Let $\mathfrak{g}$ be a simple Lie algebra over $\mathbb{C}$, corresponding to an $n \times n$ Cartan matrix, and let $\mathfrak{h}$ be the subalgebra of $\mathfrak{g}$ defined in Section 4.3. A *Chevalley basis* for $\mathfrak{g}$ is by definition a basis of $\mathfrak{g}$ of the form

$$\{h_i : 1 \leq i \leq n\} \,\dot{\cup}\, \{x_\alpha : \alpha \in \Phi\},$$

where Φ is the set of roots of $\mathfrak{g}$, satisfying the conditions

$$\begin{aligned}
[h, x_\alpha] &= \alpha(h)x_\alpha \text{ for all } h \in \mathfrak{h} \text{ and } \alpha \in \Phi, \\
[x_\alpha, x_{-\alpha}] &= -h_{\alpha^\vee}, \\
[x_\alpha, x_\beta] &= 0 \text{ if } \alpha, \beta \in \Phi,\ \alpha + \beta \notin \Phi \text{ and } \alpha + \beta \neq 0, \\
[x_\alpha, x_\beta] &= N_{\alpha,\beta}x_{\alpha+\beta} \text{ if } \alpha, \beta, \alpha + \beta \in \Phi, \text{ where} \\
N_{\alpha,\beta} &= N_{-\alpha,-\beta} = \pm(p+1),
\end{aligned}$$

where $p \geq 0$ is the greatest integer such that $\alpha - p\beta \in \Phi$, and h_α and $h_{\alpha^\vee}$ are as in Section 4.3.

It turns out that the choice of signs above is somewhat delicate, but as we shall show in Chapter 7, the theory of full heaps can be used to construct explicit Chevalley bases for simple Lie algebras over $\mathbb{C}$, except in types E_8, F_4 and G_2, where no corresponding full heap exists.

7.1 Kac's asymmetry function

In Section 7.1, we assume that the Weyl group W is associated with a simply laced Cartan matrix; that is, a generalized Cartan matrix A of finite type. Recall from Theorem 4.2.3 (iii) and (iv) that, in this case, all roots have the same length.

Proposition 7.1.1 *Let A be a simply laced Cartan matrix, and let α and β be two positive roots associated with A, normalized such that all roots γ satisfy $(\gamma, \gamma) = 2$. Then precisely one of the following situations occurs:*

(i) $(\alpha, \beta) = 2$ *and* $\alpha = \beta$;
(ii) $(\alpha, \beta) = 1$, $\alpha - \beta$ *is a root and* $\alpha + \beta$ *is not a root;*
(iii) $(\alpha, \beta) = 0$ *and neither of* $\alpha \pm \beta$ *is a root;*
(iv) $(\alpha, \beta) = -1$, $\alpha + \beta$ *is a root and* $\alpha - \beta$ *is not a root;*
(v) $(\alpha, \beta) = -2$ *and* $\alpha = -\beta$.

Proof Choose an element w in the Weyl group W such that $w(\beta) = \alpha_i$ is a fundamental root. We may replace α in the argument with $w(\alpha)$ and β by $w(\beta) = \alpha_i$, which means that it suffices to prove the statement in the case where β is a fundamental root.

Recall from Theorem 4.2.3 (i) that

$$s_\beta(\alpha) = \alpha - 2\frac{(\alpha, \beta)}{(\beta, \beta)}\beta.$$

Since $s_\beta(\alpha)$ is a real root and a linear combination (with integer coefficients) of fundamental roots, it must be the case that (α, β) is of the form $k(\beta, \beta)$ for a half-integer k. By Theorem 4.2.3 (iv), the form $(,)$ is positive definite, which means that k must lie in the set $\{1, 1/2, 0, -1/2, -1\}$. These are the five cases of the statement; the angle between α and β is equal to $0, \pi/3, \pi/2, 2\pi/3$ or π, respectively.

It is clear that if $k = 1$ (respectively, $k = -1$), we have $\alpha = \beta$ (respectively, $\alpha = -\beta$). If $k = 0$, then α and β are orthogonal, and the two vectors $\alpha \pm \beta$ both have squared length $2\sqrt{2}$, which means that neither can be a root by Theorem 4.2.3 (iii). This deals with cases (i), (iii) and (iv).

Suppose that $k = -1/2$. In this case, $\alpha - \beta$ does not have squared length 2, so cannot be a root. However, $s_\beta(\alpha) = \alpha + \beta$ is a root. This deals with case (iv), and case (ii) is handled by a similar argument. □

Corollary 7.1.2 *If α and β are distinct positive roots associated with a simply laced Cartan matrix, then it cannot be the case that $\alpha + \beta$ and $\alpha + 2\beta$ are both roots.*

Proof By Proposition 7.1.1 (ii), if $\alpha + \beta$ is a root, then $(\alpha, \beta) = -1$ (with the usual normalization). By bilinearity, we have

$$(\alpha + \beta, \beta) = (\alpha, \beta) + (\beta, \beta) = -1 + 2 = 1,$$

and then Proposition 7.1.1 (iv) shows that $\alpha + 2\beta$ is not a root. □

In order to better understand the structure of the representations of Lie algebras arising from heaps, it is convenient to break the symmetry of the bilinear form $(,)$ on V_Γ.

Definition 7.1.3 Let A be a simply laced generalized Cartan matrix of finite type. Let Q be the $\mathbb{Z}$-span of the fundamental roots α_p, and restrict the bilinear form $(,)$ to Q, normalized so that $(\alpha_p, \alpha_p) = 1$ for all p. (This is well-defined by Proposition 7.1.1.)

Choose an arbitrary total order, $<$, on the fundamental roots α_p. We define the *asymmetry function*, $\varepsilon : Q \times Q \to \{\pm 1\}$, by

$$\varepsilon(\alpha_i, \alpha_j) = \begin{cases} (-1)^{\frac{1}{2}(\alpha_i, \alpha_i)} & \text{if } i = j, \\ (-1)^{(\alpha_i, \alpha_j)} & \text{if } i < j, \\ 1 & \text{if } i > j \end{cases}$$

and extending by bimultiplicativity, stipulating that

$$\varepsilon(\alpha + \alpha', \beta) = \varepsilon(\alpha, \beta)\varepsilon(\alpha', \beta)$$

and

$$\varepsilon(\alpha, \beta + \beta') = \varepsilon(\alpha, \beta)\varepsilon(\alpha, \beta')$$

for all $\alpha, \alpha', \beta, \beta' \in Q$.

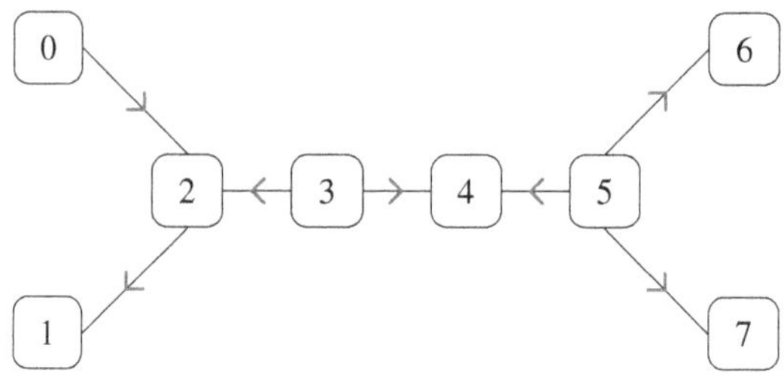

Figure 7.1 The orientation of Example 7.1.6

Definition 7.1.3 allows us to associate an asymmetry function with any orientation of the edges of the Dynkin diagram Γ corresponding to A. (The orientation is not to be confused with any directed arrows inherent in the Dynkin diagram.)

Definition 7.1.4 Let A be a simply laced generalized Cartan matrix of finite type and let Γ be the associated Dynkin diagram. Fix an orientation of Γ. We define a total order, $<$, on the vertices of Γ by defining $i > j$ if there is an arrow pointing from j to i in the orientation, and extending this relation to any total order. The asymmetry function associated with the orientation is by definition the asymmetry function associated with $<$ by Definition 7.1.3; in other words, we have

$$\varepsilon(\alpha_i, \alpha_j) = \begin{cases} -1 & \text{if } i = j \text{ or } i \to j, \\ 1 & \text{if } j \to i \text{ or } i \text{ and } j \text{ are nonadjacent.} \end{cases}$$

Remark 7.1.5 The reason the relation in Definition 7.1.4 can be extended to a total order is that the graph Γ contains no circuits. If α_i and α_j are nonadjacent vertices in Γ in then we have $\varepsilon(\alpha_i, \alpha_j) = 1$ because $(\alpha_i, \alpha_j) = 0$. For this reason, the definition of the asymmetry function associated with the orientation is independent of the choice of total order.

Example 7.1.6 Type $D_7^{(1)}$ is an example of a simply laced untwisted affine Dynkin diagram. An example of an orientation on the diagram is given in Figure 7.1. With respect to this orientation, we have $\varepsilon(\alpha_2, \alpha_3) = 1$, $\varepsilon(\alpha_5, \alpha_7) = -1$ and $\varepsilon(\alpha_1, \alpha_1) = 1$.

Proposition 7.1.7 *Let A be a simply laced generalized Cartan matrix of finite type and let Γ be the associated Dynkin diagram. Let ε be the asymmetry function associated with some fixed orientation of Γ by Definition 7.1.4. Let α and β be roots such that $\alpha \neq \pm\beta$, and let γ be an arbitrary element of Q as defined in Definition 7.1.3. Then we have*

(i) $\varepsilon(\gamma, \gamma) = (-1)^{\frac{1}{2}(\gamma,\gamma)}$;
(ii) $\varepsilon(\alpha, \alpha) = -1$;
(iii) $\varepsilon(\alpha, -\alpha) = -1$;
(iv) $\varepsilon(\alpha, \beta) = -\varepsilon(\beta, \alpha)$ *if either* $\alpha + \beta$ *or* $\alpha - \beta$ *is a root, in other words, if* $(\alpha, \beta) = \pm 1$;
(v) $\varepsilon(\alpha, \beta) = \varepsilon(\beta, \alpha)$ *if neither* $\alpha + \beta$ *or* $\alpha - \beta$ *is a root, in other words, if* $(\alpha, \beta) = 0$.

Proof Expanding $\varepsilon(\gamma, \gamma)$ by bimultiplicativity, we find that

$$\varepsilon(\gamma, \gamma) = (-1)^{\frac{1}{2}(\gamma,\gamma)},$$

which proves (i). Part (ii) follows from the fact that $(\alpha, \alpha) = 2$ for any root α, and part (iii) follows similarly.

Now let $\gamma = \alpha + \beta$ and apply (i) to obtain

$$\begin{aligned}\varepsilon(\alpha + \beta, \alpha + \beta) &= (-1)^{\frac{1}{2}(\alpha+\beta,\alpha+\beta)} \\ &= (-1)^{\frac{1}{2}(\alpha,\alpha)}(-1)^{(\alpha,\beta)}(-1)^{\frac{1}{2}(\beta,\beta)}.\end{aligned}$$

Since we have

$$\varepsilon(\alpha + \beta, \alpha + \beta) = \varepsilon(\alpha, \alpha)\varepsilon(\alpha, \beta)\varepsilon(\beta, \alpha)\varepsilon(\beta, \beta)$$

by bimultiplicativity, applying (i) again shows that

$$\varepsilon(\alpha, \beta)\varepsilon(\beta, \alpha) = (-1)^{(\alpha,\beta)}.$$

Parts (iv) and (v) now follow from Proposition 7.1.1. □

For the rest of Section 7.1, we will be interested in the situation described in Definition 5.2.13, in which F is a parabolic subheap corresponding to a simply laced Lie algebra, that is, a simple Lie algebra $\mathfrak{g}$ with a simply laced Dynkin diagram Γ^0, and $\varepsilon : E \to \Gamma$ is the associated full heap.

Proposition 7.1.8 *Let $\varepsilon : E \to \Gamma$ be a full heap with generalized Cartan matrix A. Let A^0 be the submatrix of A described in Proposition 4.3.18, and let Γ^0 be the corresponding subdiagram of Γ. Assume that Γ^0 is simply laced. Let $\mathfrak{g}$ be the simple Lie algebra associated to A^0, and let α, β be distinct positive roots of $\mathfrak{g}$. Suppose that $I, I' \in \mathcal{B}(E)$ satisfy $I \prec_\alpha I'$.*

(i) *Suppose that neither of $\alpha \pm \beta$ is a root. Then both I and I' lie in the same set K, where K is one of $J_\beta^+(E)$, $J_\beta^0(E)$, $J_\beta^-(E)$.*
(ii) *Suppose that $\alpha + \beta$ is a root. Then either (a) $I \in J_\beta^0(E)$ and $I' \in J_\beta^-(E)$ or (b) $I \in J_\beta^+(E)$ and $I' \in J_\beta^0(E)$.*
(iii) *Suppose that $\alpha - \beta$ is a root. Then either (a) $I \in J_\beta^0(E)$ and $I' \in J_\beta^+(E)$ or (b) $I \in J_\beta^-(E)$ and $I' \in J_\beta^0(E)$.*

Proof By hypothesis, we have $\chi(I, I') = \alpha$. In the situation of (i), α and β are orthogonal by Proposition 7.1.1 (iii), so that $s_\beta(\alpha) = \alpha$. Now Theorem 3.2.21 shows that

$$\chi(s_\beta(I), s_\beta(I')) = s_\beta(\chi(I, I')) = \chi(I, I') = \alpha,$$

and the assertion of (i) is a restatement of this observation.

In the situation of (ii), we have $-2(\alpha, \beta) = (\alpha, \alpha)$ by Proposition 7.1.1 (iv), and $s_\beta(\alpha) = \alpha + \beta$. Using Theorem 3.2.21 again, we find that

$$\chi(s_\beta(I), s_\beta(I')) = \alpha + \beta.$$

The two cases in the assertion of (ii) are the only ones compatible with this equation: either we have $s_\beta(I) = I$ and $s_\beta(I') \succ_\beta I'$, which is case (a), or we have $s_\beta(I) \prec_\alpha I$ and $s_\beta(I') = I'$, which is case (b).

The proof of (iii) follows similar lines to that of (ii), this time using the fact that $2(\alpha, \beta) = (\alpha, \alpha)$ by Proposition 7.1.1 (ii) and the fact that $s_\beta(\alpha) = \alpha - \beta$. □

A natural question is to try to describe a basis for the Lie algebra generated by the operators X_p, Y_p and H_p. One might guess that such a basis might include elements of the form

$$\sum_{L \in \mathcal{L}_\alpha(E)} X_L$$

for suitable α, but in general these elements do not even lie in the algebra.

However, it does turn out to be possible to construct a Chevalley basis all of whose elements are of the form

$$\sum_{L \in \mathcal{L}_\alpha(E)} \pm X_L$$

for suitable choices of signs. In order to explain the sign choices, we need to introduce the notion of the parity of a heap. Parity is the heap-theoretic analogue of Kac's asymmetry function, and it is also defined in the simply laced case in terms of orientations of Dynkin diagrams.

Definition 7.1.9 Let $\varepsilon : E \to \Gamma$ be a full heap over a simply laced Dynkin diagram. Fix an orientation of Γ, and write $p \to p'$ if there is an arrow from vertex p to vertex p'.

Let F be a finite convex subheap of E, and let $\kappa(F)$ be the number of pairs $(\alpha, \beta) \in F \times F$ such that both

(i) $\alpha > \beta$ and
(ii) either $\varepsilon(\alpha) = \varepsilon(\beta)$ or $\varepsilon(\alpha) \to \varepsilon(\beta)$.

The *parity*, $\varepsilon(F)$, of the heap F is defined to be $(-1)^{\kappa(F)}$.

Let F' be an ideal of F and let $F'' = F \backslash F'$ be the corresponding filter. We define the *relative parity*, $\varepsilon(F'', F')$, of F' and F'' to be

$$\varepsilon(F'', F') = \varepsilon(F')\varepsilon(F'')\varepsilon(F).$$

Example 7.1.10 Figure 7.2 shows the full heap $\varepsilon : E = \text{FH}(D_7^{(1)}(7)) \to D_7^{(1)}$, and a finite convex subheap F of E in which F has an ideal F' (the lower dashed region) of content α and a filter F'' (the upper dashed region) of content β.

There are two downward pointing arrows in the F' region. This corresponds to the fact that $\varepsilon(F') = (-1)^2 = 1$.

There are no downward pointing arrows in the F'' region, so $\varepsilon(F'') = (-1)^0 = 1$.

There are five downward pointing arrows overall, so $\varepsilon(F) = (-1)^5 = -1$.

There are three downward pointing arrows crossing from the F'' to the F' region. This means that

$$\varepsilon(F'', F') = \varepsilon(F')\varepsilon(F'')\varepsilon(F) = (-1)^3 = -1.$$

Exercise 7.1.11 Show that in the notation of Example 7.1.10, α, β and $\alpha + \beta$ are roots. Deduce from Example 7.1.10 that $\varepsilon(\beta, \alpha) = -1$, and prove without calculation that $\varepsilon(\alpha, \beta) = 1$.

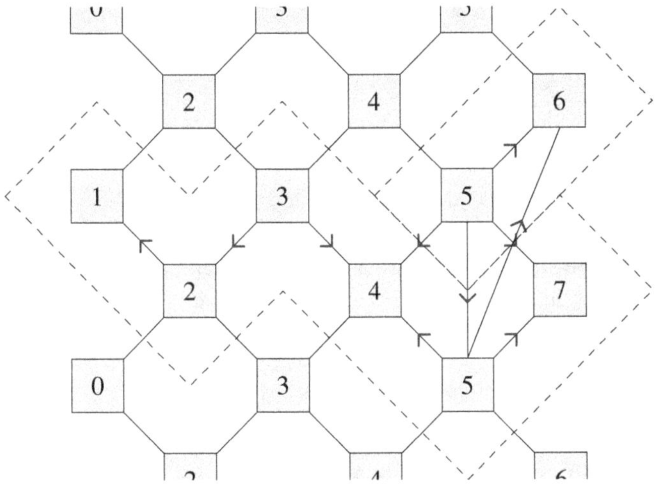

Figure 7.2 The full heap $\varepsilon : E = \mathrm{FH}(D_7^{(1)}(7)) \rightarrow D_7^{(1)}$ as in Example 7.1.10

Relative parity has the following key property.

Proposition 7.1.12 *Let F, F', F'' be as in Definition 7.1.9. We have*

$$\varepsilon(F'', F') = \varepsilon(\chi(F''), \chi(F')),$$

where the ε on the left is relative parity and the ε on the right is Kac's asymmetry function.

Proof This follows by comparing Definition 7.1.9 with Definition 7.1.4. □

7.2 Relations in simply laced simple Lie algebras

Definition 7.2.1 Let $\varepsilon : E \rightarrow \Gamma$ be a full heap over a simply laced Dynkin diagram, and let α be a positive real root. We define the linear operators

$$X_\alpha = \sum_{L \in \mathcal{L}_\alpha(E)} \varepsilon(L) X_L,$$

$$Y_\alpha = \sum_{L \in \mathcal{L}_\alpha(E)} \varepsilon(L) Y_L$$

and

$$H_\alpha = \sum_{L \in \mathcal{L}_\alpha(E)} H_L$$

on V_E.

If F is a parabolic subheap of E corresponding to a simple Lie algebra $\mathfrak{g}$, then we can define linear operators X_α, Y_α and H_α on V_F by restricting the above sums to the indexing set $\mathcal{L}_\alpha(F)$.

Remark 7.2.2 Notice that if $\alpha = \alpha_i$ is a fundamental root, then the operators X_α, Y_α and H_α on V_E are simply X_i, Y_i and H_i respectively, because the parity of a singleton heap is 1.

Remark 7.2.3 It follows from Exercise 5.4.7 that, given a fixed $I \in \mathcal{B}(E)$, at most one term X_L in the expansion of X_α satisfies $X_L(v_I) \neq 0$. It follows that, despite the infinite sum in its definition, X_α is a well-defined linear operator on V_E, and the same argument applies to Y_α and H_α.

For the rest of Section 7.2, we will consider X_α, Y_α and H_α to be operators on V_F, where F is a parabolic subheap corresponding to a simply laced simple Lie algebra $\mathfrak{g}$. The asymmetry function and parity allow us to derive some relations involving the Lie brackets of operators X_α and Y_α. The following results are still true, with the same proofs, when regarded as statements about operators on V_E, with the proviso that the roots involved should be real roots associated with the subalgebra $\mathfrak{g}$.

Lemma 7.2.4 *With the above assumptions, we have*

$$[X_\alpha, X_\beta] = \begin{cases} \varepsilon(\alpha, \beta)X_{\alpha+\beta} & \text{if } \alpha + \beta \text{ is a root,} \\ 0 & \text{otherwise.} \end{cases}$$

Proof If $\alpha + \beta$ is not representable, then it follows that $X_\alpha \circ X_\beta$ and $X_\beta \circ X_\alpha$ both act as zero on any basis element v_I, and also that $\alpha + \beta$ is not a root, by Theorem 3.2.30. This proves the result in this case, so we may assume that $\alpha + \beta$ is representable.

Let I be an ideal. If there does not exist I' with $I \prec_{\alpha+\beta} I'$, then all of the operators $X_{\alpha+\beta}$, $X_\alpha \circ X_\beta$ and $X_\beta \circ X_\alpha$ act as zero on v_I, so we may assume that such an I' exists; in particular, $\chi(I, I') = \alpha + \beta$. (By hypothesis, such an I does exist.)

First observe that we have

$$X_\alpha \circ X_\beta(v_I) = \begin{cases} \varepsilon(\alpha, \beta)\varepsilon(I' \backslash I)v_{I'} & \text{if } I \in J_\beta^-; \\ 0 & \text{otherwise} \end{cases}$$

and

$$X_\alpha \circ X_\beta(v_I) = \begin{cases} \varepsilon(\alpha, \beta)\varepsilon(I' \backslash I)v_{I'} & \text{if } I \in J_\alpha^-; \\ 0 & \text{otherwise.} \end{cases}$$

If $I \in J_\beta^-$ we define K to be the ideal for which $I \prec_\beta K$. We assume from now on that $I \in J_\beta^-$; the other case is similar. There are three subcases to consider: (a) $I' \in J_\beta^+$; (b) $I' \in J_\beta^-$ and (c) $I' \in J_\beta^0$.

The hypotheses of subcase (a) imply that $I \in J_\alpha^-$; define K' to be the ideal for which $I \prec_\alpha K'$. In this case, Theorem 3.2.21 shows that

$$s_\alpha(\beta) = s_\alpha(\chi(I, K)) = \chi(S_\alpha(I), S_\alpha(K)) = \chi(K', I') = \beta,$$

so $(\alpha, \beta) = 0$ and $\alpha + \beta$ is not a root by Proposition 7.1.1. By Proposition 7.1.7 (v), we have $\varepsilon(\alpha, \beta) = \varepsilon(\beta, \alpha)$. It follows that $[X_\alpha, X_\beta]$ acts as zero on v_I, as required.

Suppose that we are in subcase (b), and define K' to be the ideal for which $\chi(I', K') = \alpha$. Theorem 3.2.21 shows that

$$s_\beta(\alpha) = s_\beta(\chi(K, I')) = \chi(S_\alpha(K), S_\alpha(I')) = \chi(I, K') = \alpha + 2\beta,$$

which contradicts Proposition 7.1.1 and disposes of this case.

Finally, assume the hypotheses of subcase (c). Theorem 3.2.21 shows that

$$s_\beta(\alpha) = s_\alpha(\chi(K, I')) = \chi(S_\alpha(K), S_\alpha(I')) = \chi(I, I') = \alpha + \beta,$$

so $\alpha + \beta$ is a root. By Proposition 7.1.7 (iv), we have $\varepsilon(\alpha, \beta) = -\varepsilon(\beta, \alpha)$. It follows that

$$[X_\alpha, X_\beta](v_I) = \varepsilon(\alpha, \beta)\varepsilon(I'\backslash I)v_{I'},$$

which completes the proof. □

Using a parallel argument to that of Lemma 7.2.4, we obtain the following

Lemma 7.2.5 *With the above assumptions, we have*

$$[Y_\alpha, Y_\beta] = \begin{cases} -\varepsilon(\alpha, \beta)Y_{\alpha+\beta} & \text{if } \alpha + \beta \text{ is a root,} \\ 0 & \text{otherwise.} \end{cases}$$ □

Lemma 7.2.6 *Under the assumptions of Lemma 7.2.4 regarding the distinct positive roots α and β, we have*

$$[X_\alpha, Y_\beta] = \begin{cases} -\varepsilon(\alpha, \beta)X_{\alpha-\beta} & \text{if } \alpha - \beta \text{ is a positive root,} \\ \varepsilon(\alpha, \beta)Y_{\beta-\alpha} & \text{if } \alpha - \beta \text{ is a negative root,} \\ 0 & \text{otherwise.} \end{cases}$$

Proof If both $\alpha - \beta$ and $\beta - \alpha$ are not representable, then it follows that $X_\alpha \circ Y_\beta$ and $Y_\beta \circ X_\alpha$ both act as zero on any basis element v_I, and also that $\alpha - \beta$ is not a root, by Theorem 3.2.30. This proves the result in this case, so we may assume that one of $\alpha - \beta$ and $\beta - \alpha$ is representable. We will deal only with the case where $\alpha - \beta > 0$; the other case is dealt with similarly.

Let I be an ideal. If there does not exist I' with $I \prec_{\alpha-\beta} I'$, then all of the operators $X_{\alpha-\beta}$, $X_\alpha \circ Y_\beta$ and $Y_\beta \circ X_\alpha$ act as zero on v_I, so we may assume that such an I' exists; in particular, $\chi(I, I') = \alpha - \beta$. (By hypothesis, such an I does exist.)

First observe that we have

$$X_\alpha \circ Y_\beta(v_I) = \begin{cases} -\varepsilon(\alpha, \beta)\varepsilon(I'\backslash I)v_{I'} & \text{if } I \in J_\beta^+; \\ 0 & \text{otherwise} \end{cases}$$

and

$$Y_\beta \circ X_\alpha(v_I) = \begin{cases} \varepsilon(\alpha, \beta)\varepsilon(I'\backslash I)v_{I'} & \text{if } I \in J_\alpha^-; \\ 0 & \text{otherwise.} \end{cases}$$

If $I \in J_\beta^+$ we define K to be the ideal for which $K \prec_\beta I$. We assume from now on that $I \in J_\beta^+$; the other case is similar. There are three subcases to consider: (a) $I' \in J_\beta^+$; (b) $I' \in J_\beta^-$ and (c) $I' \in J_\beta^0$.

The hypotheses of subcase (a) imply the existence of an ideal K' for which

$$K \prec_\beta I \prec_{\alpha-2\beta} K' \prec_\beta I'.$$

In this case, Theorem 3.2.21 shows that

$$s_\beta(\alpha) = s_\beta(\chi(K, I')) = \chi(S_\beta(K), S_\beta(I')) = \chi(I, K') = \alpha - 2\beta,$$

which contradicts Proposition 7.1.1.

Suppose that we are in subcase (b), and define K' to be the ideal for which $\chi(I', K') = \beta$. Theorem 3.2.21 shows that

$$s_\beta(\alpha) = s_\beta(\chi(K, I')) = \chi(S_\alpha(K), S_\alpha(I')) = \chi(I, K') = \alpha,$$

which implies that $(\alpha, \beta) = 0$. Proposition 7.1.1 shows that $\alpha - \beta$ is not a root, and Proposition 7.1.7 (v) shows that $\varepsilon(\alpha, \beta) = \varepsilon(\beta, \alpha)$. It follows that $[X_\alpha, Y_\beta]$ acts as zero on v_I, as required.

Finally, assume the hypotheses of subcase (c); in this case, we have $Y_\beta \circ X_\alpha(v_I) = 0$. Theorem 3.2.21 shows that

$$s_\beta(\alpha) = s_\alpha(\chi(K, I')) = \chi(S_\alpha(K), S_\alpha(I')) = \chi(I, I') = \alpha - \beta,$$

so $\alpha - \beta$ is a root. By Proposition 7.1.7 (iv), we have $\varepsilon(\alpha, \beta) = -\varepsilon(\beta, \alpha)$. It follows that

$$\begin{aligned}
[X_\alpha, Y_\beta](v_I) &= \varepsilon(I\backslash K)\varepsilon(I'\backslash K)v_{I'} \\
&= \varepsilon(I\backslash K)^2\varepsilon(I'\backslash I)\varepsilon(\alpha - \beta, \beta)v_{I'} \\
&= \varepsilon(I'\backslash I)\varepsilon(\alpha, \beta)\varepsilon(\beta, \beta)v_{I'} \\
&= -\varepsilon(\alpha, \beta)\varepsilon(I'\backslash I)v_{I'},
\end{aligned}$$

which proves that $[X_\alpha, Y_\beta]v_I = X_{\alpha-\beta}v_I$, completing the proof. □

Lemma 7.2.7 *Maintain the assumptions of Lemma 7.2.4 regarding the distinct positive roots α and β, and suppose that $\alpha + \beta$ is a root. Then we have*

$$H_\alpha + H_\beta = H_{\alpha+\beta}.$$

Proof We define $\gamma = \alpha + \beta$. Let us consider the action of each operator on a fixed basis element, v_I.

Suppose first that $H_\gamma(v_I) = -v_I$; in other words, that there exists an ideal I' with $I \prec_\gamma I'$. Since $\gamma - \beta = \alpha$ is a root, Proposition 7.1.8 (iii) applies and gives two cases to consider.

In the first case, we have $I \in J_\beta^0(E)$ and $I' \in J_\beta^+(E)$, which means that $H_\beta(v_I) = 0$. Theorem 3.2.21 shows that

$$\beta = s_\alpha(\gamma) = s_\alpha(\chi(I, I')) = \chi(S_\alpha(I), S_\alpha(I')).$$

Using Proposition 7.1.8 (iii) again, we see that either $I \in J_\alpha^0(E)$ and $I' \in J_\alpha^+(E)$ or $I \in J_\alpha^-(E)$ and $I' \in J_\alpha^0(E)$, but the former possibility is a contradiction because it implies that $s_\alpha(I') \succ_\beta I$. It follows that $H_\alpha(v_I) = -v_I$, which proves that $H_\alpha + H_\beta$ and H_γ act the same on v_I.

The second case is that $I \in J_\beta^-(E)$ and $I' \in J_\beta^0(E)$. In this case, we can argue as in the last paragraph to show that $H_\alpha(v_I) = 0$, $H_\beta(v_I) = -v_I$, which leads to the desired conclusion and completes the analysis of the case $H_\gamma(v_I) = -v_I$.

We next suppose that $H_\gamma(v_I) = v_I$. The analysis proceeds similarly to the case of $H_\gamma(v_I) = -v_I$, except now we find that one of H_α and H_β acts as the scalar 0 and the other acts as the scalar 1.

Finally, suppose that $H_\gamma(v_I) = 0$. If both of H_α and H_β act as zero on v_I then we are done, so suppose without loss of generality that H_α acts as a nonzero scalar. We will assume from now on that $H_\alpha(v_I) = -v_I$; the case where $H_\alpha(v_I) = v_I$ is dealt with by a similar argument.

In the situation being considered, we have $I \in J_\alpha^-$, but $I \notin J_\gamma^-$; this implies that $I' \notin J_\beta^-$. Since $\alpha + \beta$ is a root, Proposition 7.1.8 (ii) shows that $I \in J_\beta^+$ and that $H_\beta(v_I) = -v_I$. This gives the desired result that

$$H_\alpha + H_\beta(v_I) = 0 = H_\gamma(v_I),$$

and completes the proof. □

The following result is immediate from the definitions. Note that the three cases given are mutually exclusive by Proposition 5.4.10.

Lemma 7.2.8 *Let $\varepsilon : E \to \Gamma$ be a full heap over a (not necessarily simply laced) Dynkin diagram and let $\mathfrak{g}$ be the simple Lie algebra corresponding to the subdiagram Γ^0. Let $I \in \mathcal{B}(E)$ and let α be a positive root associated with $\mathfrak{g}$. Then we have*

(i) $H_\alpha(v_I) = v_I$ *if and only if* $I \in J_\alpha^+(E)$;
(ii) $H_\alpha(v_I) = -v_I$ *if and only if* $I \in J_\alpha^-(E)$;
(iii) $H_\alpha(v_I) = 0$ *if and only if* $I \in J_\alpha^0(E)$. □

Lemma 7.2.9 *Let α be a positive root satisfying the assumptions of Lemma 7.2.4 Then we have*

$$[X_\alpha, Y_\alpha] = H_\alpha.$$

Proof We consider the action of each operator on a typical basis element, v_I. There are three cases to consider, according as $I \in J_\alpha^+$, $I \in J_\alpha^-$ or $I \in J_\alpha^0$. For example, if $I \in J_\alpha^+$ then we have $H_\alpha(v_I) = v_I$, $X_\alpha \circ Y_\alpha(v_I) = v_I$ and $Y_\alpha \circ X_\alpha(v_I) = 0$. The other cases are similar or easier. □

Lemma 7.2.10 *Let α and β be positive roots (not necessarily distinct) satisfying the assumptions of Lemma 7.2.4. Then we have*

$$[H_\alpha, X_\beta] = (\alpha^\vee, \beta)X_\beta$$

and

$$[H_\alpha, Y_\beta] = -(\alpha^\vee, \beta)Y_\beta.$$

Proof We prove the statement involving X_β; the other statement follows by a parallel argument.

Consider the action of each operator on a typical basis element, v_I. We may assume that $I \in J_\beta^-$, otherwise both operators act as zero. Let I' be an ideal satisfying $I \prec_\beta I'$.

Define the integers c_1 and c_2 by $H_\alpha(v_I) = c_1$ and $H_\alpha(v_{I'}) = c_2$. It follows that

$$[H_\alpha, X_\beta] = (H_a \circ X_\beta - X_\beta \circ H_\alpha)(v_I) = (c_2 - c_1)v_{I'}.$$

By Theorem 3.2.21 and Lemma 7.2.8, we have

$$s_\alpha(\beta) = s_\alpha(\chi(I, I')) = \chi(S_\alpha(I), S_\alpha(I')) = \beta + (c_1 - c_2)\alpha.$$

Recall that the action of the reflection s_α is given by

$$s_\alpha(\beta) = \beta - (\beta, \alpha^\vee)\alpha,$$

from which it follows that $(\beta, \alpha^\vee) = c_2 - c_1$. The result follows by combining these observations. □

Lemma 7.2.11 *Let F be a parabolic subheap corresponding to a simply laced simple Lie algebra $\mathfrak{g}$, and let α be a positive root associated with $\mathfrak{g}$. If α is not a fundamental root, then there exists a fundamental root α_i such that $s_i(\alpha) = \alpha - \alpha_i$ is a root.*

Proof By Corollary 5.3.12, there exists a generator s_i for which $0 < s_i(\alpha) < \alpha$. With the conventions of Proposition 7.1.1, we have $(\alpha, \alpha_i) = +1$ and $s_i(\alpha) = \alpha - \alpha_i$. It follows that $\alpha - \alpha_i$ is a root. □

Lemma 7.2.12 *Let F be a parabolic subheap corresponding to a simply laced simple Lie algebra $\mathfrak{g}$, and let $\alpha_1, \ldots, \alpha_n$ be the fundamental roots associated with $\mathfrak{g}$. Let $\mathfrak{h}$ denote the span of the operators $\{H_\alpha : \alpha \in \Phi^+\}$ indexed by the positive roots of $\mathfrak{g}$. Then the set $\{H_1, H_2, \ldots, H_n\}$ is a basis for $\mathfrak{h}$.*

Proof We first prove that the operators H_i are linearly independent. Suppose that

$$H = \sum_{i=1}^{n} \lambda_i H_i = 0.$$

By Lemma 7.2.10, we have

$$[H, X_\beta] = 0 = \left(\sum_{i=1}^{n} \lambda_i \alpha_i^\vee, \beta\right) X_\beta$$

for every positive root β. Since the positive roots span V_Γ, and since the form $(,)$ is nondegenerate by Theorem 4.2.3 (iv), it must be the case that $\lambda_i = 0$ for all i, from which it follows that $H = 0$, as desired.

To complete the proof, it suffices to prove that if α is a positive root associated with $\mathfrak{g}$, then H_α is a linear combination of the H_i. We write $\alpha = \sum_{i=1}^{n} \lambda_i \alpha_i$ and recall that the height of α is given by $h = \sum_{i=1}^{n} \lambda_i$. The proof is by induction on h. The case $h = 1$ is when α is a fundamental root, in which case there is nothing to prove, so assume that $h > 1$. By Lemma 7.2.11, there exists a fundamental root α_i such that $\alpha - \alpha_i$ is a positive root, and by Lemma 7.2.7, we have

$$H_\alpha = H_i + H_{\alpha - \alpha_i}.$$

The proof is completed by induction, because the height of $\alpha - \alpha_i$ is less than the height of α. □

Lemma 7.2.13 *Let F be a parabolic subheap corresponding to a simply laced simple Lie algebra $\mathfrak{g}$, and let X_α and Y_α be the usual operators on V_F.*

(i) *The operator X_α is nonzero and lies in the Lie subalgebra of $\mathfrak{g}$ generated by the X_i.*
(ii) *The operator Y_α is nonzero and lies in the Lie subalgebra of $\mathfrak{g}$ generated by the Y_i.*

Proof We first prove (i) by induction on the height, h, of α. If $h = 1$, then α is a fundamental root, and there is nothing to prove, so we may assume that $h > 1$. By Lemma 7.2.11, there exists a fundamental root α_i such that $\alpha - \alpha_i$ is a positive root. By Lemma 7.2.4, we have

$$[X_{\alpha-\alpha_i}, X_i] = \pm X_\alpha.$$

By the inductive hypothesis, $X_{\alpha-\alpha_i}$ lies in the subalgebra generated by the X_i, and this completes the proof.

The proof of (ii) follows the same line of argument, using Lemma 7.2.5 instead of Lemma 7.2.4. □

In order to state the main result of Section 7.2, it is convenient to introduce the following notation.

Definition 7.2.14 Let F be a parabolic subheap corresponding to a simply laced simple Lie algebra $\mathfrak{g}$. Let X_α and Y_α be the usual operators on V_F associated with positive roots $\alpha \in \Phi^+$. Given an arbitrary root $\gamma \in \Phi$ associated with $\mathfrak{g}$, we define the operator

$$E_\alpha = \begin{cases} X_\alpha & \text{if } \alpha > 0, \\ -Y_{-\alpha} & \text{if } \alpha < 0. \end{cases}$$

Theorem 7.2.15 *Let $\mathfrak{g}$ be a simply laced simple Lie algebra over $\mathbb{C}$ associated with a parabolic subheap F of a full heap, equipped with its usual generating set $\{H_i, X_i, Y_i : 1 \leq i \leq n\}$. Let $\alpha_1, \ldots, \alpha_n$ be the fundamental roots associated with $\mathfrak{g}$.*

(i) *The operators H_α and E_β on V_F satisfy the following relations:*

$$\left[H_\alpha, H_\beta\right] = 0 \textit{ for } \alpha, \beta \in \Phi^+; \tag{7.1}$$
$$\left[H_\alpha, E_\beta\right] = (\alpha^\vee, \beta)E_\beta \textit{ for } \alpha \in \Phi^+ \textit{ and } \beta \in \Phi; \tag{7.2}$$
$$[E_\alpha, E_{-\alpha}] = -H_\alpha \textit{ for } \alpha \in \Phi^+; \tag{7.3}$$
$$\left[E_\alpha, E_\beta\right] = 0 \textit{ if } \alpha, \beta \in \Phi \textit{ and } \alpha + \beta \notin \Phi \cup \{0\}; \tag{7.4}$$
$$\left[E_\alpha, E_\beta\right] = \varepsilon(\alpha, \beta)E_{\alpha+\beta} \textit{ if } \alpha, \beta, \alpha + \beta \in \Phi. \tag{7.5}$$

(ii) *A basis for $\mathfrak{g}$ is given by the set*

$$\{H_i : 1 \leq i \leq n\} \cup \{E_\alpha : \alpha \in \Phi\};$$

the associated structure constants are given by (i).

Proof Equation (7.1) holds because the operators H_α are simultaneously diagonalizable with respect to the usual basis of elements v_I. Equation (7.2) is a restatement of Lemma 7.2.10, and Equation (7.3) is a restatement of Lemma 7.2.9. Equations (7.4) and (7.5) follow by combining Lemmas 7.2.4, 7.2.5 and 7.2.6, and this completes the proof of (i).

To prove (ii), we first note that the elements listed in the statement do lie in the Lie algebra generated by the H_i, X_i and Y_i by Lemma 7.2.13. By part (i), we see that the linear span of the operators in the statement of (ii) is closed under the Lie bracket, so it remains to show that these operators are linearly independent. Suppose that

$$H = \sum_{i=1}^{n} \lambda_i H_i + \sum_{\alpha \in \Phi^+} \mu_\alpha X_\alpha + \sum_{\beta \in \Phi^+} \nu_\beta Y_\beta = 0.$$

Suppose first that $\mu_\alpha \neq 0$ for some $\alpha \in \Phi^+$. By Lemma 7.2.13 (i), there exists v_I such that $X_\alpha(v_I) \neq 0$; more precisely, we have $X_\alpha(v_I) = cv_{I'}$ for some ideal I' with $I \prec_\alpha I'$ and $c = \pm 1$. Since none of the other terms appearing in $H(v_I)$ contribute to the coefficient of $v_{I'}$, the coefficient of $v_{I'}$ in $H(v_I)$ is $\mu_\alpha c$. Since $H = 0$, we must have $\mu_\alpha = 0$ for all α.

A similar argument based on Lemma 7.2.13 (ii) shows that all the scalars ν_β are equal to zero. We deduce from Lemma 7.2.12 that all the λ_i are also zero, and this completes the proof of (ii). □

7.3 Folding

Kac also introduced a version of the asymmetry function ε for non-simply laced Lie algebras. In this case, the definition is much less transparent, but it is still possible to understand what is going on in terms of heaps.

Definition 7.3.1 Let $\varepsilon : E \to \Gamma$ be a full heap over a simply laced Dynkin diagram, and let $\mu : \Gamma \to \Gamma$ be a graph automorphism of order 2 with the property that for any vertex p of Γ, p and $\mu(p)$ are not adjacent vertices.

We then construct a new graph $\overline{\Gamma}$, whose vertices are the orbits of vertices of Γ under the action of $\{\text{id}, \mu\}$.

The vertices $\bar{p}$ and $\bar{q}$ are joined if p and q are adjacent. If q is adjacent to both the distinct vertices p and $\mu(p)$, we install a double edge between q and p with an arrow pointing towards $\bar{p}$. (It is possible for this procedure to result in a double edge with two arrows in opposite directions.)

We say that Γ *folds* to $\overline{\Gamma}$. We denote the corresponding projection map by π and call it a *folding*.

Example 7.3.2 The diagram Γ of type $D_7^{(1)}$ of Figure 7.1 has an automorphism that exchanges vertices 6 and 7 and fixes all other vertices. Under this automorphism, the diagram of type $D_7^{(1)}$ folds to the diagram $\overline{\Gamma}$ of type $B_6^{(1)}$, which was shown in Figure 5.1. There is a double arrow pointing towards $\pi(6)$ because 5 is adjacent to the distinct vertices 6 and $\mu(6)$.

Definition 7.3.3 Let $\varepsilon : E \to \Gamma$ be a full heap over a simply laced Dynkin diagram and $\pi : \Gamma \to \overline{\Gamma}$ be a folding associated with thc automorphism μ. We say that E is *compatible with* π if whenever $\alpha, \beta \in E$ satisfy $\mu(\alpha) = \mu(\beta)$, then α and β are comparable in E.

Lemma 7.3.4 *If* $\varepsilon : E \to \Gamma$ *is compatible with the folding* $\pi : \Gamma \to \overline{\Gamma}$, *then* $\varepsilon^{-1}(p) \cup \varepsilon^{-1}(\mu(p))$ *is isomorphic, as a partially ordered set, to* $\mathbb{Z}$.

Proof We may assume that $p \neq \mu(p)$, or the claim follows from Axiom (H1) of heaps. We write $C' = \varepsilon^{-1}(p)$, $C'' = \varepsilon^{-1}(\mu(p))$ and $C = C' \cup C''$. By Definition 7.3.3, C is a chain, and C is unbounded above and below because each of C' and C'' is unbounded above and below.

We will be done if we can show that C is locally finite. Let $\alpha, \beta \in C$ satisfy $\alpha < \beta$; we will prove that the interval $[\alpha, \beta]$ is finite. By Lemma 2.3.2 (ii), each of α and β has an upper and a lower bound in each of the chains C' and C''. In turn, it follows that each of the sets $I' = [\alpha, \beta] \cap C'$ and $I'' = [\alpha, \beta] \cap C''$ is bounded, and because $C' \cong C'' \cong \mathbb{Z}$, it must be the case that I' and I'' are finite. This completes the proof, because $[\alpha, \beta] = I' \cup I''$. □

Lemma 7.3.5 *Suppose that $\varepsilon : E \to \Gamma$ is a full heap associated with a generalized Cartan matrix $A = (a_{ij})$, and that E is compatible with the folding $\pi : \Gamma \to \overline{\Gamma}$. Let $p \neq \mu(p)$ be vertices of Γ. Suppose also that $\alpha \in C' = \varepsilon^{-1}(p)$ and $\beta \in C'' = \varepsilon^{-1}(\mu(p))$ are such that*

$$(\alpha, \beta) \cap \left(C' \cup C''\right) = \emptyset.$$

Then one of the following two situations must occur:

(i) *the open interval (α, β) contains a single element $\gamma \in \varepsilon^{-1}(q)$ such that $\varepsilon(\gamma)$ is adjacent to both p and $\mu(p)$; furthermore, $\varepsilon(\gamma)$ is fixed by μ if $a_{qp} = -1$, and is moved by μ if $a_{qp} = -2$;*
(ii) *the open interval (α, β) contains two distinct elements γ_1, γ_2, each of whose labels is adjacent to precisely one of p or $\mu(p)$, and each of the labels of γ_1 and γ_2 are moved by μ.*

Proof Since E is compatible with π, the elements $\alpha = E(p, z)$ and β must be comparable in E. If $\alpha' = E(p, z+1)$ is the lowest element of the chain C' that is greater than α, then α' and β must be comparable; in fact, we must have $\alpha < \beta < \alpha'$. By Definition 7.3.3, we know that $\alpha < \beta$ and $\beta < \alpha'$ cannot be covering relations. It follows that we have

$$\alpha < \delta < \beta < \delta' < \alpha'$$

for some covering relations $\alpha < \delta$ and $\delta' < \alpha'$. Applying Axiom (F3) to the interval $[\alpha, \alpha']$, we see that the elements δ and δ' are the unique elements with this property, and that there is precisely one element γ in each of the intervals (α, β) and (β, α') such that $\varepsilon(\gamma)$ is adjacent to p.

Reversing the roles of p and $\mu(p)$ in the above argument, we also see that there is precisely one element γ' in each of the intervals (α, β) and (β, α') such that $\varepsilon(\gamma')$ is adjacent to $\mu(p)$. Suppose that $\gamma = \gamma'$, from which the first assertion of (i) follows. There are two possibilities: the first is that $\mu(\varepsilon(\gamma))$ is distinct from $\varepsilon(\gamma)$ but is adjacent to both p and $\mu(p)$, in which case $a_{qp} = -2$. The other possibility is that $\mu(\varepsilon(\gamma)) = \varepsilon(\gamma)$; in this case, the definition of folding shows that there is a single arrow pointing from q to p in the Dynkin diagram, which means that $a_{qp} = -1$.

The other possibility is that $\gamma \neq \gamma'$, in which case $\varepsilon(\gamma')$ cannot be fixed by μ, or $\varepsilon(\gamma') = \mu(\varepsilon(\gamma'))$ would be adjacent to $p = \mu(\mu(p))$, contrary to hypothesis. A similar argument shows that $\varepsilon(\gamma)$ is not fixed by μ, and statement (ii) holds. □

Lemma 7.3.6 *Suppose that $\varepsilon : E \to \Gamma$ is compatible with the folding $\pi : \Gamma \to \overline{\Gamma}$. Suppose that p and q are adjacent vertices of Γ, and suppose that $p \neq \mu(p)$. If $\alpha, \beta \in E$ satisfy $\varepsilon(\alpha) = \mu(p)$ and $\varepsilon(\beta) = q$, then α and β are comparable in E.*

Proof Suppose that α and β are in the statement, so that, in particular, $\mu(p)$ and q are not adjacent vertices. By Axiom (F2), there exists an element $\gamma = E(p, z) \in \varepsilon^{-1}(p)$ such that either γ covers β or β covers γ. We will treat the case where γ covers β; the other case is dealt with by a parallel argument.

Let $\gamma' = E(p, z-1)$ be the greatest element of the chain $\varepsilon^{-1}(p)$ that is less than γ. Consideration of the edge chain $\varepsilon^{-1}\{p, q\}$ shows that $\gamma' < \beta$. Since E is compatible with π, both γ and γ' must be comparable with α (because $\varepsilon(\alpha) = \mu(\varepsilon(\gamma))$) and also with β (because $\varepsilon(\beta)$ and $\varepsilon(\gamma)$ are adjacent). Since α and β are not comparable, we must have $\gamma' < \alpha < \gamma$. The definition of folding means that p and $\mu(p)$ are not adjacent, so neither of the relations $\gamma' < \alpha$ or $\alpha < \gamma$ corresponds to a covering relation. We must therefore have $\delta, \delta' \in E$ such that

$$\gamma' < \delta' < \alpha < \delta < \gamma$$

for which $\gamma' < \delta'$ and $\delta < \gamma$ are covering relations. Since α and β are incomparable, the elements $\{\delta', \beta, \delta\}$ are distinct elements in the p-interval (γ, γ') whose labels are adjacent to p. This contradicts Axiom (F3). □

Theorem 7.3.7 *If $\varepsilon : E \to \Gamma$ is a full heap over a simply laced untwisted affine Dynkin diagram and $\pi : \Gamma \to \overline{\Gamma}$ is a folding such that E is compatible with π, then $\overline{\varepsilon} := \pi \circ \varepsilon : E \to \overline{\Gamma}$ is a full heap.*

Proof We first prove that $\overline{E} = \pi \circ \varepsilon : E \to \overline{\Gamma}$ is a heap. The vertex chain requirement of Axiom (H1) holds by the definition of compatibility, and the edge chain requirement follows from Lemma 7.3.6. Axiom (H2) holds because the underlying partially ordered set, E, is unchanged by the folding.

Axiom (F1) for full heaps holds by Lemma 7.3.4. Axiom (F2) for $\overline{E}$ is inherited from the corresponding property of E, so it remains to prove Axiom (F3). Consider a closed $\bar{p}$-interval $[x, y]$ in $\overline{E}$. There are two cases, according as $\varepsilon(x) = \varepsilon(y)$ or $\varepsilon(x) = \mu(\varepsilon(y))$. If $\varepsilon(x) = \mu(\varepsilon(y))$, Axiom (F3) holds by Lemma 7.3.5, so we assume from now on that $\varepsilon(x) = \varepsilon(y)$.

There are two elements $z, z' \in [x, y]$ with labels adjacent to $p = \varepsilon(x)$. If $q = \varepsilon(z)$ is adjacent to $\mu(p)$ in Γ, then there is an element a of E such that either a covers z or z covers a; we will only deal with the first possibility because the argument for the second is similar. We now have $x < z < y$ and $x < z < a$. Since a and y are comparable by the definition of compatibility and $z < a$ is a covering relation, we must have $x < a < y$, and this contradicts the assumption that $[x, y]$ is a $\bar{p}$-interval in $\overline{E}$. We conclude that neither of z or z' is adjacent to $\mu(p)$, from which it follows that $\bar{p}$ and $\bar{q}$ are connected by a simple edge in $\overline{\Gamma}$ and Axiom (F3) holds. □

All known full heaps over non simply laced Dynkin diagrams can be constructed from simply laced examples using Proposition 7.3.7.

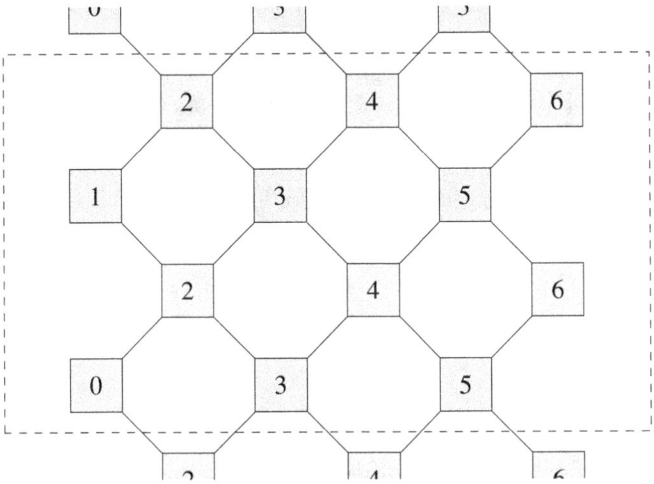

Figure 7.3 The full heap of Example 7.3.8

Example 7.3.8 Consider the folding π described in Example 7.3.2. This folding is compatible with the full heap $\varepsilon : E \to D_7^{(1)}$ given in Example 7.1.10, because every element of E labelled 6 is comparable with every element of E labelled 7. It follows from Theorem 7.3.7 that $\pi \circ \varepsilon : E \to B_6^{(1)}$ is also a full heap; it is shown in Figure 7.3.

Proposition 7.3.9 *Suppose that $\varepsilon : E \to \Gamma$ is a full heap over a simply laced untwisted affine Dynkin diagram and $\pi : \Gamma \to \overline{\Gamma}$ is a folding such that E is compatible with π. Suppose furthermore that $\overline{\Gamma}$ is an affine Dynkin diagram. Then there is a unique injective homomorphism of groups groups $\iota : W(\overline{\Gamma}) \to W(\Gamma)$ such that*

$$\iota(s_{\overline{a}}) = \prod_{i \in \pi^{-1}(a)} s_i .$$

(The order in the product is immaterial.)

Proof Let us fix a as in the statement. Note that the vertices in $\pi^{-1}(a)$ are pairwise nonadjacent, which shows that the product can be taken in any order, as claimed. The definition of folding shows that at most one of the operators S_i (as i ranges over $\pi^{-1}(a)$) can act nontrivially on any given proper ideal $I \in \mathcal{B}(E)$. Identifying the underlying posets of the folded and unfolded heaps, it then follows that the operator $S_{\overline{a}}$ on the folded heap acts in the same way on proper ideals as the operator $\prod_{i \in \pi^{-1}(a)} S_i$ on the unfolded heap. The assertions now follow from Theorem 3.2.27 (ii). □

Proposition 7.3.10 *Suppose that $\varepsilon : E \to \Gamma$ is a full heap over a simply laced untwisted affine Dynkin diagram and $\pi : \Gamma \to \overline{\Gamma}$ is a folding such that E is compatible with π. Suppose further that Γ is an affine Dynkin diagram, that π sends the vertex 0 of Γ to the vertex $0_{\overline{\Gamma}}$ of $\overline{\Gamma}$, and that vertex 0 of Γ is the only element in $\pi^{-1}(\pi(0))$.*

(i) *If* $\varepsilon_F : F \to \Gamma_0$ *is the principal subheap of* $\varepsilon : E \to \Gamma$, *then* $\pi \circ \varepsilon_F : F \to \overline{\Gamma}_0$ *is the principal subheap of* $\pi \circ \varepsilon : E \to \overline{\Gamma}$.
(ii) *The injective homomorphism of groups* $\iota : W(\overline{\Gamma}) \to W(\Gamma)$ *of Proposition 7.3.9 restricts to an injective homomorphism between the corresponding finite Weyl groups.*
(iii) *Let* $\mathfrak{g}$ *and* $\overline{\mathfrak{g}}$ *be the simple Lie algebras corresponding to the affine Dynkin diagrams* Γ *and* $\overline{\Gamma}$ *respectively. Then there is a unique injective homomorphism of Lie algebras* $\iota : \overline{\mathfrak{g}} \to \mathfrak{g}$ *satisfying*

$$\begin{aligned} \iota(X_{\overline{a}}) &= \textstyle\sum_{i \in \pi^{-1}(a)} X_i, \\ \iota(Y_{\overline{a}}) &= \textstyle\sum_{i \in \pi^{-1}(a)} Y_i, \\ \iota(H_{\overline{a}}) &= \textstyle\sum_{i \in \pi^{-1}(a)} H_i. \end{aligned}$$

Proof The hypotheses ensure that π induces a bijection between the vertex chain $\varepsilon^{-1}(0_\Gamma)$ of E and the vertex chain $(\pi \circ \varepsilon)^{-1}(0_{\overline{\Gamma}})$ of $\overline{E}$. Part (i) follows from this observation and the definition of principal subheap. Part (ii) then follows from Proposition 7.3.9. Part (iii) follows by adapting the argument of Proposition 7.3.9 to the operators X_i, Y_i and H_i, where the role of Theorem 3.2.27 (ii) is played by the simplicity of the Lie algebras. □

Example 7.3.11 Recall the full heap $\varepsilon : E \to D_7^{(1)}$ of Example 7.3.8, which folds to a full heap $\pi \circ \varepsilon : E \to B_6^{(1)}$. This folding has the property that the only element mapping to vertex 0 of $B_6^{(1)}$ is vertex 0 of $D_7^{(1)}$.

Proposition 7.3.9 shows that the subgroup of $W(D_7^{(1)})$ generated by the set

$$\{s_0, s_1, s_2, s_3, s_4, s_5, s_6s_7\}$$

is naturally isomorphic to $W(B_6^{(1)})$.

Proposition 7.3.10 (ii) shows that the finite Weyl group $W(D_7)$, generated by

$$\{s_1, s_2, s_3, s_4, s_5, s_6, s_7\},$$

has a subgroup isomorphic to $W(B_6)$, generated by

$$\{s_1, s_2, s_3, s_4, s_5, s_6s_7\}.$$

Proposition 7.3.10 (iii) shows that the subalgebra of the simple Lie algebra of type D_7 generated by the set

$$\{X_i, Y_i, H_i : 1 \leq i \leq 5\} \cup \{X_6 + X_7, Y_6 + Y_7, H_6 + H_7\}$$

is isomorphic to the Lie algebra of type B_6. (Furthermore, as in the Weyl group cases above, the isomorphism identifies the canonical generating sets with each other.)

Exercise 7.3.12 Show that the results of Example 7.3.11 generalize to describe a relationship between types D_l and B_{l-1} for general l.

Exercise 7.3.13 Construct a folding $\pi : A_{2l-1}^{(1)} \to C_l^{(1)}$ that is (a) compatible with the full heap $\mathrm{FH}(A_{2l-1}^{(1)}(1))$ and (b) sends vertex 0 in type A to vertex 0 in type C. Use the folding to prove the following:

(i) The subgroup of the affine Weyl group $W(A_{2l-1}^{(1)})$ generated by the set

$$\{s_0, s_1s_{2l-1}, s_2s_{2l-2}, \ldots, s_{l-1}s_{l+1}, s_l\}$$

is isomorphic to the affine Weyl group $W(C_l^{(1)})$.

(ii) The subgroup of the finite Weyl group $W(A_{2l-1})$ generated by the set

$$\{s_1s_{2l-1}, s_2s_{2l-2}, \ldots, s_{l-1}s_{l+1}, s_l\}$$

is isomorphic to the finite Weyl group $W(C_l)$.

(iii) The subalgebra of the simple Lie algebra of type A_{2l-1} generated by the subset

$$\{X_i + X_{2l-i}, Y_i + Y_{2l-i}, H_i + H_{2l-i} : 1 \leq i < l\} \cup \{X_l, Y_l, H_l\}$$

is isomorphic to the simple Lie algebra of type C_l.

Exercise 7.3.14 Construct a folding $\pi : A_{2l-1}^{(1)} \to C_l^{(1)}$ that is compatible with the full heap $\mathrm{FH}(A_{2l-1}^{(1)}(1))$ but that does not send vertex 0 to vertex 0.

Exercise 7.3.15 Construct a folding $\pi : E_6^{(1)} \to F_4^{(1)}$ that sends vertex 0 to vertex 0. Show that the folding is not compatible with either of the full heaps over $E_6^{(1)}$.

Exercise 7.3.16 Construct a folding $\pi : A_3^{(1)} \to A_1^{(1)}$ that is compatible with the full heap $\mathrm{FH}(A_3^{(1)}(1))$. Use the folding to prove that the subgroup of the affine Weyl group $W(A_3^{(1)})$ generated by the two elements s_0s_2 and s_1s_3 is isomorphic to the affine Weyl group $W(A_1^{(1)})$.

Exercise 7.3.17 Construct a folding $\pi : D_l^{(1)} \to D_{l-1}^{(2)}$ that is compatible with the full heap $\mathrm{FH}(D_l^{(1)}(l))$.

(i) Use the folding to prove that the subgroup of the affine Weyl group $W(D_l^{(1)})$ generated by the set

$$\{s_0s_1, s_2, s_3, \ldots, s_{l-2}, s_{l-1}s_l\}$$

is isomorphic to the affine Weyl group $W(D_{l-1}^{(2)})$.

(ii) State a version of this result for the full heap $\mathrm{FH}(D_l^{(1)}(l-1))$.

Exercise 7.3.18 Construct a folding $\pi : D_{2l}^{(1)} \to A_{2l-1}^{(2)}$ that is compatible with the full heap $\mathrm{FH}(D_{2l}^{(1)}(1))$. Use the folding to prove that the subgroup of the affine Weyl group $W(D_{2l}^{(1)})$ generated by the set

$$\{s_0s_l, s_1s_{l-1}, \ldots, s_{l-1}s_{l+1}, s_l\}$$

is isomorphic to the affine Weyl group $W(A_{2l-1}^{(2)})$.

7.4 Long and short roots

Definition 7.4.1 Let $\varepsilon : E \to \Gamma$ be a full heap over a simply laced Dynkin diagram, and let $\mu : \Gamma \to \Gamma$ be a graph automorphism of order 2 with the property that p and $\mu(p)$ are never adjacent vertices. Denote the corresponding automorphism of $W(\Gamma)$ by μ_W, and the corresponding automorphism of V_Γ by μ_{V_Γ}. Suppose that μ induces a folding $\pi : \Gamma \to \overline{\Gamma}$ and that π is compatible with the folding. Then μ induces a linear map $\mu_V : V_\Gamma \to V_{\overline{\Gamma}}$ such that

$$\mu_V(\alpha_i) = \alpha_{\overline{i}}$$

for each vertex i of Γ, and μ induces a function from the generators of $W(\overline{\Gamma})$ to $W(\Gamma)$ such that

$$\mu_S(S_{\overline{i}}) = \prod_{j:\overline{j}=\overline{i}} S_j.$$

If α is a real root of Γ, then the root $\mu_V(\alpha)$ of $\overline{\Gamma}$ is said to be *long* if $\mu_{V_\Gamma}(\alpha) = \alpha$, and *short* otherwise. We also define a linear map $\mu_\pi : V_\Gamma \to V_\Gamma$ by

$$\mu_\pi(\alpha) := \frac{\alpha + \mu_{V_\Gamma}(\alpha)}{2}.$$

Proposition 7.4.2 *In the notation of Definition 7.4.1, the function μ_S extends to an injective homomorphism of groups from $W(\overline{\Gamma})$ to $W(\Gamma)$.*

Proof Note that by Definition 7.3.3, at most one term in the product

$$\prod_{j:\overline{j}=\overline{i}} S_j$$

can act nontrivially on any given basis element $v_I \in V_\Gamma$. Furthermore, identifying the underlying partially ordered sets of the original heap E and the folded heap $\overline{E}$ in the usual way, we see that this product corresponds to the generator $S_{\overline{i}}$ of $W(\overline{\Gamma})$. The other assertions now follow. □

Proposition 7.4.3 *Maintain the above notation, and suppose that $\overline{\Gamma}$ is a Dynkin diagram of affine type.*

(i) *Let β be a real root of $\overline{\Gamma}$ and suppose that $I, I' \in \mathcal{B}(\overline{E})$ satisfy $\chi(I, I') = \beta$. If J and J' are the corresponding ideals of E, then we have $\chi(J, J') = \alpha$ for a real root α of Γ. In particular, every real root of $\overline{\Gamma}$ is of the form $\mu_V(\alpha)$ for some real root α of Γ.*
(ii) *Suppose that β is a real root of $\overline{\Gamma}$ and α is a real root of Γ such that $\mu_V(\alpha) = \beta$, then the only real roots γ of Γ such that $\mu_V(\gamma) = \beta$ are $\gamma = \alpha$ and $\gamma = \mu_{V_\Gamma}(\alpha)$.*
(iii) *Let α be a real root of Γ. If $\mu_V(\alpha)$ is short, then $(\alpha, \mu(\alpha)) = 0$.*
(iv) *If, furthermore, the fundamental root α_0 of $\overline{\Gamma}$ corresponds to a singleton orbit of μ, then $\mu_V(\alpha)$ is a real (respectively, imaginary) root of $\overline{\Gamma}$ if and only if α is a real (respectively, imaginary) root of Γ.*

Proof By linearity, we may assume that $\beta > 0$. Since β is a real root, the pair (I, I') can be trivialized to a singleton heap by Theorem 5.3.16, say by the sequence $S_{k_1} \cdots S_{k_l}$ in $W(\overline{\Gamma})$. Replacing each generator S_k in this product by the product

$$\prod_{j:\overline{j}=k} S_j$$

in $W(\Gamma)$, we obtain by Proposition 7.4.2 a sequence of generators of $W(\Gamma)$ trivializing the pair (I, I') in $\mathcal{B}(E)$, also to a singleton heap. The first assertion of (i) now follows, and the second assertion is a consequence of Theorem 3.2.30.

To prove (ii), it is enough to consider the case $\beta > 0$; the other case follows as a consequence. If β is not a fundamental root, then the argument in the previous paragraph applies to reduce the argument to the case where β is a fundamental root. In the latter case, α and γ are also fundamental roots, and the assertions of (ii) follow from the definitions.

It is enough to prove (iii) in the case where α is a fundamental root, and in this case, the result follows from the assumption that α and $\mu(\alpha)$ correspond to nonadjacent vertices.

To prove (iv), let δ and δ' be the lowest positive imaginary roots of Γ and Γ', respectively. By hypothesis, the fundamental root α_0 in $\overline{\Gamma}$ corresponds to a fundamental root, α_0', of Γ. Since Γ is simply laced, the coefficient of α_0 in δ is 1 by Proposition 4.3.18 (ii), and in turn, the coefficient of α_0' in δ' is 1. By Theorem 5.3.13 (iv), E has period δ and $\overline{E}$ has period δ'. It follows that the period, δ', of $\overline{E}$ satisfies $\mu(\delta) = k\delta'$ for some integer k. Since the coefficient of α_0 in $\mu(\delta)$ is the same as the coefficient of α_0' in δ, namely 1, we must have $k = 1$. It now follows from Definition 4.2.14 that μ_V identifies the imaginary roots of Γ with those of $\overline{\Gamma}$.

Now let α be a real root of Γ. We can use (i) to complete the proof of (ii) if we can show that $\mu_V(\alpha)$ is a real root of Γ'. Let (I, I') be ideals of E with $\chi(I, I') = \alpha$. Regarded as ideals of $\overline{E}$, we have $\chi(I, I') = \mu_V(\alpha)$. If the pair (I, I') were not trivializable in $\overline{E}$, then by Theorem 5.3.16, $\mu_V(\alpha)$ would be a multiple of δ', and the argument used above to prove (i) rules this out. It follows that (I, I') is trivializable in $\overline{E}$. This induces a trivialization of (I, I') in E, and arguing as in the proof of (i) above, we see that (I, I') is trivializable in $\overline{E}$ to a singleton heap. It follows that $\mu_V(\alpha)$ is a real root, which completes the proof. □

Exercise 7.4.4 Use the folding $\pi : A_{2l-1}^{(1)} \to C_l^{(1)}$ in Exercise 7.3.13 to show that the fundamental roots in type C_l may be identified with the roots

$$\alpha_1 + \alpha_{2l-1},\ \alpha_2 + \alpha_{2k-2},\ \ldots,\ \alpha_{l-1} + \alpha_{l+1},\ \alpha_l$$

in type A_{2l-1}. Show that this correspondence extends linearly to produce an isometric copy of the root system of type C_l within the span of the roots of type A_{2l-1}, with respect to the inner product in type A_{2l-1}.

Exercise 7.4.5 Use the folding $\pi : D_l^{(1)} \to B_{l-1}^{(1)}$ in Example 7.3.11 and Exercise 7.3.12 to show that the fundamental roots in type B_{l-1} may be identified with the roots

$$\alpha_1,\ \alpha_2,\ \ldots,\ \alpha_{l-2},\ \alpha_{l-1} + \alpha_l$$

in type D_l. Show that this correspondence extends linearly to produce an isometric copy of the root system of type B_{l-1} within the span of the roots of type D_l, with respect to the inner product in type D_l.

Proposition 7.4.3 (iv) is not true without the hypothesis that α_0 belong to a singleton orbit, as the next exercise shows.

Exercise 7.4.6 Show that in each of the foldings $\pi : \Gamma_1 \to \Gamma_2$ below, there is a real root of Γ_1 that maps to the lowest positive imaginary root of Γ_2:

(i) the folding $\pi : D_l^{(1)} \to D_{l-1}^{(2)}$ of Exercise 7.3.17;
(ii) the folding $\pi : D_{2l}^{(1)} \to A_{2l-1}^{(2)}$ of Exercise 7.3.18.

Lemma 7.4.7 *Maintain the notation of Definition 7.4.1, and let $[x, y]$ be a p-interval in the folded heap $\overline{E}$.*

(i) *If the open interval (x, y) contains a single element z with a label q adjacent to p, then we have $a_{pq} = -2$ and the fundamental root α_p is short. The fundamental root α_q is short if $a_{qp} = -2$, and long if $a_{qp} = -1$.*
(ii) *If the open interval (x, y) contains two elements z_1, z_2 with (respective) labels q_1 and q_2 adjacent to p, then we have $a_{pq_1} = a_{pq_2} = -1$ and either (a) the fundamental root α_p is long or (b) all the fundamental roots α_p, α_{q_1} and α_{q_2} are short.*

Proof Suppose that we have an element z satisfying the hypothesis of (i). Let x', y' be the elements of the unfolded heap E corresponding to x and y, respectively. If (x', y') were a p'-interval for some p' in Γ, it would violate Axiom (F3) for E, so we must have $\varepsilon(x') = \mu(\varepsilon(y'))$. Since $\varepsilon(p') \neq \mu(\varepsilon(p'))$, it follows that α_p is short, which proves the first assertion of (i). The rest of (i) follows from Lemma 7.3.5 (i).

Suppose that we have elements z_1 and z_2 satisfying the hypotheses of (ii). By Axiom (F3), we must have $a_{pq_1} = a_{pq_2} = -1$. If that α_p is short, then α_{q_1} and α_{q_2} are also short, by Lemma 7.3.5 (ii). □

Lemma 7.4.8 *Maintain the hypotheses of Proposition 7.4.3 (iii), let β be a real root of $\overline{\Gamma}$ and let $w \in W(\overline{\Gamma})$.*

(i) *The root $w(\beta)$ is long (respectively, short) if and only if β is long (respectively, short).*
(ii) *The root β is long if and only if β is conjugate to a long fundamental root of $\overline{\Gamma}$.*

Proof By Proposition 7.4.3 (iii), there exists a real root α of β such that $\mu_V(\alpha) = \beta$. Let $w' = \mu_S(w)$ be the element of $W(\Gamma)$ given by Proposition 7.4.2. It follows from the definition of μ_S that w' is fixed by μ_W.

Observe that $\mu_{V_\Gamma}(w'(\alpha)) = w(\mu_{V_\Gamma}(\alpha))$. It follows from this that $\mu_V(w'(\alpha)) = w(\mu_V(\alpha)) = w(\beta)$ is long if and only if $\mu_V(\alpha) = \beta$ is long, which completes the proof of (i). Part (ii) is an immediate consequence of (i). □

The next result shows that the definitions of "long" and "short" roots in Definition 7.4.1 are compatible with the inner product $(,)$ on $V_{\overline{\Gamma}}$.

Proposition 7.4.9 *Maintain the notation of Definition 7.4.1. Suppose that every short real root of $\overline{\Gamma}$ has length m, and that every long real root of $\overline{\Gamma}$ has length M. If both long and short roots exist, then we have $M^2 = 2m^2$. In the latter situation, if α_p and α_q are adjacent fundamental roots in $\overline{\Gamma}$, then α_p and α_q have distinct lengths if and only if they are connected by an edge with a single arrow pointing towards the short root.*

Proof It follows from Lemma 7.4.8 and Theorem 4.2.3 (ii) that every long (respectively, short) real root has the same length with respect to $(,)$ and is W-conjugate to a long (respectively, short) fundamental root. Now suppose that α_p and α_q are adjacent fundamental roots, so that each of a_{pq} and a_{qp} comes from the set $\{-1, -2\}$.

Choose a p-interval $[x, y]$ containing an element z labelled q (which is possible by Exercise 2.2.8).

Suppose first that $a_{pq} = -2$, meaning that z is the only element in the open interval (x, y). Lemma 7.4.7 shows that α_p is short, and also that α_q is short (respectively, long) if $a_{qp} = -2$ (respectively, $a_{qp} = -1$).

It remains to deal with the case where $a_{pq} = -1$, in which there are two elements $z = z_1, z_2$ in (x, y) with labels adjacent to p. If $a_{qp} = -2$, we can argue as in the previous paragraph, so suppose that $a_{qp} = a_{pq} = -1$. Let z' be the element for which $[z, z']$ is a q-interval, and let y' be the unique element, other than u, in (z, z') such that $\overline{\varepsilon}(y')$ is adjacent to q. Applying Lemma 7.4.7 (ii) to the p-interval $[x, y]$, which contains $z = z_1$, we find that if α_p is short, then α_q is short. Similarly, applying Lemma 7.4.7 (ii) to the q-interval $[z, z']$, which contains y, we find that if α_q is short, then so is α_p. We conclude that α_p and α_q have the same length.

Recall from Theorem 4.2.3 (iii) that

$$\frac{(\alpha_q, \alpha_q)}{(\alpha_p, \alpha_p)} = \frac{a_{pq}}{a_{qp}}.$$

All the assertions now follow from this observation coupled with the above case analysis and the fact that the Dynkin diagram is connected. □

Lemma 7.4.10 *Maintain the notation of Definition 7.4.1. Let $\mathfrak{g}$ be the simple Lie algebra associated with the unfolded heap E, and let $\overline{\mathfrak{g}}$ be the simple Lie algebra associated with the folded heap $\overline{E}$. Let $(,)_{\mathfrak{g}}$ (respectively, $(,)_{\overline{\mathfrak{g}}}$) be a bilinear form associated with $\mathfrak{g}$ (respectively, $\overline{\mathfrak{g}}$) in the sense of Theorem 4.2.3.*

(i) *There is a nonzero constant λ such that for all roots α, β of $\mathfrak{g}$, we have*

$$(\mu_\pi(\alpha), \mu_\pi(\beta))_{\overline{\mathfrak{g}}} = \lambda(\alpha, \beta)_{\mathfrak{g}}.$$

(ii) *If α and β are short positive roots of $\mathfrak{g}$ with $(\alpha, \beta)_{\mathfrak{g}} = (\alpha, \mu_{V_\Gamma}(\beta))_{\mathfrak{g}} = 0$, then $\mu_V(\alpha + \beta)$ is not a root of $\overline{\mathfrak{g}}$.*

Proof Let us normalize the form $(,)_{\mathfrak{g}}$ such that $(\alpha, \alpha) = 2$ for all roots α, which is possible by Theorem 4.2.3 (iii).

If α is short, then the hypotheses on μ show that $(\alpha, \mu_{V_\Gamma}(\alpha))_{\mathfrak{g}} = 0$. It follows from this that

$$(\mu_\pi(\alpha), \mu_\pi(\alpha))_{\mathfrak{g}} = \frac{(\alpha, \alpha)_{V_\Gamma} + 2(\alpha, \mu_{V_\Gamma}(\alpha)) + (\mu_{V_\Gamma}(\alpha), \mu_{V_\Gamma}(\alpha))}{4} = 1.$$

If α is long, then we have $(\alpha, \mu_{V_\Gamma}(\alpha))_{\mathfrak{g}} = 2$. It follows from this that $(\mu_\pi(\alpha), \mu_\pi(\alpha))_{\mathfrak{g}} = 2$.

Suppose that $\mathfrak{g}$ has both long and short roots. By Proposition 7.4.9, the squared length of the long roots is twice the squared length of the short roots; this is the same ratio obtained in the previous two paragraphs. By Theorem 4.2.3 (iii), these ratios characterize the bilinear form $(,)_{\overline{\mathfrak{g}}}$ up to multiplication by nonzero scalars, which proves part (i). Let us choose $\lambda = 1$ for the rest of the proof.

Suppose for a contradiction that α and β are short positive roots of $\mathfrak{g}$ with $(\alpha, \beta)_{\mathfrak{g}} = 0$, but that $\gamma' = \mu_V(\alpha + \beta)$ is a root of $\overline{\mathfrak{g}}$. Because the bilinear form $(,)_{\mathfrak{g}}$ is μ_{V_Γ}-invariant, we must also have

$$(\mu_{V_\Gamma}(\alpha), \mu_{V_\Gamma}(\beta))_{\mathfrak{g}} = (\mu_{V_\Gamma}(\alpha), \beta) = 0.$$

The hypotheses and Proposition 7.4.3 (iii) show that $(\gamma', \gamma')_{\overline{\mathfrak{g}}} = 2$ and that γ' is long. It follows from Proposition 7.4.3 (ii) that there is a unique root γ of $\mathfrak{g}$ such that $\mu_V(\gamma) = \gamma'$; in turn, we have $\mu_\pi(\gamma) = \gamma$, and

$$\gamma = \frac{\alpha + \mu(\alpha) + \beta + \mu(\beta)}{2}.$$

Since α and β are positive, so are γ and γ'. The above formula for γ shows that $s_\alpha(\gamma) = \gamma - \alpha$, which implies that $\gamma - \alpha$ is a root and also that

$$\mu_V(\gamma - \alpha) = \mu_V(\beta).$$

Proposition 7.4.3 (ii) now implies that either $\gamma = \alpha + \beta$, or $\gamma = \alpha + \mu(\beta)$. The first possibility, together with Proposition 7.1.1, contradicts the assumption that $(\alpha, \beta) = 0$, and the second possibility contradicts the assumption that $(\alpha, \mu(\beta)) = 0$. This completes the proof of (ii). □

Lemma 7.4.11 *Maintain the notation of Definition 7.4.1, and suppose that the root* α_0 *satisfies the hypotheses of Proposition 7.4.3 (iv). If* α *and* $\alpha + 2\beta$ *are both roots associated with the simple Lie algebra* $\overline{\mathfrak{g}}$*, then* $\alpha + \beta$ *is another such root.*

Proof Let us use the normalization $\lambda = 1$ of Lemma 7.4.10, so that long (respectively, short) roots have squared length 2 (respectively, 1).

Theorem 4.2.3 (i) shows that the Weyl group W stabilizes the root lattice (i.e., the $\mathbb{Z}$-span of the roots). It follows from the formula for reflections in Remark 4.2.15 that whenever α and β are roots, (α, β) must be an integer or half-integer. By symmetry of the form $(,)$, we have

$$(\alpha + 2\beta, \alpha + 2\beta) = (\alpha, \alpha) + 4(\alpha, \beta) + 4(\beta, \beta).$$

The above remarks show that $4(\alpha, \beta)$ and $4(\beta, \beta)$ are both even integers.

If α is long, then $(\alpha, \alpha) = 2$; the right hand side of the equation is even, and therefore equal to 2 by Proposition 7.4.9. It follows that $(\alpha, \beta) = -(\beta, \beta)$ and that $s_\beta(\alpha) = \alpha + 2\beta$.

On the other hand, if α is short, then $(\alpha, \alpha) = 1$; the right hand side of the equation is odd, and therefore equal to 1 by Proposition 7.4.9. We conclude as before that $s_\beta(\alpha) = \alpha + 2\beta$.

Theorem 3.2.30 guarantees the existence of ideals I, I' with $I \prec_\alpha I'$. Since $s_\beta(\alpha) = \alpha + 2\beta$, Theorem 3.2.21 shows that $I \in J_\beta^+$ and $I' \in J_\beta^-$; in other words, there exist ideals K, K' such that

$$K \prec_\beta I \prec_\alpha I' \prec_\beta K'.$$

Let J, J', L, L' be the ideals of the unfolded heap corresponding to I, I', K, K', respectively. We then have

$$L \prec_{\beta'} J \prec_{\alpha'} J' \prec_{\beta''} L',$$

where α', β' and β'' are roots of $\mathfrak{g}$ that project to α, β and β respectively under μ_V. We must have $\beta' \neq \beta''$, because otherwise Theorem 3.2.21 would show that $s_{\beta'}(\alpha') = \alpha' + 2\beta'$, and this contradicts Proposition 7.1.1. It follows that $s_{\beta'}(\alpha') = \alpha' + \beta'$, and that $\alpha' + \beta'$ is a root. By Proposition 7.4.3 (iv), $\mu_V(\alpha' + \beta') = \alpha + \beta$ is also a root, as required. □

Remark 7.4.12 It is not true in general that if α and $\alpha + 2\beta$ are real roots, then $\alpha + \beta$ is a real root. For example, in the situation of Proposition 4.3.18 (iii), $\theta = \delta - \alpha_0$ and $\theta + 2\alpha_0 = \delta + \alpha_0$ are both real roots, but $\theta + \alpha_0 = \delta$ is not.

Lemma 7.4.13 *Maintain the notation of Definition 7.4.1. Suppose that α, β and $\gamma = \alpha + \beta$ are (pairwise distinct) positive real roots of $\overline{\Gamma}$. Let α'' be a root of Γ such that $\mu_V(\alpha'') = \alpha$; such a root exists by Proposition 7.4.3 (i).*

(i) *If α is long, then γ is long if and only if β is long.*
(ii) *It is possible to find real roots α', β', γ' in Γ such that $\mu_V(\alpha') = \alpha$, $\mu_V(\beta') = \beta$, $\mu_V(\gamma') = \gamma$ and $\alpha' + \beta' = \gamma'$. Furthermore, it is possible to find such roots such that $\alpha' = \alpha''$.*

Proof Let us normalize the bilinear form on the roots of $\overline{\Gamma}$ in the usual way, so that the short roots have squared length 1, and the long roots have squared length 2.

To prove (i), it suffices to observe that if α is fixed by μ_{V_Γ}, then $\alpha + \beta$ is fixed by μ_{V_Γ} if and only if β is as well.

If α and β are both long, then γ is long by (i). The existence of α', β' and γ' in (ii) then follows from the definition of long roots and Proposition 7.4.3 (i).

The last assertion of (ii) follows from the first assertion by Proposition 7.4.3 (ii): if $\alpha' \neq \alpha''$, then we have $\mu(\alpha') = \alpha''$ and we can replace the equation $\alpha' + \beta' = \gamma'$ by $\mu(\alpha') + \mu(\beta') = \mu(\gamma')$.

Suppose now that one of α is long and the other is short; without loss of generality, suppose that α is long. Arguing as in the previous paragraph, there exists a root α' of Γ such that $\mu_V(\alpha') = \alpha$ and $\mu_\pi(\alpha') = \alpha$; there also exists a root β' of Γ such that $\mu_V(\beta') = \beta$ and

$$\mu_\pi(\beta') = \frac{\beta' + \mu_{V_\Gamma}(\beta')}{2}.$$

Since γ is short by (i), Lemma 7.4.10 (i) implies that $(\mu_\pi(\alpha' + \beta'), \mu_\pi(\alpha' + \beta')) = 1$. The assumptions, together with Lemma 7.4.10 (ii), imply the following identities:

$$\begin{aligned} (\alpha', \mu_\pi(\alpha')) &= 2; \\ (\beta', \mu_\pi(\beta')) &= 0; \\ (\alpha', \beta') &= (\mu_\pi(\alpha'), \mu_\pi(\beta')); \\ (\alpha', \mu(\beta')) &= (\mu_\pi(\alpha'), \beta'). \end{aligned}$$

By Proposition 7.1.1, the four quantities in the last two of the above equations can only be equal to -1 or to 0, and the fact that γ is short forces them all to be equal to -1. Further applications of Proposition 7.1.1 show that $\alpha' + \beta'$ and $\alpha' + \mu_\pi(\beta')$ are both roots of Γ. We can now satisfy the conditions of (ii) by taking $\gamma' = \alpha' + \beta'$.

Finally, suppose that both of α and β are short. Arguing as before, there exist roots α' and β' of Γ such that $\mu_V(\alpha') = \alpha$, $\mu_V(\beta') = \beta$,

$$\mu_\pi(\alpha') = \frac{\alpha' + \mu_{V_\Gamma}(\alpha')}{2}$$

and

$$\mu_\pi(\beta') = \frac{\beta' + \mu_{V_\Gamma}(\beta')}{2}.$$

We now have

$$\begin{aligned} (\alpha', \mu_\pi(\alpha')) &= 0; \\ (\beta', \mu_\pi(\beta')) &= 0; \\ (\alpha', \beta') &= (\mu_\pi(\alpha'), \mu_\pi(\beta')); \\ (\alpha', \mu(\beta')) &= (\mu_\pi(\alpha'), \beta'). \end{aligned}$$

If γ is short, then we must have $(\mu_\pi(\alpha' + \beta'), \mu_\pi(\alpha' + \beta')) = 1$. This is only compatible with the preceding equations if α' is orthogonal to both of β' and $\mu(\beta')$, and this contradicts Lemma 7.4.10 (ii). It follows that γ is long, and that $(\mu_\pi(\alpha' + \beta'), \mu_\pi(\alpha' + \beta')) = 2$. This implies that α' is orthogonal to precisely one of β' and $\mu_\pi(\beta')$; without loss of generality, assume that α' is orthogonal to $\mu_\pi(\beta')$, and that $(\alpha', \beta') = -1$. It follows that $\gamma' = \alpha' + \beta'$ is a root, and that it satisfies the conclusion of (ii). □

Lemma 7.4.14 *Maintain the notation of Definition 7.4.1. Suppose that α, β are distinct positive real roots of $\overline{\Gamma}$ and that $\alpha - \beta$ is also a root. Let α'' be a root of Γ such that $\mu_V(\alpha'') = \alpha$; such a root exists by Proposition 7.4.3 (i).*

(i) *If α is long, then γ is long if and only if β is long.*
(ii) *It is possible to find real roots α', β', γ' in Γ such that $\mu_V(\alpha') = \alpha$, $\mu_V(\beta') = \beta$, $\mu_V(\gamma') = \gamma$ and $\alpha' - \beta' = \gamma'$. Furthermore, it is possible to find such roots such that $\alpha' = \alpha''$.*

Proof Part (i) is proved using the argument of Lemma 7.4.13 (i).

The last assertion of (ii) follows from the first by copying the argument of Lemma 7.4.13 (ii). As for the first assertion, if $\alpha - \beta$ is positive, the conclusion of (ii) follows from Lemma 7.4.13 (ii) applied to the triple $\alpha - \beta, \beta, \alpha$. On the other hand, if $\alpha - \beta$ is negative, we apply Lemma 7.4.13 (ii) to the triple $\beta - \alpha, \alpha, \beta$, and the conclusion of (ii) follows. □

Lemma 7.4.15 *Maintain the notation of Definition 7.4.1, and let β be a (short) root of $\overline{\Gamma}$. If β' and β'' are distinct roots of Γ for which $\mu_V(\beta') = \mu_V(\beta'') = \beta$, then the sets*

$$\{J^+_{\beta'}, J^-_{\beta'}, J^+_{\beta''}, J^-_{\beta''}\}$$

are pairwise disjoint.

Proof This follows by applying Proposition 5.4.10 to the folded heap $\overline{E}$. □

Lemma 7.4.16 *Maintain the notation of Definition 7.4.1, and let α, β be distinct positive roots of $\overline{E}$. Let I, I' be proper ideals of $\overline{E}$ such that $I \prec_\alpha I'$, and let J, J' be the corresponding ideals of the unfolded heap, E.*

(i) *If $\alpha + \beta$ is a root, then either $I \in J^+_\beta(\overline{E})$ or $I' \in J^-_\beta(\overline{E})$ (or both).*
(ii) *If $\alpha - \beta$ is a root, then either $I \in J^-_\beta(\overline{E})$ or $I' \in J^+_\beta(\overline{E})$ (or both).*

Proof By Proposition 7.4.3 (ii) we have $J \prec_{\alpha'} J'$ for a root α' of Γ.

By Lemma 7.4.13 (ii), there exists a root β' of Γ such that $\alpha' + \beta' = \gamma'$ is a root of Γ for which $\mu_V(\gamma') = \alpha + \beta$. By Proposition 7.1.8 (ii), we either have $J \in J^+_{\beta'}(E)$ or $J' \in J^-_{\beta'}(E)$, and (i) follows.

The proof of (ii) follows similar lines, but uses Lemma 7.4.14 (ii) and Proposition 7.1.8 (iii). □

Unfortunately, there is no good analogue of Proposition 7.1.1 in the non-simply laced case, because if α and β are orthogonal roots for a non-simply laced simple Lie algebra, it may or may not be the case that $\alpha + \beta$ is a root. However, the following result is the analogue of Proposition 7.1.8 for non-simply laced Lie algebras; note that the six conditions given are mutually exclusive and exhaustive.

Proposition 7.4.17 *Maintain the notation of Definition 7.4.1, and suppose that the root α_0 satisfies the hypotheses of Proposition 7.4.3 (iv). Let A be the generalized Cartan matrix corresponding to the folded heap $\overline{E}$, let A^0 be the submatrix of A described in Proposition 4.3.18, and let $\overline{\Gamma}^0$ be the corresponding subdiagram of $\overline{\Gamma}$. Let $\mathfrak{g}$ be the simple Lie algebra associated with A^0, and let α, β be distinct positive roots of $\mathfrak{g}$. Suppose that $I, I' \in \mathcal{B}(\overline{E})$ satisfy $I \prec_\alpha I'$.*

(i) *If $I, I' \in J^0_\beta(\overline{E})$ then neither of $\alpha \pm \beta$ is a root.*
(ii) *If I and I' lie in the same set K, where K is one of $\{J^+_\beta(\overline{E}), J^-_\beta(\overline{E})\}$, then $(\alpha, \beta) = 0$ and either (a) neither of $\alpha \pm \beta$ is a root, or (b) both roots α and β are short, and both of $\alpha \pm \beta$ are long roots.*
(iii) *If either $I \in J^0_\beta(\overline{E})$ and $I' \in J^-_\beta(\overline{E})$ or $I \in J^+_\beta(\overline{E})$ and $I' \in J^0_\beta(\overline{E})$, then $\alpha + \beta$ is a root, but $\alpha - \beta$ and $\alpha + 2\beta$ are not.*
(iv) *If either (a) $I \in J^0_\beta(\overline{E})$ and $I' \in J^+_\beta(\overline{E})$ or (b) $I \in J^-_\beta(\overline{E})$ and $I' \in J^0_\beta(\overline{E})$, then $\alpha - \beta$ is a root, but $\alpha + \beta$ and $\alpha - 2\beta$ are not.*
(v) *If $I \in J^+_\beta(\overline{E})$ and $I' \in J^-_\beta(\overline{E})$, then α is a long root, β is a short root, $\alpha + \beta$ is a short root, $\alpha + 2\beta$ is a long root, and neither $\alpha - \beta$ nor $\alpha + 3\beta$ is a root.*
(vi) *If $I \in J^-_\beta(\overline{E})$ and $I' \in J^+_\beta(\overline{E})$, then α is a long root, β is a short root, $\alpha - \beta$ is a short root, $\alpha - 2\beta$ is a long root, and neither $\alpha + \beta$ nor $\alpha - 3\beta$ is a root.*

Proof Observe first that the six conditions given are mutually exclusive and exhaustive.

Part (i) is immediate from Lemma 7.4.16.

Let J and J' be the ideals of the unfolded heap E corresponding to I and I' respectively. It follows that we have $J \prec_{\alpha'} J'$ for some root α' of Γ with $\mu_V(\alpha') = \alpha$.

Suppose that $\alpha + \beta$ is a root. By Lemma 7.4.13, there exist roots α'' and β' of Γ with $\mu_V(\alpha'') = \alpha$, $\mu_V(\beta') = \beta$ and $\mu_V(\alpha'' + \beta') = \alpha + \beta$. By replacing α'' by $\mu_\pi(\alpha'')$ if necessary, we may assume that $\alpha'' = \alpha'$. By Proposition 7.1.8 applied to the pair (J, J'), we see that it is not possible for J and J' both to lie in $J^0_{\beta'}(E)$; it follows that it is not possible for I and I' both to lie in $J^0_\beta(\overline{E})$.

A similar argument, using Lemma 7.4.14, shows that if $\alpha - \beta$ is a root, then it is not possible for I and I' both to lie in $J^0_\beta(\overline{E})$. This proves (i).

The hypotheses of (ii) imply that

$$\chi(s_\beta(I), s_\beta(I')) = \chi(I, I'),$$

and Theorem 3.2.21 now shows that $s_\beta(\alpha) = \alpha$, which implies that $(\alpha, \beta) = 0$. It follows that $s_\beta(\alpha + \beta) = \alpha - \beta$, so that if one of $\alpha \pm \beta$ is a root, then so is the other. Since the squared length of $\alpha + \beta$ is the sum of the squared lengths of α and β, the only way $\gamma = \alpha \pm \beta$ can be a root is if α and β are both short and γ is long, by Proposition 7.4.9. To prove (ii), it therefore suffices by Lemma 7.4.8 (ii) to prove that one of $\alpha \pm \beta$ is indeed a root.

Suppose for now that $I, I' \in J^-_\beta(\overline{E})$. It follows that we have $J \in J^-_{\beta'}(E)$ and $J' \in J^-_{\beta''}(E)$, where β' and β'' are roots of Γ by Proposition 7.4.3 (i). Assume that $\beta' = \beta''$, and set $\delta = \mu_\pi(\beta')$. It follows that $J, J' \in J^0_\delta(E)$, and in turn that $s_{\beta'}(\alpha') = s_\delta(\alpha') = \alpha'$. By Lemma 7.4.10 (ii), $\alpha + \beta$ cannot be a root, and neither can $\alpha - \beta$. We may therefore assume that $\beta'' = \mu_\pi(\beta') \neq \beta'$, from which it follows that $s_{\beta''}(\alpha') = \alpha' + \beta''$. Since $\alpha' + \beta''$ is a root, Proposition 7.4.3 (iv) shows that $\alpha + \beta$ is also a root.

The case where $I, I' \in J^+_\beta(\overline{E})$ follows similar lines, and argues that in this case, $\alpha - \beta$ must be a root. This completes the proof of (ii).

Under the hypotheses of (iii), Theorem 3.2.21 shows that $s_\beta(\alpha) = \alpha + \beta$, so $\alpha + \beta$ must be a root. Using lemmas 7.4.15 and 7.4.16 and the hypotheses, we see that it is not possible for either of $\alpha - \beta$ or $\alpha + 2\beta$ to be roots, and this completes the proof of (iii).

The proof of (iv) follows similar lines to the proof of (iii).

Now assume the hypotheses of (v), and let the ideals K, K' be those satisfying

$$K \prec_\beta I \prec_\alpha I' \prec_\beta K'.$$

Denote the corresponding ideals of E by

$$L \prec_{\beta'} J \prec_{\alpha'} J' \prec_{\beta''} L'.$$

Note that we must have $\beta' \neq \beta''$, because it is not possible for $s_{\beta'}(\alpha') = \alpha' + 2\beta'$, by Proposition 7.1.1; it follows that β is short. In turn, we have $s_{\beta'}(\alpha') = \alpha' + \beta'$, so that $\alpha' + \beta'$ is a real root of Γ and $\mu(\alpha' + \beta') = \alpha + \beta$ is a real root of $\overline{\Gamma}$. Since $s_\beta(\alpha) = \alpha + 2\beta$, it follows that $\alpha + 2\beta$ is also a real root of $\overline{\Gamma}$. We cannot have $\alpha - \beta$ be a root by Lemma 7.4.16 (ii), and $\alpha + 3\beta$ cannot be a root by Lemma 7.4.16 (i) applied to the sum of roots $(\alpha + 2\beta) + \beta$.

Part (vi) can be proved by modifying the statement of (v), in this case by considering a chain of ideals

$$I \prec_\beta K \prec_{\alpha-2\beta} K' \prec_\beta I'. \qquad \square$$

Corollary 7.4.18 *Maintain the notation of Definition 7.4.1, and let α, β be real roots associated with $\overline{\Gamma}$. Suppose that for some positive integer k, the roots*

$$\alpha, \alpha + \beta, \ldots, \alpha + k\beta$$

are all roots. Then we have $k \leq 2$.

Proof Suppose that

$$\alpha - \beta, \alpha, \alpha + \beta$$

are all roots. We will deal with the case where $\alpha > 0$; the other case follows using a similar argument. By symmetry, we may assume that $\beta > 0$.

The only case of Proposition 7.4.17 compatible with this situation is part (ii), and this shows that α is a short root and $\alpha \pm \beta$ are both long roots. Since $\alpha + \beta$ is long, it cannot be the case that $(\alpha + \beta) \pm \beta$ are both roots, again by Proposition 7.4.17 (ii). Since α is a root, $\alpha + 2\beta$ is not a root. A similar argument shows that $\alpha - 2\beta$ is not a root, and this completes the proof. □

7.5 Relations in non-simply laced simple Lie algebras

In Section 7.5, we will consider X_α, Y_α and H_α to be operators on V_F, where F is a parabolic subheap corresponding to a non-simply laced simple Lie algebra $\mathfrak{g}$ obtained from folding a simply laced Lie algebra. We assume in addition that the root α_0 of the folded Lie algebra satisfies the hypotheses of Proposition 7.4.3 (iv). As in Section 7.2, the following results are still true, with the same proofs, when regarded as statements about operators on V_E, with the proviso that the roots involved should be real roots associated with the subalgebra $\mathfrak{g}$.

Lemma 7.5.1 *With the above assumptions, we have*

$$[X_\alpha, X_\beta] = \begin{cases} \varepsilon(\alpha, \beta) X_{\alpha+\beta} & \text{if } \alpha + \beta \text{ is a root but } \alpha - \beta \text{ is not a root,} \\ 2\varepsilon(\alpha', \beta') X_{\alpha+\beta} & \text{if } \alpha \pm \beta \text{ are both roots,} \\ 0 & \text{if } \alpha + \beta \text{ is not a root.} \end{cases}$$

Here, the roots α' and β' are (any) roots of the unfolded root system such that $\mu_V(\alpha') = \alpha$ and $\mu_V(\beta') = \beta$.

Proof The proof is identical to the proof of Lemma 7.2.4, except for subcases (a) and (b), which proceed as follows.

Suppose that I, I', K are as in the proof of Lemma 7.2.4, and assume the hypotheses of subcase (a). As before, define K' to be the ideal for which $I \prec_\alpha K'$. Let J, J', L, L' be the ideals of the unfolded heap corresponding to I, I', K, K' respectively. By Proposition 7.4.3 (i), we have

$$J \prec_{\beta'} L \prec_{\alpha'-\beta'} L' \prec_{\beta''} J',$$

where α', β', β'' are real roots of the unfolded heap, $\mu_V(\alpha' - \beta') = \alpha - \beta$, and $\mu_V(\beta') = \mu_V(\beta'') = \beta$. Because it is not possible for $s_{\beta'}(\alpha' - \beta') = \alpha' + 2\beta'$ by Proposition 7.1.1, Theorem 3.2.21 shows that $\beta' \neq \beta''$, $\mu_{V_\Gamma}(\beta') = \beta''$,

$s_{\beta'}(\alpha' - \beta') = \alpha'$ and

$$s_{\beta'} s_{\beta''}(\alpha' - \beta') = \alpha' + \beta''.$$

This shows that $\alpha' - \beta'$ and $\alpha' + \beta''$ are both roots, which by Proposition 7.4.3 (iv) implies that $\alpha - \beta$ and $\alpha + \beta$ are both roots. By Definition 7.3.3, we have $\varepsilon(\alpha', \beta') = \varepsilon(\alpha', \beta'')$ and $\varepsilon(\beta', \alpha') = \varepsilon(\beta'', \alpha')$. We now have

$$X_\alpha \circ X_\beta(v_I) = \varepsilon(\alpha', \beta')\varepsilon(I' \backslash I) v_{I'}$$

and

$$X_\alpha \circ X_\beta(v_I) = \varepsilon(\beta'', \alpha')\varepsilon(I' \backslash I) v_{I'}.$$

By Proposition 7.1.7 (iv), we have $\varepsilon(\alpha', \beta'') = -\varepsilon(\beta'', \alpha)$; it follows that

$$[X_\alpha, X_\beta](v_I) = 2\varepsilon(\alpha', \beta')\varepsilon(I' \backslash I) v_{I''},$$

and hence that $[X_\alpha, X_\beta] = 2\varepsilon(\alpha', \beta') X_{\alpha+\beta}$, which resolves subcase (a).

Now suppose that we are in subcase (b). As in subcase (a), we have

$$J \prec_{\beta'} L \prec_{\alpha'} J' \prec_{\beta''} L'$$

for real roots α', β' and β'' with $\mu_V(\alpha') = \alpha$, $\mu_V(\beta') = \mu_V(\beta'') = \beta$, and $\beta' \neq \beta'' = \mu_{V_\Gamma}(\beta')$. The argument used to prove subcase (a) shows that $\alpha + \beta$ and $\alpha + 2\beta$ are both roots; Corollary 7.4.18 then shows that $\alpha - \beta$ cannot be a root. Note that there is no ideal K' with $I \prec_\alpha K'$, because we would have $K' \prec_\beta I'$, contrary to assumption. It follows that $X_\beta \circ X_\alpha(v_I) = 0$ and that

$$X_\alpha \circ X_\beta(v_I) = \varepsilon(\alpha', \beta')\varepsilon(I' \backslash I) v_{I'}.$$

This gives the desired conclusion, that

$$[X_\alpha, X_\beta] = \varepsilon(\alpha', \beta') X_{\alpha+\beta}.$$

□

Arguing as in Lemma 7.5.1, we can also prove the following result.

Lemma 7.5.2 *With the above assumptions, we have*

$$[Y_\alpha, Y_\beta] = \begin{cases} -\varepsilon(\alpha', \beta') Y_{\alpha+\beta} & \text{if } \alpha + \beta \text{ is a root but } \alpha - \beta \text{ is not a root,} \\ 2\varepsilon(\alpha', \beta') Y_{\alpha+\beta} & \text{if } \alpha \pm \beta \text{ are both roots,} \\ 0 & \text{if } \alpha + \beta \text{ is not a root.} \end{cases}$$

Here, the roots α' and β' are (any) roots of the unfolded root system such that $\mu_V(\alpha') = \alpha$ and $\mu_V(\beta') = \beta$. □

Lemma 7.5.3 *With the above assumptions, we have*

$$[X_\alpha, Y_\beta] = \begin{cases} -\varepsilon(\alpha', \beta') X_{\alpha-\beta} & \text{if } \alpha - \beta \text{ is a positive root but } \alpha + \beta \text{ is not a root,} \\ \varepsilon(\alpha', \beta') Y_{\beta-\alpha} & \text{if } \alpha - \beta \text{ is a negative root but } \alpha + \beta \text{ is not a root,} \\ -2\varepsilon(\alpha', \beta') X_{\alpha-\beta} & \text{if } \alpha - \beta \text{ is a positive root and } \alpha + \beta \text{ is also a root,} \\ 2\varepsilon(\alpha', \beta') Y_{\beta-\alpha} & \text{if } \alpha - \beta \text{ is a negative root and } \alpha + \beta \text{ is also a root,} \\ 0 & \text{if } \alpha - \beta \text{ is not a root.} \end{cases}$$

Here, the roots α' and β' are (any) roots of the unfolded root system such that $\mu_V(\alpha') = \alpha$ and $\mu_V(\beta') = \beta$.

Proof As in the proof of Lemma 7.2.6, we can reduce to the case where $\alpha - \beta > 0$. The proof is essentially identical to the proof of Lemma 7.2.6, except for subcases (a) and (b), which proceed as follows.

Let I, I', K, K' be as in subcase (a) of the proof of Lemma 7.2.6, and let J, J', L, L' be the corresponding ideals of the unfolded heap, so that we have

$$K \prec_\beta I \prec_{\alpha-2\beta} K' \prec_\beta I'.$$

Applying Proposition 7.4.17 (vi) to the pair (K, I'), we find that $\alpha + \beta$ is not a root. Since $I' \notin J_\beta^-$, the operator $Y_\beta \circ X_\alpha$ acts as zero on v_I. Arguing as in Lemma 7.2.6, we find that $\alpha - 2\beta$ is a root, and also that we must have

$$L \prec_{\beta'} J \prec_{\alpha'-\beta'-\beta''} L' \prec_{\beta''} J',$$

where $\beta'' = \mu_{V_\Gamma}(\beta')$, $\beta' \neq \beta''$, $\mu_V(\alpha') = \alpha$ and $\mu_V(\beta') = \mu_V(\beta'') = \beta$. The same argument shows that $\alpha' - \beta'$ and $\alpha' - \beta''$ are roots, which means by Proposition 7.4.3 (iv) that $\alpha - \beta$ is a root. The analysis now proceeds similarly to subcase (c), and we conclude that

$$[X_\alpha, Y_\beta] = -\varepsilon(\alpha, \beta) X_{\alpha-\beta},$$

as required.

Now let I, I', K, K' be as in subcase (b) of the proof of Lemma 7.2.6, and let J, J', L, L' be the corresponding ideals of the unfolded heap, so that we have

$$K \prec_\beta I \prec_{\alpha-\beta} I' \prec_\beta K'.$$

As before, we find that

$$L \prec_{\beta'} J \prec_{\alpha'-\beta'} J' \prec_{\beta''} L',$$

$\mu_{V_\Gamma}(\beta') = \beta''$, $\beta' \neq \beta''$, $\mu(\alpha') = \alpha$ and $\mu(\beta') = \beta$. Theorem 3.2.21 shows that

$$s_{\beta'}(\alpha') = s_{\beta'}(\chi(K, I')) = \chi(S_{\beta'}(K), S_{\beta'}(I')) = \chi(I, I') = \alpha' - \beta',$$

so $\alpha' - \beta'$ is a root. Proposition 7.4.3 (iv) now shows that $\alpha - \beta$ is a root, and Proposition 7.4.17 (ii) shows that both of $\alpha \pm \beta$ are roots. The observations of Lemma 7.2.6 show that

$$X_\alpha \circ Y_\beta(v_I) = -\varepsilon(\alpha, \beta)\varepsilon(I' \backslash I) v_{I'}$$

and also that

$$Y_\beta \circ X_\alpha(v_I) = \varepsilon(\alpha, \beta)\varepsilon(I' \backslash I) v_{I'}.$$

It follows that

$$[X_\alpha, Y_\beta](v_I) = -2\varepsilon(\alpha, \beta) X_{\alpha-\beta} v_{I'},$$

as required. □

The following result, which is the non-simply laced analogue of Lemma 7.2.8, is immediate from the definitions.

Lemma 7.5.4 *Maintain the above assumptions regarding the distinct positive roots α and β, and suppose that $\gamma = \alpha + \beta$ is a root. Then we have*

$$(\alpha, \alpha)H_\alpha + (\beta, \beta)H_\beta = (\gamma, \gamma)H_\gamma.$$

Proof Define c_1, c_2, c_3 to be the elements of $\{-1, 0, +1\}$ such that $H_\alpha(v_I) = c_1 v_I$, $H_\beta(v_I) = c_2 v_I$ and $H_\gamma(v_I) = c_3(v_I)$. Let I, I' be ideals for which $\chi(I, I') = \gamma$; these exist by Theorem 3.2.30. By Lemma 7.4.16 (i), we must either have $I \in J_\alpha^-$ or $I \in J_\beta^-$ (or both). It follows that if both $c_1 = 0$ and $c_2 = 0$, then we also have $c_3 = 0$, as required. We may therefore reduce to the case where at least one of c_1 or c_2 is nonzero; without loss of generality, we may assume that c_1 is nonzero. By considering the dual heap if necessary, we may assume that $c_1 = -1$.

Since $\alpha + \beta$ is a root and $I \prec_\alpha I'$, Proposition 7.4.17 there are five cases to consider: two from part (ii), two from part (iii), and one from part (v) of that result. The cases are as follows:

(a) $I \in J_\beta^+$ and $I' \in J_\beta^+$, so that $c_2 = 1$;
(b) $I \in J_\beta^-$ and $I' \in J_\beta^-$, so that $c_2 = -1$;
(c) $I \in J_\beta^0$ and $I' \in J_\beta^-$, so that $c_2 = 0$;
(d) $I \in J_\beta^+$ and $I' \in J_\beta^0$, so that $c_2 = 1$;
(e) $I \in J_\beta^+$ and $I' \in J_\beta^-$, so that $c_2 = -1$.

In case (a), α and β are short but γ is long, and $(\alpha, \beta) = 0$. We then have $(\gamma, \gamma) = 2(\alpha, \alpha) = 2(\beta, \beta)$ by Proposition 7.4.9. By hypothesis, there exists an ideal I'' with $I'' \prec_\beta I$. Since $s_\alpha(\beta) = \alpha$, Theorem 3.2.21 shows that

$$\beta = s_\alpha(\beta) = s_\alpha(\chi(I'', I)) = \chi(S_\alpha(I''), S_\alpha(I)) = \chi(S_\alpha(I''), I').$$

Since $I'' \prec_{\alpha+\beta} I'$, we have $I'' \in J_\alpha^-$. The argument of the first paragraph of the proof now shows that $c_3 = 0$; it follows that

$$c_1(\alpha, \alpha) + c_2(\beta, \beta) = c_3(\gamma, \gamma),$$

satisying the assertion in this case.

The hypotheses of cases (b), (c) and (e) all imply that $c_3 = -1$, so we treat these cases together. The conclusion in case (b) follows from a calculation similar to the one in the previous paragraph. In case (c), the proof of Proposition 7.4.17 (iii) shows that we have $s_\beta(\alpha) = \alpha + \beta = \gamma$, which proves that $(\alpha, \alpha) = (\gamma, \gamma)$ and completes the proof in this case. In case (e), we know from Proposition 7.4.17 (v) that α is long and that β and γ are short, which shows that

$$(\alpha, \alpha) = 2(\beta, \beta) = 2(\gamma, \gamma),$$

and the conclusion follows.

In case (d), as in case (a), there exists an ideal I'' with $I'' \prec_\beta I$. There are three subcases, according as

(1) $I'' \in J_\alpha^-$,
(2) $I'' \in J_\alpha^0$, or
(3) $I'' \in J_\alpha^+$.

Subcase (1) is reminiscent of case (b): both roots α and β are short, and an argument like that of the first paragraph shows that $c_3 = 0$. Subcase (2) is reminiscent of case (c). In this situation, we also have $c_3 = 0$, and we can apply Theorem 3.2.21 to the pairs (I'', I) and (I, I') to show that $s_\alpha(\beta) = \gamma = s_\beta(\alpha)$. It follows that $(\alpha, \alpha) = (\beta, \beta)$. Finally, subcase (3) is reminiscent of case (e), and $c_3 = 1$. Here, β is long and α, γ are short. In all three subcases, the proof is completed by a short calculation. □

Remark 7.5.5 Note that the operators H_α (where α is a positive root) can be identified with the coroots $\alpha^\vee$, because they satisfy the same linear relations:

$$(\alpha, \alpha)\alpha^\vee + (\beta, \beta)\beta^\vee = (\gamma, \gamma)\gamma^\vee.$$

The same is true in Lemma 7.2.7, but, in the simply laced case, there is no distinction between roots and coroots.

Lemma 7.5.6 *Maintain the above assumptions, let α be a positive root, and let β be a (not necessarily distinct) positive root. Then we have*

$$\begin{aligned} [X_\alpha, Y_\alpha] &= H_\alpha; \\ [H_\alpha, X_\beta] &= (\alpha^\vee, \beta)X_\beta; \\ [H_\alpha, Y_\beta] &= -(\alpha^\vee, \beta)Y_\beta. \end{aligned}$$

Proof The proof is essentially the same as the proofs of Lemmas 7.2.9 and 7.2.10. □

Lemma 7.5.7 *Maintain the notation of Definition 7.4.1, and suppose that the root α_0 satisfies the hypotheses of Proposition 7.4.3 (iv). Let F be a parabolic subheap corresponding to a folded non-simply laced simple Lie algebra $\mathfrak{g}$, and let α be a positive root associated with $\mathfrak{g}$. If α is not a fundamental root, then there exists a fundamental root α_i such that $\alpha - \alpha_i$ is a root.*

Proof By Corollary 5.3.12, there exists a generator s_i for which $0 < s_i(\alpha) < \alpha$. By Theorem 3.2.30, there exist ideals I, I' with $I \prec_\alpha I'$. Applying Theorem 3.2.21 to the pair (I, I'), we see that either $s_i(\alpha) = \alpha - \alpha_i$ or $s_i(\alpha) = \alpha - 2\alpha_i$. In the first case, the proof is completed by copying the argument of Lemma 7.2.11. In the second case, Lemma 7.4.11 shows that $\alpha - \alpha_i$ is a root, as required. □

Lemma 7.5.8 *Maintain the notation of Definition 7.4.1, and suppose that the root α_0 satisfies the hypotheses of Proposition 7.4.3 (iv). Let F be a parabolic subheap corresponding to a folded non-simply laced simple Lie algebra $\mathfrak{g}$, and let $\alpha_1, \ldots, \alpha_n$ be the fundamental roots associated with $\mathfrak{g}$. Let $\mathfrak{h}$ denote the span of the operators $\{H_\alpha : \alpha \in \Phi^+\}$ indexed by the positive roots of $\mathfrak{g}$. Then the set $\{H_1, H_2, \ldots, H_n\}$ is a basis for $\mathfrak{h}$.*

Proof This is proved in the same way as Lemma 7.2.12, using Lemmas 7.5.6, 7.5.7 and 7.5.4. □

Exercise 7.5.9 Let $\varepsilon : E \to \Gamma$ be a full heap over an affine Dynkin diagram. Suppose that Γ has $n + 1$ vertices, numbered $0, 1, \ldots, n$ in such a way that 0 is the additional vertex relative to the corresponding Dynkin diagram Γ_0 of finite type. Let $\{H_i : 0 \leq i \leq n\}$ be the corresponding set of diagonal operators on $\mathcal{B}$, and let θ be the highest root of Γ_0. Use Theorem 5.3.13 and Lemma 7.5.4 to show that we have $H_0 = -H_\theta$. Deduce that the operators H_i are not linearly independent.

Lemma 7.5.10 *Maintain the notation of Definition 7.4.1, and suppose that the root α_0 satisfies the hypotheses of Proposition 7.4.3 (iv). Let F be a parabolic subheap corresponding to a folded non-simply laced simple Lie algebra $\mathfrak{g}$, and let X_α and Y_α be the usual operators on V_F.*

(i) *The operator X_α is nonzero and lies in the Lie subalgebra of $\mathfrak{g}$ generated by the X_i.*
(ii) *The operator Y_α is nonzero and lies in the Lie subalgebra of $\mathfrak{g}$ generated by the Y_i.*

Proof The proof is nearly identical to the proof of Lemma 7.2.13, but using Lemmas 7.5.7, 7.5.1 and 7.5.2. The only other change is that we have

$$[X_{\alpha-\alpha_i}, X_i] = cX_\alpha$$

for some $c \in \{\pm 1, \pm 2\}$. □

The next definition is the direct analogue of Definition 7.2.14.

Definition 7.5.11 Let F be a parabolic subheap corresponding to a folded non-simply laced simple Lie algebra $\mathfrak{g}$. Let X_α and Y_α be the usual operators on V_F associated with positive roots $\alpha \in \Phi^+$. Given an arbitrary root $\gamma \in \Phi$ associated with $\mathfrak{g}$, we define the operator

$$E_\alpha = \begin{cases} X_\alpha & \text{if } \alpha > 0, \\ -Y_{-\alpha} & \text{if } \alpha < 0. \end{cases}$$

We can now state and prove the non-simply laced analogue of Theorem 7.2.15.

Theorem 7.5.12 *Let $\overline{\mathfrak{g}}$ be a folded non-simply laced simple Lie algebra over $\mathbb{C}$ associated with a parabolic subheap F of a full heap, equipped with its usual generating set $\{H_i, X_i, Y_i : 1 \leq i \leq n\}$. Let $\alpha_1, \dots, \alpha_n$ be the fundamental roots associated with $\overline{\mathfrak{g}}$.*

(i) *The operators H_α and E_β on V_F satisfy the following relations:*

$$\begin{aligned} \left[H_\alpha, H_\beta\right] &= 0 \textit{ for } \alpha, \beta \in \Phi^+; && (7.6) \\ \left[H_\alpha, E_\beta\right] &= (\alpha^\vee, \beta)E_\beta \textit{ for } \alpha \in \Phi^+ \textit{ and } \beta \in \Phi; && (7.7) \\ [E_\alpha, E_{-\alpha}] &= -H_\alpha \textit{ for } \alpha \in \Phi^+; && (7.8) \\ \left[E_\alpha, E_\beta\right] &= 0 \textit{ if } \alpha, \beta \in \Phi \textit{ and } \alpha + \beta \notin \Phi \cup \{0\}; && (7.9) \\ \left[E_\alpha, E_\beta\right] &= \varepsilon(\alpha, \beta)E_{\alpha+\beta} \textit{ if } \alpha, \beta, \alpha + \beta \in \Phi \textit{ but } \alpha - \beta \notin \Phi; && (7.10) \\ \left[E_\alpha, E_\beta\right] &= 2\varepsilon(\alpha, \beta)E_{\alpha+\beta} \textit{ if } \alpha, \beta, \alpha + \beta, \alpha - \beta \in \Phi. && (7.11) \end{aligned}$$

(ii) *A basis for $\mathfrak{g}$ is given by the set*

$$\{H_i : 1 \leq i \leq n\} \cup \{E_\alpha : \alpha \in \Phi\};$$

the associated structure constants are given by (i).

Proof The proof follows the same lines as the proof of Theorem 7.2.15, this time using Lemmas 7.5.6, 7.5.1, 7.5.2 and 7.5.3 to prove (i), and Lemmas 7.5.10 and 7.5.8 to prove (ii). □

7.6 Notes and references

1 Our definition of Chevalley bases is taken from Borel's paper [6]. Borel states that the existence of bases satisfying all the conditions except for the last one ($N_{\alpha,\beta} = N_{-\alpha,-\beta} = \pm(p+1)$) is a classical result due to Weil. The last condition comes from Chevalley's paper [14]. Another early use of the term "Chevalley basis" appears in Carter's survey paper [10]. However, the definition in [10] differs slightly from the one here: the signs may be chosen arbitrarily, and we have $[x_\alpha, x_{-\alpha}] = h_{\alpha^\vee}$ (rather than $-h_{\alpha^\vee}$).

2 The span of the operators H_i in Theorem 7.2.15 and Theorem 7.5.12 is called the *Cartan subalgebra* of the associated simple Lie algebra, and is denoted by $\mathfrak{h}$.

3 Kac's asymmetry function $\varepsilon(\alpha, \beta)$ of Definition 7.1.3 appears in [37, section 7.8]. Kac uses it to construct Chevalley bases in the simply laced case in [37, (7.8.5)]. Adjusting for notation, Kac's definition is as follows:

$$\begin{aligned} [h, h'] &= 0 \text{ if } h, h' \in \mathfrak{h}, \\ [h, E_\alpha] &= \alpha(h)E_\alpha \text{ if } h \in \mathfrak{h} \text{ and } \alpha \in \Phi, \\ [E_\alpha, E_{-\alpha}] &= -h_{\alpha^\vee}, \\ [E_\alpha, E_\beta] &= 0 \text{ if } \alpha, \beta \in \Phi,\ \alpha+\beta \notin \Phi \text{ and } \alpha+\beta \neq 0, \\ [E_\alpha, E_\beta] &= \varepsilon(\alpha,\beta)E_{\alpha+\beta} \text{ if } \alpha, \beta, \alpha+\beta \in \Phi. \end{aligned}$$

In this case, the number p in the definition of Chevalley basis is equal to zero, by Proposition 7.1.1 (iv).

4 It follows from Proposition 5.4.10 and the constructions of this chapter that if α is a positive real root, then the squares of the operators X_α and Y_α are zero. The corresponding property for irreducible representations of simple Lie algebras over $\mathbb{C}$ is a characterization of minuscule representations; see Bourbaki [8, VIII section 7.3, proposition 7].

5 Most of the definitions and results pertaining to principal subheaps in Sections 7.1 and 7.2 are due to Wildberger [90]. In particular, the definition of parity of a heap (Definition 7.1.9) and the operators in Definition 7.2.1 are due to Wildberger. Wildberger uses the term "minuscule heap" to refer to principal subheaps in the simply laced case; full heaps do not appear in [90]. The main result of [90] is a result similar to Theorem 7.2.15 (ii), although no proof appears in the paper. A result similar to Lemma 7.2.7 also appears, again with no proof.

6 The Chevalley bases of [37] corresponding to simple Lie algebras with minuscule cases are reconstructed using heaps in Green [28, theorem 6.7]; this result incorporates Theorems 7.2.15 and 7.5.12.

7 Theorem 7.3.7 is a streamlined version of [28, proposition 6.1].

8 Suppose that $\varepsilon : E \to \Gamma$ is a full heap over an untwisted affine Lie algebra, and let $\mathfrak{g}$ be the simple Lie algebra corresponding to the Dynkin diagram Γ_0 of finite type. In this case, the Lie algebra generated by all operators X_p, Y_p, H_p (including the case $p = 0$) turns out to be isomorphic to the infinite-dimensional loop algebra corresponding to $\mathfrak{g}$. It is possible to add an additional diagonal heap operator D so as to generate a Lie algebra isomorphic to the affine Kac–Moody algebra modulo its one-dimensional centre; see [28, theorem 7.10] for details.

9 Kac proves in [37, Lemma 8.6] that loop algebras have no nontrivial ideals with finite dimensional quotients. This result can be used to prove that the Lie algebra generated

by the operators X_p, Y_p, H_p gives a faithful representation of the loop algebra in general (even in the twisted cases).

10 Kac constructs Chevalley bases in the non-simply laced case using the asymmetry function in the corresponding simply laced case. Kac uses the notation α' and β' for simply laced roots that fold to the non-simply laced roots α and β, respectively. The construction appears in [37, (7.9.3)]; adjusting for notation, it is as follows:

$$\begin{aligned}
[h, h'] &= 0 \text{ if } h, h' \in \mathfrak{h}, \\
[h, E_\alpha] &= \alpha'(h) E_\alpha \text{ if } h \in \mathfrak{h} \text{ and } \alpha \in \Phi, \\
[E_\alpha, E_{-\alpha}] &= -c_\alpha h_\alpha \text{ for } \alpha \in \Phi, \\
[E_\alpha, E_\beta] &= 0 \text{ if } \alpha, \beta \in \Phi,\ \alpha + \beta \notin \Phi \text{ and } \alpha + \beta \neq 0, \\
[E_\alpha, E_\beta] &= \varepsilon(\alpha, \beta) E_{\alpha+\beta} \text{ if } \alpha, \beta, \alpha + \beta \in \Phi.
\end{aligned}$$

Here, the number c_α is equal to 1 if α is long, and equal to r (the order of μ) if α is short. In type G_2, it is necessary to consider automorphisms of order 3 as well as 2, so r can be equal to 2 or 3.

11 The construction of Theorem 7.2.15 produces a Chevalley basis for all simply laced simple Lie algebras over $\mathbb{C}$. This turns out to be true even in type E_8, where there is no available minuscule representation or full heap (see Theorem 6.6.2).

12 The construction of Theorem 7.5.12 produces a Chevalley basis for all non-simply laced simple Lie algebras over $\mathbb{C}$. It turns out that analogous constructions work even in types F_4 and G_2, where there is no available minuscule representation or full heap (see Theorem 6.6.2). However, it does not seem to be possible to construct these Chevalley bases without some reference to the corresponding simply laced case.

13 It is also possible to construct Chevalley bases for the simple Lie algebras of types F_4 and G_2 in terms of the Lie algebras of types E_6 and D_4, the latter are described explicitly by Theorem 7.2.15. In the case of type E_6, one can use the unique nonidentity graph automorphism (identifying vertex 1 with 5, and 2 with 4) to construct the algebra of type F_4, provided a compatible orientation of E_6 is taken. The construction works even though the folding does not respect the full heap structure. In the case of type G_2, one needs to use an automorphism of order 3 that identifies all three end vertices in type D_4.

14 There are many other approaches to constructing Chevalley bases for Lie algebras. For example, Vavilov [87] gives a combinatorial construction of the simple Lie algebras of types E_6, E_7 and E_8, using the minuscule representations in the first two cases and the adjoint representation for E_8. The construction produces Chevalley bases for the Lie algebras, as well as the associated Chevalley groups. The introduction of [87] provides an extensive list of references to other approaches to the problem.

8

Combinatorics of Weyl groups

The weights of a minuscule representation may be regarded as points in Euclidean space. As we shall see in Chapter 8, the convex hull of these points forms a polytope with interesting combinatorial properties, and the action of the Weyl group on the polytope gives additional insight into the nature of minuscule representations.

Section 8.1 introduces the notion of a minuscule system. This provides a convenient way to describe explicit coordinates for the weights of all minuscule representations. This is useful for later purposes when concrete constructions are required.

Section 8.2 describes the action of the Weyl group as a permutation group on the weights of a minuscule representation. It is well-known that this action is transitive, but we go further and describe the W-orbits on ordered pairs of weights. This turns out to be important for some later combinatorial constructions.

Section 8.3 describes the remarkable relationship between the weights of a minuscule representation of weight ω_p and the positive roots of the Weyl group in which α_p appears with nonzero coefficient.

Section 8.4 introduces the weight polytopes of minuscule representations; that is, the convex hull of the set of weights of a minuscule representation. Section 8.5 analyses the combinatorics of the faces of the weight polytope. Finally, Section 8.6 shows how to associate families of graphs with the weight polytopes. These graphs come equipped with an action of the Weyl group, and include several families of graphs that are of independent interest.

8.1 Minuscule systems

In order to understand minuscule modules better, it is convenient to introduce the notion of a minuscule system, which is an abstraction of the set of weights of a minuscule module.

Definition 8.1.1 Let W be a finite Weyl group with root system Φ, acting by reflections on Euclidean space $\mathbb{R}^n$, where n may be larger than the number of standard generators of W. We call a subset $\Psi \subset \mathbb{R}^n$ a *minuscule system* for W if the following conditions hold:

(i) $W.\Psi = \Psi$;
(ii) W acts transitively on Ψ;

(iii) for all $\alpha \in \Phi$ and $\lambda \in \Psi$, we have $2\lambda.\alpha = c\alpha.\alpha$ for some $c \in \{+1, 0, -1\}$ (depending on λ and α).

Remark 8.1.2 In condition (iii) in Definition 8.1.1, the hypothesis $\alpha \in \Phi$ can be replaced by $\alpha \in \Delta$, by combining condition (ii) with the fact that every root of W is W-conjugate to a fundamental root. We will freely use this equivalent condition below, because it reduces the amount of checking required to identify minuscule systems.

Proposition 8.1.3 *Let $\mathfrak{g}$ be a simple Lie algebra over $\mathbb{C}$, let W be the Weyl group of $\mathfrak{g}$ and let V_F be a minuscule module for $\mathfrak{g}$ with highest weight ω_p. Suppose that W acts on $\mathbb{R}^N$ by reflections and that $\Psi \subset \mathbb{R}^N$ is a minuscule system for W with respect to this action.*

(i) *There is a unique element $\psi \in \Psi$ such that $\psi - s_i(\psi)$ is a nonnegative combination of fundamental roots for all generators s_i.*
(ii) *If ψ is fixed by all generators s_i other than s_p, but $s_p(\psi) \neq \psi$, then Ψ is an isometric copy of the set of weights for the minuscule module V_F.*
(iii) *The action of W on the weights of the minuscule module V_F is equivalent to the action of W on the cosets of the subgroup H generated by all generators s_i other than s_p.*

Proof By Definition 8.1.1 (ii), the set Ψ is the orbit under the reflection action of W on any one element of Ψ. By replacing $\mathbb{R}^n$ by the real span of Ψ, we now obtain part (i) as a consequence of Theorem 4.3.10 (iii) and (vii).

The hypothesis of (ii) together with Theorem 4.3.10 (iv) shows that the stabilizer of ψ is precisely the subgroup H of W generated by the set $\{s_i : i \neq p\}$. The same hypothesis shows that $s_p(\psi) = \psi - c'\alpha_p$ for some positive scalar c'. Definition 8.1.1 (iii) then shows that $s_p(\psi) = \psi - \alpha_p$, meaning that $c = 1$ in the notation of Definition 8.1.1. It follows that $\lambda.\alpha = (\alpha.\alpha)/2$.

By Corollary 6.6.3, there exists a principal subheap F of a full heap corresponding to the minuscule module V_F. Let Ψ' be the set of weights of V_F, regarded as vectors in the real span of Ψ'. By Lemma 5.5.1, the unique maximal element of F is labelled p. Using the action of W on the ideals of F, it follows that if v is a highest weight vector for V_F, then we have $s_i(v) = v$ for all $i \neq p$, and $s_p(v) = v - \alpha_p$. It follows that v lies in the fundamental chamber, in the sense of Theorem 4.3.10. By Theorem 4.3.10 (iv), the stabilizer of v is precisely the group H. This shows that Ψ and Ψ' are isomorphic as W-sets, since they are both equivalent to the action of W on the cosets of H. Furthermore, the fact that $s_p(v) = v - \alpha_p$ shows that $v.a = (\alpha.\alpha)/2$, from which it follows that Ψ and Ψ' are isometric, proving (ii) and (iii). □

Exercise 8.1.4 Let W and H be as in Proposition 8.1.3 (iii). Show that no nontrivial normal subgroup N of W is contained in H.

Example 8.1.5 Let $\varepsilon_1, \varepsilon_2, \ldots, \varepsilon_n$ be an orthonormal basis of $\mathbb{R}^n$. Let W be a finite Weyl group of type B_n, acting on $\mathbb{R}^n$, with fundamental roots

$$\alpha_1 = 4(\varepsilon_1 - \varepsilon_2), \quad \alpha_2 = 4(\varepsilon_2 - \varepsilon_3), \quad \ldots, \quad \alpha_{n-1} = 4(\varepsilon_{n-1} - \varepsilon_n) \quad \text{and } \alpha_n = 4\varepsilon_n,$$

and let Ψ be the set of 2^n vectors in $\mathbb{R}^n$, all of whose entries are equal to ± 2.

It is not hard to see that the vectors α_i above are linearly independent. Since their inner products are compatible with the Cartan matrix of type B in the sense of Theorem 4.2.3 (ii), it follows that we may regard these as the fundamental roots of the Weyl group.

Note that for $i < n$, s_i acts on Ψ by exchanging the i-th and $(i+1)$-th coordinates, and s_n acts on Ψ by changing the sign of the n-th coordinate; it follows that Ψ satisfies part (i) of Definition 8.1.1. The description just given in terms of signed permutations shows that the action of W on Ψ is transitive, establishing part (ii). To prove part (iii), we recall from Remark 8.1.2 that it is enough to check elements of Ψ against fundamental roots α. If $i < n$ and $\lambda \in \Psi$, we see that $\lambda.\alpha_i \in \{16, 0, -16\}$, whereas $\alpha_i.\alpha_i = 32$, as required. We also observe that the short fundamental root α_n satisfies $\lambda.\alpha_n \in \{8, 0, -8\}$, whereas $\alpha_n.\alpha_n = 16$. This completes the proof that Ψ is a minuscule system in type B_n.

Taking $\psi = (+2, +2, \ldots, +2)$, we see that ψ is fixed by all the s_i for $i < n$, but that $s_n(\psi) = \psi - \alpha_n$. It follows by Proposition 8.1.3 that Ψ is an isometric copy of the set of weights of the spin representation of the simple Lie algebra of type B_n.

The element $\psi' = (-2, -2, \ldots, -2)$ has the property that for all i, $s_i(\psi') - \psi'$ is a nonnegative linear combination of positive roots. It follows that ψ' corresponds to the lowest weight of the spin representation, and ψ to the highest. Since s_n is the only generator not fixing ψ', it follows that the lowest weight of the spin representation is $-\omega_n$.

From the data in Appendix B, we see that the highest root in type B_n is

$$\alpha_1 + 2(\alpha_2 + \alpha_3 + \cdots + \alpha_n),$$

and we showed in Exercise 6.4.9 that the highest short root in type B_n is

$$\alpha_1 + \alpha_2 + \cdots + \alpha_n.$$

It follows that, in the coordinate system above, the highest root is $4\varepsilon_1 + 4\varepsilon_2$, and the highest short root is $4\varepsilon_1$.

Example 8.1.6 Let $n \geq 4$, let $\alpha_1, \ldots, \alpha_{n-1}$ be the vectors in Example 8.1.5, and let $\alpha'_n = 4(\varepsilon_{n-1} + \varepsilon_n)$. Arguing as in Example 8.1.5, we can prove that these vectors may be regarded as the fundamental roots of type D_n.

Let Ψ be the set of $2n$ vectors $\{\pm 4\varepsilon_i : 1 \leq i \leq n\}$. In this case, the generators s_i (for $i < n$) act by exchanging the i-th and $(i+1)$-th coordinates. The generator s_n acts by exchanging the $(n-1)$-th and n-th coordinates, followed by changing the sign in both the $(n-1)$-th and n-th positions. It follows from this that $W(D_n)$ acts transitively on Ψ. To complete the verification of Definition 8.1.1, it remains to check condition (iii) for the fundamental roots. If α is a fundamental root and $\lambda \in \Psi$, we see that $\lambda.\alpha \in \{16, 0, -16\}$ and $\alpha.\alpha = 32$, as required. This completes the proof that Ψ is a minuscule system in type D_n.

Taking $\psi = 4\varepsilon_1$, we see that ψ is fixed by all the s_i except s_1, and that $s_1(\psi) = \psi - \alpha_1$. It follows by Proposition 8.1.3 that Ψ is an isometric copy of the set of weights of the natural representation of the simple Lie algebra of type D_n, with highest weight ω_1.

A similar argument shows that the vector $-\psi$ is the lowest weight vector of the natural representation; it has weight $-\omega_1$.

Example 8.1.7 Let $\varepsilon_0, \varepsilon_1, \ldots, \varepsilon_7$ be an orthonormal basis for $\mathbb{R}^8$. If $0 \leq i < j \leq 7$, we define

$$v_{i,j} = v_{j,i} := 4(\varepsilon_i + \varepsilon_j) - \left(\sum_{i=0}^{7} \varepsilon_i \right);$$

for example, $v_{0,1} = (3, 3, -1, -1, -1, -1, -1, -1)$. Let $\Psi^+ = \{v_{i,j} : 0 \leq i < j \leq 7\}$ and Ψ be the set of 56 vectors given by $\Psi^+ \cup -(\Psi^+)$. For $1 \leq i \leq 6$, define $\alpha_i := 4(\varepsilon_i - \varepsilon_{i+1})$, and define

$$\alpha_7 = (-2, -2, -2, -2, 2, 2, 2, 2).$$

We also define $K_0 = \{0, 1, 2, 3\}$ and $K_7 = \{4, 5, 6, 7\}$.

It is by now a familiar observation that the vectors $\alpha_1, \ldots, \alpha_6$ are linearly independent, and the vector α_7 is not a linear combination of them because it does not lie in the hyperplane $x_0 = 0$. By taking scalar products, we see that the vectors $\alpha_1, \ldots, \alpha_7$ may be regarded as the fundamental roots in type E_7.

In the Weyl group $W(E_7)$, the generators s_i (for $1 \leq i \leq 6$) act by exchanging the i-th and $(i+1)$-th coordinates. The action of the generator s_7 is as follows. If $1 \leq i < j \leq 7$ and $\{i, j\} \subset K \in \{K_0, K_7\}$, then we have $s_7(v_{i,j}) = -v_{k,l}$, where $K = \{i, j, k, l\}$. If these conditions are not met, then $s_7(v_{i,j}) = v_{i,j}$. For example, we have $s_7(v_{0,1}) = -v_{2,3}$ and $s_7(v_{3,4}) = v_{3,4}$; note that it is always true that $w(-v_{i,j}) = -w(v_{i,j})$.

It follows from the action described above and the list of elements in Ψ that any two elements of Ψ with the same ε_0-entry are conjugate under the action of $W(E_7)$. There are four possible values for the ε_0-coordinate, namely 3, 1, -1 and -3. Since $s_7(v_{0,1}) = -v_{2,3}$, $s_7(-v_{6,7}) = v_{4,5}$ and $s_7(v_{2,3}) = -v_{0,1}$, it follows (by considering the ε_0-entries of the vectors involved) that $W(E_7)$ acts transitively on Ψ. This establishes conditions (i) and (ii) of Definition 8.1.1. To prove condition (iii), we first note that for any fundamental root α_i, we have $\alpha_i.\alpha_i = 32$. If we have $1 \leq i \leq 6$ and $\lambda \in \Psi$, a case analysis shows that $\lambda.\alpha_i \in \{16, 0, -16\}$ as required, and a separate case analysis shows that $\lambda.\alpha_7 \in \{16, 0, -16\}$ as well. It follows that Ψ is a minuscule system for $W(E_7)$.

The vector $-v_{0,7}$ is fixed under the action of all the s_i except s_6; in this case, we have $s_6(-v_{0,7}) = -v_{0,7} - \alpha_6$. It follows that the corresponding minuscule representation (with highest weight ω_6) has highest weight vector corresponding to $-v_{0,7}$, and a similar argument shows that the lowest weight vector corresponds to $v_{0,7}$, of weight $-\omega_6$.

Exercise 8.1.8 Let $\alpha_1, \ldots, \alpha_{n-1}$ be the vectors in Example 8.1.5, and let $W(A_{n-1})$ be the Weyl group generated by the corresponding reflections. Let Ψ be the set of $\binom{n}{k}$ vectors in $\mathbb{R}^n$ such that each entry of each vector is equal to $+2$ or -2, and there is a total of k entries equal to $+2$.

(i) Show that the vectors α_i may be regarded as the fundamental roots in a root system of type A_{n-1}.
(ii) Show that Ψ is a minuscule system in type A_{n-1}.

(iii) Show that Ψ is an isometric copy of the set of weights of a minuscule representation of the simple Lie algebra $W(A_{n-1})$.
(iv) Show that the representation in (iii) has highest weight ω_k, and that the element of Ψ corresponding to the highest weight vector is $(\underbrace{+2, +2, \ldots, +2}_{k}, -2, \ldots, -2)$.
(v) Show that the representation in (iii) has lowest weight $-\omega_{n-k}$, and that the element of Ψ corresponding to the lowest weight vector is $(-2, \ldots, -2, \underbrace{+2, +2, \ldots, +2}_{k})$.

Exercise 8.1.9 Let $n \geq 4$ and let $\alpha_1, \ldots, \alpha_{n-1}, \alpha'_n \in \mathbb{R}^n$ be the vectors in Example 8.1.6 corresponding to the fundamental roots in type D_n. Let $W(D_n)$ be the Weyl group generated by the corresponding reflections. Let Ψ be the set of 2^{n-1} vectors in $\mathbb{R}^n$ such that the total number of entries equal to -2 is even, and let Ψ' be the set of 2^{n-1} vectors in $\mathbb{R}^n$ such that the total number of entries equal to -2 is odd.

(i) Show that each of Ψ and Ψ' is a minuscule system in type D_n.
(ii) Show that each of Ψ and Ψ' is an isometric copy of the set of weights of a minuscule representation of the simple Lie algebra $W(D_n)$.
(iii) Show that the representation corresponding to Ψ has highest weight ω_n, and that the element of Ψ corresponding to the highest weight vector is $(+2, +2, \ldots, +2)$.
(iv) Show that if n is even, then the representation corresponding to Ψ has lowest weight $-\omega_n$ and that the lowest weight vector corresponds to $(-2, -2, \ldots, -2)$.
(v) Show that if n is odd, then the representation corresponding to Ψ has lowest weight $-\omega_{n-1}$ and that the lowest weight vector corresponds to $(-2, -2, \ldots, -2, +2)$.
(vi) Show that the representation corresponding to Ψ' has highest weight ω_{n-1}, and that the element of Ψ' corresponding to the highest weight vector is $(+2, +2, \ldots, +2, -2)$.
(vii) Show that if n is odd, then the representation corresponding to Ψ has lowest weight $-\omega_n$ and that the lowest weight vector corresponds to $(-2, -2, \ldots, -2)$.
(viii) Show that if n is even, then the representation corresponding to Ψ' has lowest weight $-\omega_{n-1}$ and that the lowest weight vector corresponds to $(-2, -2, \ldots, -2, +2)$.

Exercise 8.1.10 Let $\alpha_1, \ldots, \alpha_{n-1}$ be the vectors in Example 8.1.5, and let $\alpha''_n = 8\varepsilon_n$. Let Ψ be the set of $2n$ vectors given in Example 8.1.6.

(i) Show that the vectors $\alpha_1, \ldots, \alpha_{n-1}, \alpha''_n$ defined above may be regarded as the fundamental roots for a root system of type C_n.
(ii) Show that the generator s_i (for $i < n$) acts by exchanging the i-th and $(i+1)$-th coordinates, and that the generator s_n acts by changing the sign of the n-th coordinate.
(iii) Show that Ψ is a minuscule system in type C_n.
(iv) Show that the representation corresponding to Ψ has highest weight ω_1, corresponding to the vector $\psi = 4\varepsilon_1$ and lowest weight $-\omega_1$, corresponding to $-\psi$.

Exercise 8.1.11 Let Ψ be the set of 56 vectors defined in Example 8.1.7, and define

$$\Psi_1 := \{\lambda \in \Psi : \lambda.v_{0,7} = 8\}$$

and

$$\Psi_6 := \{\lambda \in \Psi : \lambda.v_{0,7} = -8\}.$$

Regard the vectors $\alpha_1, \alpha_2, \alpha_3, \alpha_4, \alpha_5, \alpha_7$ defined in Example 8.1.7 as the fundamental roots for a root system of type E_6.

(i) Show that Ψ_1 consists precisely of the elements

$$\{v_{i,7} : 1 \leq i \leq 6\} \cup \{-v_{i,j} : 1 \leq i < j \leq 6\} \cup \{v_{0,i} : 1 \leq i \leq 6\},$$

and that $\Psi_6 = -\Psi_1$.

(ii) Show that Ψ_1 and Ψ_6 are each minuscule systems in type E_6.

(iii) Show that the representation corresponding to Ψ_1 has highest weight ω_1 and highest weight vector corresponding to $v_{1,7}$.

(iv) Show that the representation corresponding to Ψ_1 has lowest weight $-\omega_5$ and lowest weight vector corresponding to $v_{0,6}$.

(v) Show that the representation corresponding to Ψ_6 has highest weight ω_5 and highest weight vector corresponding to $-v_{0,6}$.

(vi) Show that the representation corresponding to Ψ_6 has lowest weight $-\omega_1$ and lowest weight vector corresponding to $-v_{1,7}$.

(vii) Deduce that the two minuscule representations in type E_6 each have dimension 27.

Exercise 8.1.12 For each minuscule module of the simple Lie algebras of types A, B, C and D, find a W-invariant bijection between (a) the elements of Ψ described in the minuscule systems above and (b) the ideals of the corresponding principal subheaps described in Chapter 6.

Remark 8.1.13 The ideals of the principal subheaps in types E_6 and E_7 were not classified in Chapter 6, but Example 8.1.7 and Exercise 8.1.11 give a nice parametrization of them. We will examine the exact nature of the correspondence later.

8.2 Weyl groups as permutation groups

A *permutation group* on a set Ω is a subgroup G of the symmetric group on Ω. If the action of G on Ω is transitive, then we call G a transitive permutation group.

If $\alpha \in \Omega$, we write G_α for the stabilizer of α in Ω. If G acts transitively on Ω and $\alpha, \beta \in \Omega$, then the stabilizers G_α and G_β are conjugate.

Suppose that G is a transitive permutation group on Ω, and fix $\alpha \in \Omega$. The group G acts diagonally on $\Omega \times \Omega$ via $g.(\beta, \gamma) = (g.\beta, g.\gamma)$. The G-orbits in $\Omega \times \Omega$ are called *orbitals*. The orbits of G_α on Ω are called *suborbits*.

Proposition 8.2.1 *Let G be a transitive permutation group on a set Ω, and let $\alpha \in \Omega$. Then there is a bijection ψ from the set of orbitals of the action to the set of suborbits* *(with respect to α) defined by*

$$\phi(\Gamma) = \Gamma(\alpha) = \{\beta : (\alpha, \beta) \in \Gamma\}. \qquad \square$$

Exercise 8.2.2 Prove Proposition 8.2.1.

Exercise 8.2.3 In the setup of Proposition 8.2.1, show that the diagonal, $\{(\beta, \beta) : \beta \in \Omega\}$, is one of the orbitals.

Definition 8.2.4 The *rank* of a transitive permutation group is defined to be the number of orbitals of the action (i.e., the number of suborbits). A permutation group of rank 2 is said to be *doubly transitive*.

Exercise 8.2.5 Show that G acts doubly transitively on Ω if and only if G acts transitively on unordered pairs of elements from Ω.

It is convenient for later purposes to introduce a shorthand notation for principal subheaps of full heaps. We will take the usual labelled Hasse diagram and rotate it clockwise by 45°; the covering relations are then represented by a shared vertical line.

Example 8.2.6 The shorthand notation for the principal subheap in Figure 6.12 is

<table>
<tr><td></td><td></td><td></td><td></td><td>6</td><td>3</td><td>4</td><td>5</td></tr>
<tr><td></td><td></td><td>5</td><td>4</td><td>3</td><td>2</td><td></td><td></td></tr>
<tr><td></td><td></td><td>4</td><td>3</td><td>2</td><td>1</td><td></td><td></td></tr>
<tr><td>1</td><td>2</td><td>3</td><td>6</td><td></td><td></td><td></td><td></td></tr>
</table>

It should be noted that there are other ways to render the principal subheap of $L(E_6, \omega_5)$ as a planar diagram, such as the one used in [84], namely

<table>
<tr><td></td><td></td><td></td><td>1</td><td>2</td><td>3</td><td>4</td><td>5</td></tr>
<tr><td></td><td></td><td></td><td>2</td><td>3</td><td>6</td><td></td><td></td></tr>
<tr><td></td><td></td><td>4</td><td>3</td><td>4</td><td></td><td></td><td></td></tr>
<tr><td>1</td><td>2</td><td>3</td><td>6</td><td>5</td><td></td><td></td><td></td></tr>
</table>

The same shorthand technique can be used for other labelled partially ordered sets.

Definition 8.2.7 Assume that F is a principal subheap of a full heap $\varepsilon : E \to \Gamma$ over an affine Dynkin diagram, and that V_F is the corresponding minuscule representation of the simple Lie algebra $\mathfrak{g}$ with Dynkin diagram Γ_0. Let p be a vertex of Γ_0, and let

$$F(p, 1) < F(p, 2) < \cdots < F(p, k)$$

be a complete list of the elements in the chain $F \cap \varepsilon^{-1}(p)$. Let $\mathfrak{g}_p$ be the subalgebra of $\mathfrak{g}$ generated by the set

$$\{e_i, f_i, h_i : i \neq p\}.$$

For each $1 \leq i \leq k$, let

$$D_{p,i} := \{x \in F : x \leq F(p, i)\}$$

and

$$U_{p,i} := \{x \in F : x \geq F(p, i)\},$$

and define $D(p, 0) = U(p, k + 1) = \emptyset$. Finally, for $1 \le i \le k + 1$, define

$$S_{p,i} := F \backslash (D_{p,i-1} \cup U_{p,i}).$$

Example 8.2.8 Let F be the principal subheap of Example 8.2.6, corresponding to the representation $L(E_6, \omega_5)$, and let $p = 3$. The elements of F labelled 3 form a chain $F(3, 1) < F(3, 2) < F(3, 3) < F(3, 4)$. The subheap $D_{3,0}$ is empty, and the subheaps $D_{3,i}$ for $1 \le i \le 4$ are

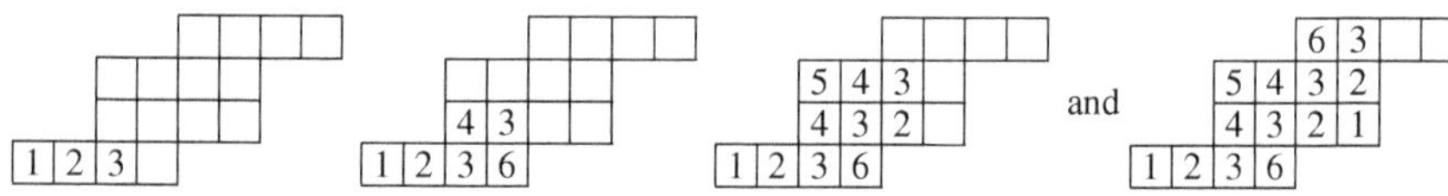

The subheap $U_{3,5}$ is empty, and the subheaps $U_{3,i}$ for $1 \le i \le 4$ are

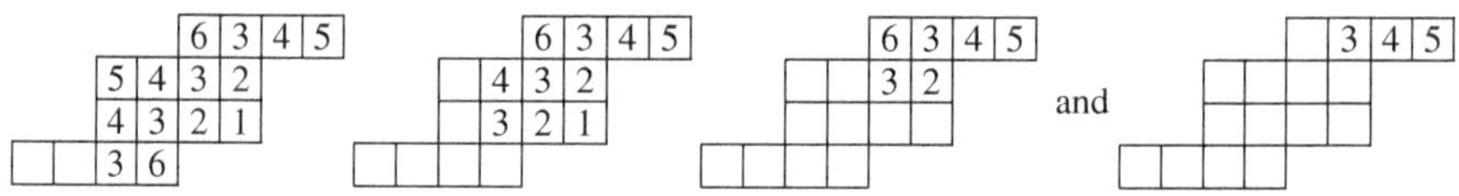

The subheaps $S_{3,i}$ for $1 \le i \le 5$ are

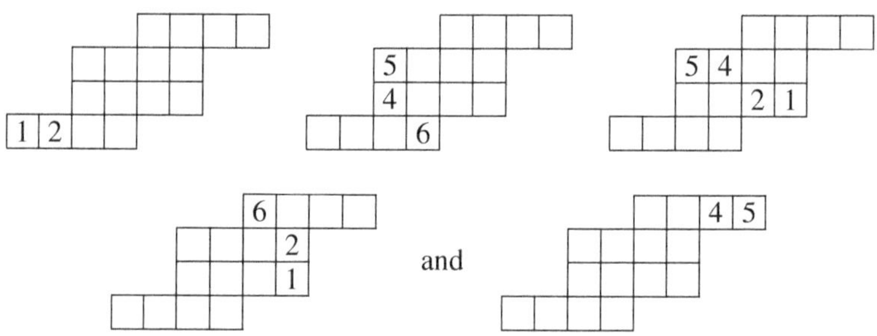

Proposition 8.2.9 *Maintain the notation of Definition 8.2.7, and let $\Gamma_{0,p} = \Gamma \backslash \{0, p\}$. Let W_0 be the Weyl group with Dynkin diagram Γ_0 and let $W_{0,p}$ be the subgroup of W_0 generated by the elements $\{s_i : i \neq p\}$. Let $\mathfrak{g}_p$ be the Lie subalgebra of $\mathfrak{g}$ corresponding to $W_{0,p}$.*

(i) *The subheap $S_{p,i}$ contains no occurrences of p, and may therefore be regarded as a heap over $\Gamma_{0,p}$, by restriction.*
(ii) *There is a bijection between ideals of $S_{p,i}$ and ideals of $U_{p,i}$ containing $D_{p,i-1}$.*
(iii) *If the faithful transitive action of W_0 on $J(F)$ given in Proposition 3.3.8 is restricted to $W_{0,p}$ then, under the identifications of (i) and (ii) the orbits are precisely the sets $J(S_{p,i})$.*
(iv) *If the $\mathfrak{g}$-module V_F is restricted to the subalgebra $\mathfrak{g}_p$, then under the identifications of (i) and (ii), the module decomposes as a direct sum*

$$\bigoplus_{i=1}^{k+1} V_{S_{p,i}}.$$

Proof Part (i) follows because the set $D_{p,i-1}$ contains all elements of the form $F(p, 0), \ldots, F(p, i-1)$ and the set $U_{p,i}$ contains all elements of the form $U_{p,i}, \ldots, U_{p,k}$.

Part (ii) follows by elementary properties of partially ordered sets, as in the proof of Lemma 3.3.3 (v).

Let I be an ideal of F, and define i to be the largest integer for which $F(p, i) \in I$, or $i = 0$ if no such integer exists. It follows that we have $D(p, i) \subset I \subset F \backslash U(p, i + 1)$. By (i), the subheap $U(p, i + 1) \cup D(p, i)$ contains all occurrences of p in F which means that $D(p, i)$, $F \backslash U(p, i + 1)$ and I are in the same $W_{0,p}$ orbit. Conversely, every ideal in the same $W_{0,p}$ orbit as I must have the same intersection with the vertex chain $\varepsilon^{-1}(p)$, which means that the $W_{0,p}$-orbit of I consists precisely of those ideals that contain $D(p, i)$ but are contained in $F \backslash U(p, i + 1)$. Part (iii) now follows.

Recall from the definitions that if $v_I \in V_F$ is a basis element for which $e_i(v_I) \neq 0$, then $e_i(v_I) = v_{s_i(I)}$, and that the same is true for the operators f_i. Conversely, if $s_i(I) \neq I$, then either $e_i(v_I) = v_{s_i(I)}$ or $f_i(v_I) = v_{s_i(I)}$. Since the Lie algebras $\mathfrak{g}$ and $\mathfrak{g}_p$ are both generated by the elements of the form e_i and f_i, as i ranges over some indexing set, (iv) is a restatement of (iii). □

Remark 8.2.10 Results such as Proposition 8.2.9 (iv), which describes how a module decomposes upon restriction to a subalgebra, are known as *branching rules*.

Example 8.2.11 Let us consider the principal subheap F of $\varepsilon : E_7^{(1)}(6) \to E_7^{(1)}$ shown in Figure 6.14, with the notational conventions of Definition 8.2.7. The subdiagram of the Dynkin diagram of type E_7 is a Dynkin diagram of type D_6. We identify the labels 2, 3, 4, 5, 6, 7 of E_7 with 5, 4, 3, 2, 1, 6, respectively.

The principal subheap in this case is given by

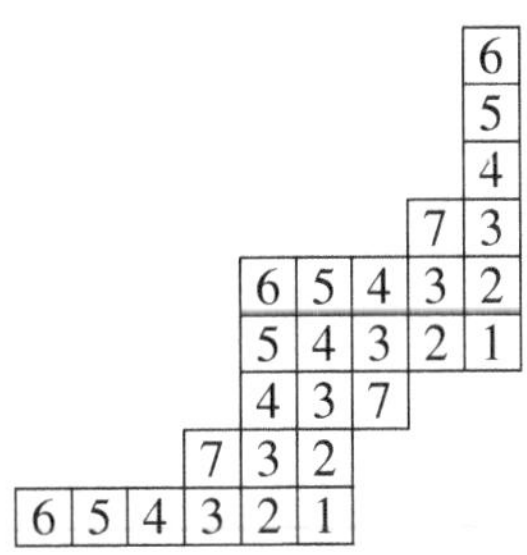

and the subheaps $S_{1,k}$ for $k \in \{1, 2, 3\}$ are given by

$S_{1,1} =$

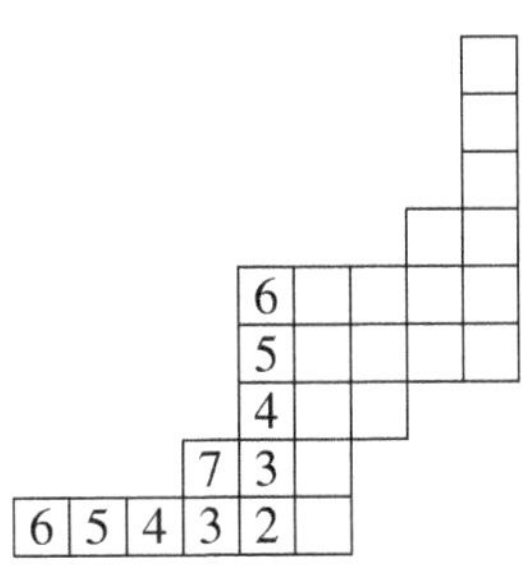

$S_{1,2} =$

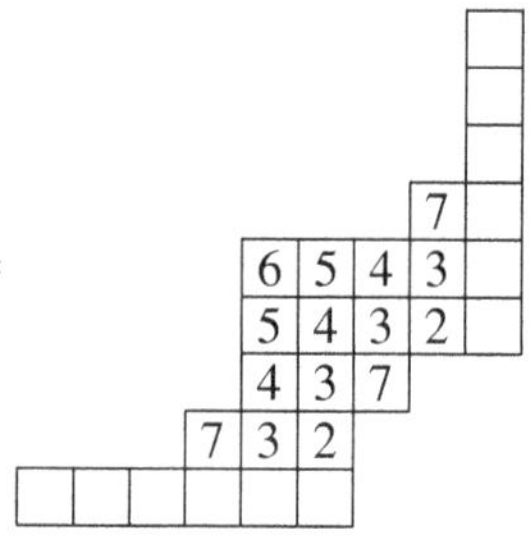

and

$S_{1,3} =$

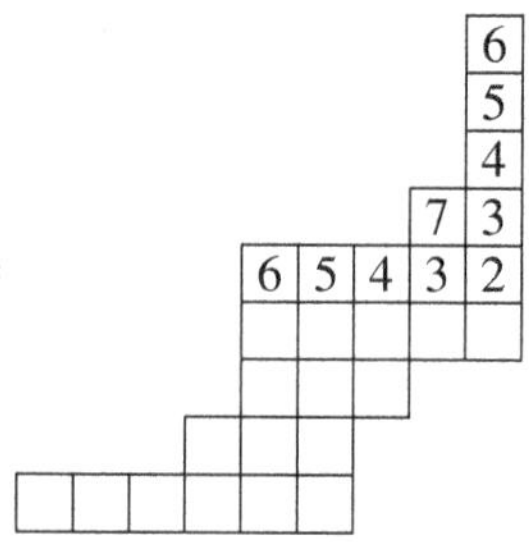

Under the identifications above, these three subheaps are isomorphic (in **Heap**, after ignoring empty boxes and unused labels) to the principal subheaps of the minuscule representations $L(D_6, \omega_1)$, $L(D_6, \omega_6)$ and $L(D_6, \omega_1)$, respectively.

By Proposition 8.2.9 (iv), it follows that with the above identifications, the restriction of $L(E_7, \omega_6)$ to the subalgebra of type D_6 is given by

$$L(D_6, \omega_1) \oplus L(D_6, \omega_6) \oplus L(D_6, \omega_1).$$

Exercise 8.2.12 Consider the Dynkin diagram of type D_6, and its type A_5 subgraph obtained by omitting the vertex labelled 5. Show that the spin representation $L(D_6, \omega_6)$ decomposes on restriction to

$$L(A_5, \omega_1) \oplus L(A_5, \omega_3) \oplus L(A_5, \omega_5).$$

Exercise 8.2.13 Consider the Dynkin diagram of type B_l, and its type A_{l-1} subgraph obtained by omitting the vertex labelled l. Show that the spin representation $L(B_l, \omega_l)$ decomposes on restriction to

$$\sum_{i=0}^{l} L(A_{l-1}, \omega_i),$$

with the convention that $L(A_{l-1}, \omega_0)$ and $L(A_{l-1}, \omega_l)$ are both equal to the one-dimensional trivial module. By counting dimensions, deduce the familiar identity

$$2^l = \sum_{i=0}^{l} \binom{l}{i}.$$

Exercise 8.2.14 Consider the Dynkin diagram of type B_l, and its type B_{l-1} subgraph obtained by omitting the vertex labelled 1. Show that the spin representation $L(B_l, \omega_l)$

decomposes on restriction to

$$L(B_{l-1}, \omega_{l-1}) \oplus L(B_{l-1}, \omega_{l-1}).$$

Exercise 8.2.15 Consider the Dynkin diagram of type D_l, and its type A_{l-1} subgraph obtained by omitting the vertex labelled l. Show that the spin representation $L(D_l, \omega_l)$ decomposes on restriction to

$$\sum_{i=0}^{\lfloor l/2 \rfloor} L(A_{l-1}, \omega_{2i}),$$

with the convention that $L(A_{l-1}, \omega_0)$ and $L(A_{l-1}, \omega_l)$ are both equal to the one-dimensional trivial module. By counting dimensions, deduce the identity

$$2^{l-1} = \sum_{i=0}^{\lfloor l/2 \rfloor} \binom{l}{2i}.$$

Exercise 8.2.16 Consider the Dynkin diagram of type D_l, and its type D_{l-1} subgraph obtained by omitting the vertex labelled 1.

(i) Show that the spin representation $L(D_l, \omega_l)$ decomposes on restriction to

$$L(D_{l-1}, \omega_{l-1}) \oplus L(D_{l-1}, \omega_{l-2}).$$

(ii) Show that the spin representation $L(D_l, \omega_{l-1})$ decomposes on restriction to

$$L(D_{l-1}, \omega_{l-1}) \oplus L(D_{l-1}, \omega_{l-2}).$$

(iii) Show that the natural representation $L(D_l, \omega_1)$ decomposes on restriction to

$$L(D_{l-1}, \omega_0) \oplus L(D_{l-1}, \omega_1) \oplus L(D_{l-1}, \omega_0),$$

where $L(D_{l-1}, \omega_0)$ denotes the trivial module.

Exercise 8.2.17 Consider the Dynkin diagram of type E_6. Consider the type D_5 subgraph obtained by omitting the vertex labelled 5. Show that the minuscule representation $L(E_6, \omega_5)$ decomposes on restriction to the direct sum of (a) a trivial representation, (b) a 16-dimensional spin representation and (c) a copy of the 10-dimensional natural representation $L(D_5, \omega_1)$. (Notice that the highest weight of the spin representation obtained depends on the exact identification of D_5 as a subgraph of E_6.)

Exercise 8.2.18 Consider the Dynkin diagram of type E_7. Consider the type E_6 subgraph obtained by omitting the vertex labelled 6. Show that the minuscule representation $L(E_7, \omega_6)$ decomposes on restriction to a direct sum

$$M \oplus L(E_6, \omega_1) \oplus L(E_6, \omega_5) \oplus M,$$

where M is the trivial module. Prove also that the weights in each summand (reading left to right) lie at distance 0, $\sqrt{32}$, $\sqrt{64}$ and $\sqrt{96}$ respectively from the lowest weight vector $v_{0,7}$ of $L(E_7, \omega_6)$. (Compare with Exercise 8.1.11.)

Exercise 8.2.19 Consider the Dynkin diagram of type E_7. Consider the type A_6 subgraph obtained by omitting the vertex labelled 7. Show that the minuscule representation $L(E_7, \omega_6)$ decomposes on restriction to a direct sum of two 28-dimensional representations, namely

$$L(A_6, \omega_2) \oplus L(A_6, \omega_6).$$

Proposition 8.2.20 *Let Ψ be the set of weights of a minuscule representation $L(\omega_p)$ of a simple Lie algebra $\mathfrak{g}$, regarded as points in Euclidean space. (Here, ω_p is the highest weight of the module, according to the numbering in Appendix B.) Let*

$$D = \{|\lambda_1 - \lambda_2| : \lambda_1, \lambda_2 \in \Psi\},$$

where $|\ |$ denotes Euclidean distance. Let N be the cardinality of the set D, and write

$$D = \{d_0, d_1, \ldots, d_{N-1}\},$$

where $0 = d_0 < d_1 < \cdots < d_{N-1}$.

(i) *We have*

$$\begin{aligned}
N &= \min(k+1, l-k+2) \text{ if } \mathfrak{g} \text{ has type } A_l;\\
N &= l+1 \text{ if } \mathfrak{g} \text{ has type } B_l;\\
N &= 3 \text{ if } \mathfrak{g} \text{ has type } C_l;\\
N &= 3 \text{ if } \mathfrak{g} \text{ has type } D_l \text{ and } k = 1;\\
N &= \lfloor l/2 \rfloor + 1 \text{ if } \mathfrak{g} \text{ has type } D_l \text{ and } k \in \{l-1, l\};\\
N &= 3 \text{ if } \mathfrak{g} \text{ has type } E_6;\\
N &= 4 \text{ if } \mathfrak{g} \text{ has type } E_7.
\end{aligned}$$

In each case, the sequence

$$d_0^2 < d_1^2 < \cdots < d_{N-1}^2$$

is an arithmetic progression.

(ii) *Let F be a principal subheap of a full heap giving rise to the representation $L(\omega_p)$. Then the number of elements of F with label p is equal to $N - 1$.*

Proof There are two methods available to prove (i): one by using the results of Section 8.1, and one by using the Trivialization Theorem.

In the first four cases listed in (i), it is perhaps easier to use the explicit descriptions of the weights given in Section 8.1.

For the last three cases listed in (i), we use a trivialization argument, as follows. Let λ_1 and λ_2 be elements of Ψ, and let F be a principal subheap of a full heap $\varepsilon : E \to \Gamma$ corresponding to the set of weights (such an F exists by Corollary 6.6.3). Let I_1 and I_2 be the ideals of F corresponding to λ_1 and λ_2, respectively. Since $\chi(I_1, I_2)$ contains α_0 with coefficient 0, we may identify I_1 and I_2 with ideals of E and apply Theorem 5.3.16 to find an element of the Weyl group W of $\mathfrak{g}$ for which $(w(I_1), w(I_2))$ is relatively trivial. Since W acts by orthogonal transformations, we have $|\lambda_1 - \lambda_2| = |w(\lambda_1) - w(\lambda_2)|$. Since the Lie algebras in question are simply laced, it follows that $N - 1$ is equal to the length of the longest antichain in F. By inspecting the heaps F, we see that the longest antichain has length $\lfloor l/2 \rfloor$ for the spin modules in type D_l, length 2 for the two 27-dimensional modules in type E_6, and length 3 for the minuscule module $L(E_7, \omega_6)$. The squares of the numbers in the set D are proportional to the sizes of the subsets of this longest antichain, which is why they form an arithmetic progression. This completes the proof of (i).

The proof of (ii) follows by a case by case count. The most complicated case is in type A_l; in this situation, the heap F corresponding to ω_p has the form of a k by $(l+1)-k$ rectangle (see Figure 6.4), so the number of elements labelled k is equal to the shorter dimension of this rectangle. □

Proposition 8.2.21 *Let Ψ be the set of weights of a minuscule representation $L(\omega_p)$ of a simple Lie algebra $\mathfrak{g}$, and let α be a root of $\mathfrak{g}$. Then there exist weights $\lambda_1, \lambda_2 \in \Psi$ such that $\alpha = \lambda_1 - \lambda_2$.*

Proof By Corollary 6.6.3, there is a principal subheap F of a full heap $\varepsilon : E \to \Gamma$ corresponding to Ψ. The reflection s_α corresponding to α is an involution in W, and Theorem 5.4.13 shows that s_α interchanges two ideals, I_1 and I_2, of F. The corresponding weights λ_1 and λ_2 satisfy $s_\alpha(\lambda_1) = \lambda_2$. Using the formula for a reflection and Corollary 6.6.6, we find that $\lambda_1 - \lambda_2 = \pm\alpha$. □

Theorem 8.2.22 *Let Ψ be the set of weights of a minuscule representation $L(\omega_p)$ of a simple Lie algebra $\mathfrak{g}$, regarded as points in Euclidean space. (Here, ω_p is the highest weight of the module, according to the numbering in Appendix B.) Let $\mathfrak{g}_p$ be the subalgebra of $\mathfrak{g}$ of Definition 8.2.7 obtained by omitting those generators indexed by the vertex p.*

(i) *The rank N of the Weyl group W of $\mathfrak{g}$ acting on the weights of $L(\omega_p)$ is given by*

$$\begin{array}{ll} \min(k+1, l-k+2) & \textit{if } \mathfrak{g} \textit{ has type } A_l; \\ l+1 & \textit{if } \mathfrak{g} \textit{ has type } B_l; \\ 3 & \textit{if } \mathfrak{g} \textit{ has type } C_l; \\ 3 & \textit{if } \mathfrak{g} \textit{ has type } D_l \textit{ and } k=1; \\ \lfloor l/2 \rfloor + 1 & \textit{if } \mathfrak{g} \textit{ has type } D_l \textit{ and } k \in \{l-1, l\}; \\ 3 & \textit{if } \mathfrak{g} \textit{ has type } E_6; \\ 4 & \textit{if } \mathfrak{g} \textit{ has type } E_7. \end{array}$$

(ii) *The pair of weights (λ_1, λ_2) is in the same W-orbit as the pair (μ_1, μ_2) if and only if the Euclidean distances $|\lambda_1 - \lambda_2|$ and $|\mu_1 - \mu_2|$ are equal.*
(iii) *Upon restriction to the subalgebra $\mathfrak{g}_p$, $L(\omega_p)$ decomposes as a direct sum of N irreducible submodules, $L_0, \ldots, L_{N-1}$, each of which is a direct sum of weight spaces of $L(\omega_p)$.*
(iv) *If λ_i is a weight of L_i, then the distance $d_i = |\omega_p - \lambda_i|$ depends only on i.*

Proof By Proposition 8.2.9 (iii), the rank of the action of W on the weights is equal to one more than the number of elements labelled k in the heap F. Parts (i) and (ii) now follow from Proposition 8.2.20.

The irreducibility assertion of (iii) follows from Proposition 5.2.12, and the rest follows from Proposition 8.2.9 (iv) and the constructions.

Part (iv) follows from the fact that the Weyl group acts by orthogonal transformations, together with Proposition 8.2.9 (iii).

The first assertion of (v) follows from the definition of N and the fact that the Weyl group acts transitively on the weights of a minuscule representation. The second assertion of (v) follows from the first. □

Exercise 8.2.23 Recall the parametrization of the set Ψ of weights of $L(B_l, \omega_l)$ in Example 8.1.5. The *Hamming distance*, $d(\psi_1, \psi_2)$, between two elements ψ_1 and ψ_2 of Ψ is defined to be the number of coordinate positions at which ψ_1 and ψ_2 disagree.

(i) Show that d is a metric on Ψ.
(ii) Show that the set of possible values of d on ψ is $\{0, 1, \ldots, l\}$.
(iii) Show that if we have $\psi_1, \psi_2, \psi_3, \psi_4 \in \Psi$, then the ordered pairs (ψ_1, ψ_2) and (ψ_3, ψ_4) are conjugate under the action of $W(B_l)$ if and only if $d(\psi_1, \psi_2) = d(\psi_3, \psi_4)$.

Exercise 8.2.24 Recall the parametrization of the set Ψ of weights of $L(D_l, \omega_l)$ in Exercise 8.1.9, and define the Hamming distance on Ψ as in Exercise 8.2.23

(i) Show that the set of possible values of d on Ψ is $\{0, 2, 4, \ldots, l'\}$, where $l' \in \{l-1, l\}$.
(ii) Show that if we have $\psi_1, \psi_2, \psi_3, \psi_4 \in \Psi$, then the ordered pairs (ψ_1, ψ_2) and (ψ_3, ψ_4) are conjugate under the action of $W(B_l)$ if and only if $d(\psi_1, \psi_2) = d(\psi_3, \psi_4)$.

Exercise 8.2.25 Give a direct proof that the number of orbits of $W(A_l) \cong S_{l+1}$ acting on the k-th exterior power of $L(\omega_1)$ is $\min(k+1, l-k+2)$.

Lemma 8.2.26 *Let V_F be a minuscule representation of a simple Lie algebra $\mathfrak{g}$ with Weyl group W. Then the longest element w_0 of W exchanges the highest weight λ_1 and lowest weight λ_2 of V_F.*

Proof Let C be the fundamental chamber of the action of W on the span of the fundamental roots, as in Theorem 4.3.10. The highest weight of V_F lies in $\overline{C}$, and the lowest weight lies in $-\overline{C}$. Let $w \in W$ be such that $w(\lambda_1) = \lambda_2$. Then $w(\overline{C}) \cap -\overline{C}$ is nonempty, which by Humphreys [36, exercise 5.13] means that $w(\lambda_1) = w_0(\lambda_1)$, as required. □

Lemma 8.2.27 *Let F be a principal subheap corresponding to a minuscule representation, V_F, and suppose that F is self-dual as a heap. Then the action of the longest element w_0 on the ideals $J(F)$ is given by $w_0(I) = *_F(F \backslash I)$.*

Proof The proof is by induction on $|I|$. If $|I| = 0$, then the result follows by Lemma 8.2.26. Because F is self-dual, we have

$$*_F(F \backslash S_i(I)) = S_i(*_F(F \backslash I)),$$

and the inductive step follows from this. □

Exercise 8.2.28 Show that the representation $L(A_5, \omega_3)$ has a self-dual principal subheap and weights that are closed under negation, and yet the Weyl group has trivial centre. (Recall that the centres of finite Weyl groups are described in Proposition 4.3.17.)

8.3 Ideals of roots

In this section, we show how the elements of a parabolic subheap corresponding to a minuscule weight ω_p can be identified (up to isomorphism of posets) with the set of positive roots in which α_p appears with nonzero coefficient.

In order to do this, we need to know more about short and long roots in types B_l and C_l. We now recall the explicit description of root systems of types A, B, C and D from Section 5.3 for easy reference.

Proposition 8.3.1 *Let $\varepsilon_1, \varepsilon_2, \ldots, \varepsilon_n$ be linearly independent vectors in $\mathbb{R}^n$.*

(i) *In type A_{n-1}, the set of positive roots is $\{\varepsilon_i - \varepsilon_j : 1 \leq i < j \leq n\}$.*
(ii) *In type B_n, the set of long positive roots is $\{\varepsilon_i \pm \varepsilon_j : 1 \leq i < j \leq n\}$, and the set of short positive roots is $\{\varepsilon_i : 1 \leq i \leq n\}$.*
(iii) *In type C_n, the set of short positive roots is $\{\varepsilon_i \pm \varepsilon_j : 1 \leq i < j \leq n\}$, and the set of long positive roots is $\{2\varepsilon_i : 1 \leq i \leq n\}$.*
(iv) *In type D_n, the set of positive roots is $\{\varepsilon_i \pm \varepsilon_j : 1 \leq i < j \leq l\}$,*

Proof See Exercises 5.3.20 (ii), 5.3.31 (i), 5.3.24 (iii) and 5.3.27 (ii), respectively. □

Proposition 8.3.2 *Let $W = W(B_n)$ and let H be the maximal parabolic subgroup of W generated by all s_i other than the generator s_n, where α_n is the short fundamental root. For each root α, let $c(\alpha)$ be the coefficient of α_n in α. There are five orbits of H in its action on the roots, as follows:*

(i) *the roots $\{\varepsilon_i + \varepsilon_j : 1 \leq i < j \leq n\}$, each of which is long and satisfies $c(\alpha) = 2$;*
(ii) *the roots $\{\varepsilon_i : 1 \leq i \leq n\}$, each of which is short and satisfies $c(\alpha) = 1$;*
(iii) *the roots $\{\varepsilon_i - \varepsilon_j : 1 \leq i \neq j \leq n\}$, each of which is long and satisfies $c(\alpha) = 0$;*
(iv) *the roots $\{-\varepsilon_i : 1 \leq i \leq n\}$, each of which is short and satisfies $c(\alpha) = -1$;*
(v) *the roots $\{-\varepsilon_i - \varepsilon_j : 1 \leq i < j \leq n\}$, each of which is long and satisfies $c(\alpha) = -2$.*

Proof The parabolic subgroup H is of type A_{n-1} and acts on the coordinate positions by permutation. The classification of orbits now follows from Proposition 8.3.1 (ii). The statements about long and short roots follow by considering lengths of vectors, and the statements about c-values follow from elementary linear algebra. □

The next result now follows immediately.

Corollary 8.3.3 *Two roots of $W(B_n)$ are in the same $W(A_{n-1})$-orbit if and only if they contain the fundamental root α_n with the same coefficient.* □

Proposition 8.3.4 *Let $W = W(C_n)$ and let H be the maximal parabolic subgroup of W generated by all s_i other than the generator s_1, which is furthest away from the long fundamental root in the diagram. For each root α, let $c(\alpha)$ be the coefficient of α_1 in α. There are six orbits of H in its action on the roots, as follows:*

(i) *the highest root, $\alpha = 2\varepsilon_1$, which is long and satisfies $c(\alpha) = 2$;*
(ii) *the roots $\{\varepsilon_1 \pm \varepsilon_i : 1 < i \leq n\}$, each of which is short and satisfies $c(\alpha) = 1$;*

(iii) *the roots* $\{\pm 2\varepsilon_i : 1 < i \leq n\}$, *each of which is long and satisfies* $c(\alpha) = 0$;
(iv) *the roots* $\{\pm\varepsilon_i \pm \varepsilon_j : 1 < i < j \leq n\}$, *each of which is short and satisfies* $c(\alpha) = 0$;
(v) *the roots* $\{-\varepsilon_1 \pm \varepsilon_i : 1 < i \leq n\}$, *each of which is short and satisfies* $c(\alpha) = -1$;
(vi) *the negative of highest root,* $\alpha = -2\varepsilon_1$, *which is long and satisfies* $c(\alpha) = -2$.

Proof The assertions about long and short roots, and about c-values, follow from linear algebra as in the proof of Proposition 8.3.2.

The parabolic subgroup H is of type C_{n-1} and acts as signed permutations on the coordinates $2, 3, \ldots, n$. Two roots in type C_n are W-conjugate if and only if they have the same length. It follows that two roots in type C_n are H-conjugate if and only if they have the same length and the same coefficient of ε_1. The classification of orbits follows from this observation and the assertions about long and short roots. □

Lemma 8.3.5 *Let* F *be the principal subheap of a full heap of type* $B_n^{(1)}$ *or* $C_n^{(1)}$, *labelled* $F(q, i)$ *as in Definition 8.2.7. Let* ω_p *and* W *be the minuscule weight and finite Weyl group associated with* F. *Let* χ *be the map sending each ideal* I *of the subheap* $F \backslash U_{p,3}$ *to its content,* $\chi(I)$. *Then* χ *is a bijection, and the nonzero elements in the image of* χ *are precisely the set of roots of* W *that contain* α_p *with nonzero coefficient.*

Proof Let H be the parabolic subgroup of W generated by all s_i other than s_p. By Proposition 8.2.9, this action has $k + 1$ orbits, where k is the number of occurrences of p in F. The action of H on the union of the three orbits corresponding to the lowest occurrences of p in F is equivalent to the action of H on the ideals of $F \backslash U_{p,3}$.

One of these three orbits consists of only the empty set, of content 0. Another orbit contains a singleton ideal with content α_p, which is a short root. In type B, the third orbit contains an ideal of content $2\alpha_p + \alpha_{p-1}$, whereas in type C, the third orbit contains the ideal F, whose content is the highest root; in both cases, these roots are long and contain α_p with nonzero coefficient. The conclusion now follows from Proposition 8.3.2 (i) and (ii) in type B, and from Proposition 8.3.4 (i) and (ii) in type C. □

Lemma 8.3.6 *Let* F *be the principal subheap of a full heap over an affine Dynkin diagram, let* ω_p *and* W *be the minuscule weight and finite Weyl group associated with* F, *and assume that* W *is of simply laced type. Let* χ *be the map sending each ideal* I *of the subheap* $F \backslash U_{p,2}$ *to its content,* $\chi(I)$. *Then* χ *is a bijection, and the nonzero elements in the image of* χ *are precisely the set of roots of* W *that contain* α_p *with nonzero coefficient (necessarily equal to* 1*).*

Proof Let H be the parabolic subgroup of W generated by all s_i other than s_p.

As in the proof of Theorem 5.1.5, the coefficient of α_p in the highest root θ is 1, and thus the coefficient of α_p in any positive root α is at most 1. By Theorem 4.2.3 (iii), α is W-conjugate to θ, and by Corollary 4.3.11, there is a sequence

$$\alpha = v_0, v_1, \ldots, v_k = \theta$$

such that for all $0 < i \leq k$, $v_i - v_{i-1}$ is a positive multiple of a fundamental root. If α_p appears in α with nonzero coefficient (i.e., coefficient 1), then none of these fundamental roots is equal to α_p, and this implies that α is H-conjugate to θ.

The rest of the argument is similar to, but easier than, the proof of Lemma 8.3.5. In this case, we consider the action of H on the union of the lowest two orbits, or equivalently, on the ideals of $F \backslash U_{p,2}$. One of these orbits consists only of the empty ideal. The preceding argument shows that the other orbit consists of a set of ideals with distinct contents, and that the resulting contents are precisely the positive roots with nonzero coefficient α_p; in other words, coefficient 1. □

Lemma 8.3.7 *Let F be any principal subheap of a full heap $\varepsilon : E \to \Gamma$ over an affine Dynkin diagram. Let ω_p and Γ_0 be the minuscule weight and finite type Dynkin diagram associated with F. Let Q be the subposet of F consisting of all join irreducible elements, as in Theorem A.1.11. Then Q is isomorphic to $F \backslash (U_{p,2} \cup D_{p,1})$ if Γ_0 is simply laced, and Q is isomorphic to $F \backslash (U_{p,3} \cup D_{p,1})$ if Γ_0 is doubly laced.*

Proof The proof is a case by case check using Corollary 6.6.3 and the classification of minuscule representations (Theorem 5.1.5).

The easiest cases are in types A and C, where the isomorphism claimed is an equality.

For the representation $L(D_n, \omega_1)$, all the elements of F are join irreducible except the bottom element and the element immediately above the nontrivial antichain. The heap $F \backslash (U_{p,2} \cup D_{p,1})$ consists of all elements of F except the top and bottom elements. These are isomorphic posets by inspection.

Consider the spin representation $L(B_n, \omega_n)$. Let F_1 be the set of lowest elements in each vertex chain. The join irreducible elements are then obtained by removing the bottom element from the union $F_1 \cup \varepsilon^{-1}(n)$. This is isomorphic to the poset $F \backslash (U_{p,3} \cup D_{p,1})$.

Consider representations $L(D_n, \omega_n)$ and $L(D_n, \omega_{n-1})$. Their principal subheaps F' are isomorphic as posets to the principal subheap of $L(B_n, \omega_n)$, and the subheap $F' \backslash (U'_{p,2} \cup D'_{p,1})$ is isomorphic to the subheap $F \backslash (U_{p,3} \cup D_{p,1})$ of the previous paragraph. The result now follows from the result in type B.

The three exceptional minuscule representations are treated by ad hoc arguments. For $L(E_6, \omega_1)$ and $L(E_6, \omega_5)$, we find that both posets are isomorphic to the principal subheap of $L(D_5, \omega_5)$. For $L(E_7, \omega_6)$, we find that the poset is isomorphic to the principal subheap of $L(E_6, \omega_5)$. □

Exercise 8.3.8 Verify the claims about E_6 and E_7 given in the proof of Lemma 8.3.7.

Example 8.3.9 The principal subheap F corresponding to the spin representation $L(B_5, \omega_5)$ is

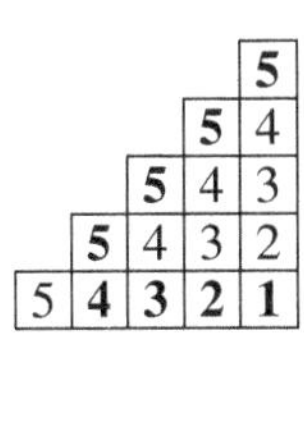

where the join irreducible elements are shown in bold. The subheap $F \backslash (U_{p,3} \cup D_{p,1})$ is given by

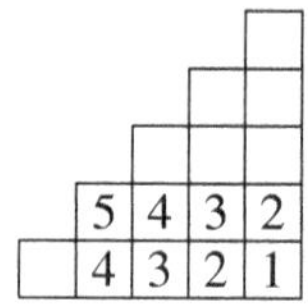

;

as an abstract poset, this subheap is isomorphic to $[2] \times [4]$

We are now ready to prove the main result of this section. Recall from Appendix A that $[m]$ is the poset consisting of the set $\{1, 2, \ldots, m\}$, equipped with the natural partial order. We will write $J^2(P)$ for $J(J(P))$, and so on.

Theorem 8.3.10 *Let Λ_p be the set of weights of a minuscule representation of a simple Lie algebra $\mathfrak{g}$ with highest weight ω_p. Let R_p be the set of (positive) roots of $\mathfrak{g}$ in which α_p appears with positive coefficient. Equip Λ_p and R_p with the usual partial order, as in Definition 5.2.7.*

(i) *The poset Λ_p is isomorphic to $J(R_p)$.*
(ii) *The poset R_p is isomorphic to the underlying poset of F, the principal subheap corresponding to $L(\mathfrak{g}, \omega_p)$.*
(iii) *A reflection s_α moves ω_p if and only if $\alpha \in R_p$.*
(iv) *The following numbers are equal:*
 (a) *the number of reflections moving any weight of $L(\mathfrak{g}, \omega_p)$;*
 (b) *$|R_p|$;*
 (c) *$|F|$;*
 (d) *$\sum_{\alpha \in \Phi^+} \langle \omega_p, \alpha^\vee \rangle$.*
(v) *The isomorphism type of R_p is given in the following table.*

Type	*p*	*isomorphism type of R_p*
A_l	k	$[k] \times [l+1-k] \cong J([k-1] \dot{\cup} [l-k])$
B_l	l	$J([2] \times [l-1])$
C_l	1	$[2l-1] \cong J([2l-2])$
D_l	1	$J^{l-3}([2] \times [2])$
D_l	$l-1, l$	$J([2] \times [l-2])$
E_6	$1, 5$	$J^2([2] \times [3])$
E_7	6	$J^3([2] \times [3])$

Proof If $\mathfrak{g}$ is simply laced, then Lemma 8.3.6 shows that we have an isomorphism of posets

$$R_p \cong J(F \backslash (U_{p,2} \cup D_{p,1})),$$

whereas in types B and C, Lemma 8.3.5 shows that

$$R_p \cong J(F \backslash (U_{p,3} \cup D_{p,1})).$$

It now follows from 8.3.7 and Theorem A.1.11 that $\Lambda_p \cong R_p$, proving (i).

It follows from Corollary 6.6.3 and the definitions that Λ_p is isomorphic as a poset to $J(F)$. Part (ii) follows from this and Theorem A.1.11.

We now prove (iii). Let $\alpha \in R_p$. By Lemmas 8.3.5 and 8.3.6, there is a filter of F of content α, and then Theorem 5.4.13 shows that s_α moves the highest weight ω_p. Conversely, suppose that s_α moves ω_p. Since the unique maximal element of F is labelled p, it follows from Theorem 5.4.13 that α_p appears in α with positive coefficient, and thus that $\alpha \in R_p$.

The equality of (b) and (c) in (iv) follows from (ii), and the equality of (a) and (b) follows from (iii) and the transitivity of the action of W on the weights. The formula for a reflection,

$$s_\alpha(\lambda) = \lambda - (\lambda, \alpha^\vee)\alpha,$$

and Theorem 5.4.13 show that $(\omega_p, \alpha^\vee)$ is equal to 1 if s_α moves ω_p, and is equal to 0 otherwise. This proves the equality of (a) and (d) and completes the proof of (iv).

Part (iv) follows from (iii) and the fact that the Weyl group acts transitively on the weights.

It remains to prove (v). In types A_l and C_l, it is clear from the constructions of principal subheaps that we have $F \cong [k] \times [l+1-k]$ in type A and $F \cong [2l-1]$ in type C. The assertions involving J follow from Theorem A.1.11 and the comments about join irreducible elements in the proof of Lemma 8.3.7.

The assertion in type B_l follows from the observation that

$$F \backslash (U_{p,3} \cup D_{p,1}) \cong [2] \times [l-1],$$

and the cases $L(D_l, \omega_{l-1})$ and $L(D_l, \omega_l)$ follow from the same observation. This also deals with the case $L(D_4, \omega_1)$, whose principal subheap is isomorphic to those of $L(D_4, \omega_3)$ and and $L(D_4, \omega_4)$.

We showed in the proof of Lemma 8.3.7 that every element in the principal subheap F_l of $L(D_l, \omega_1)$ is join irreducible, except the top and bottom elements. It follows that $F_l \cong J(F_{l-1})$ for $l > 4$. The case of $L(D_l, \omega_1)$ for $l > 4$ now follows by induction, since the base case $l = 4$ has already been proved.

It follows by inspection and the result for $L(D_5, \omega_5)$ that in type E_6 we have

$$F \backslash (U_{p,2} \cup D_{p,1}) \cong J([2] \times [3]),$$

which proves the result in type E_6.

Finally, it follows by inspection and the result for $L(E_6, \omega_5)$ that in type E_7 we have

$$F \backslash (U_{p,2} \cup D_{p,1}) \cong J^2([2] \times [3]),$$

and this proves the result in type E_7. □

The following corollary is now immediate.

Corollary 8.3.11 *Suppose that a minuscule representation of a simple Lie algebra* $\mathfrak{g}$ *has highest weight* ω_p*, and let* R_p *be the set of (positive) roots of* $\mathfrak{g}$ *in which* α_p *appears with positive coefficient, equipped with the usual total order. Then* R_p *is a distributive lattice.* □

Exercise 8.3.12 Give a direct proof that in type A_l, the poset of roots that contain the fundamental root α_k with coefficient 1 is isomorphic to $[k] \times [l+1-k]$.

8.4 Weight polytopes

Definition 8.4.1 If V is a vector space over $\mathbb{R}$ and $X \subset V$, then the *affine hull*, $\mathrm{Aff}(X)$, of X in V is the set of affine combinations of finite subsets of points in X; that is, the set of all vectors

$$\left\{ \sum_{i=1}^{k} \lambda_i x_i : x_i \in X,\ \lambda_i \in \mathbb{R},\ \sum_{i=1}^{k} \lambda_i = 1,\ k \in \mathbb{N} \right\}.$$

The *dimension* of the affine hull of X is the dimension of the vector space of all differences of vectors in X; that is, the dimension of the space

$$\left\{ \sum_{i=1}^{k} \lambda_i x_i : x_i \in X,\ \lambda_i \in \mathbb{R},\ \sum_{i=1}^{k} \lambda_i = 0,\ k \in \mathbb{N} \right\}.$$

The *convex hull*, $\mathrm{Conv}(X)$, of X in V is the set of convex combinations of finite subsets of points in X; that is, the subset of $\mathrm{Aff}(X)$ given by the set

$$\left\{ \sum_{i=1}^{k} \lambda_i x_i : x_i \in X,\ \lambda_i \in \mathbb{R},\ \lambda_i \geq 0,\ \sum_{i=1}^{k} \lambda_i = 1,\ k \in \mathbb{N} \right\}.$$

The following is a fundamental result in metric geometry.

Theorem 8.4.2 *A subset $P \subseteq \mathbb{R}^d$ is the convex hull of a finite set of points if and only if it is a bounded intersection of finitely many closed half spaces.* □

Definition 8.4.3 A subset P satisfying either of the conditions of Theorem 8.4.2 is called a *polytope*. If P is the convex hull of the finite set V, then we write $P = \Pi(V)$. The *dimension* of a polytope P is the dimension of its affine hull. An *automorphism* of a polytope $P \subset \mathbb{R}^d$ is an isometry of $\mathrm{Aff}(P)$ fixing P setwise. The group of automorphisms of P (under composition of maps) is denoted by $\mathrm{Aut}(P)$.

The usual definition of the term "polytope" is more general and does not include convexity as a necessary requirement. However, for our purposes. it is more convenient to adopt Definition 8.4.3.

Definition 8.4.4 The *vertices* of a polytope P are the points x of P for which $\mathrm{Conv}(P \backslash \{x\}) \neq P$.

Example 8.4.5 Recall the 2^n points $(\pm 2, \pm 2, \ldots, \pm 2)$, which appeared in Example 8.1.5 as the weights of the spin representation of the simple Lie algebra of type B_n. The convex hull of these points is the n-dimensional *hypercube* (of side 4). It can also be described as the set of points $(x_1, x_2, \ldots, x_n)$ lying in the intersection of the half spaces

$$x_1 \geq -2, \quad x_2 \geq -2, \quad \ldots, \quad x_n \geq -2, \quad x_1 \leq 2, \quad x_2 \leq 2, \quad \ldots, \quad x_n \leq 2.$$

The vertices of the hypercube are the points $(\pm 2, \pm 2, \ldots, \pm 2)$ themselves; this is proved in Proposition 8.4.7 below.

In general, it is much more difficult than in Example 8.4.5 to pass between two characterizations of a polytope given by Theorem 8.4.2.

Lemma 8.4.6 *If all the points of $X \subset \mathbb{R}^n$ lie on the surface of a fixed sphere S of some dimension, then* $\text{Conv}(X) \cap S = X$.

Proof This follows by induction on k using Definition 8.4.1. □

Proposition 8.4.7 *Let Ψ be the set of weights of a minuscule representation.*

(i) *The vertices of the polytope* $\text{Conv}(\Psi)$ *are precisely the elements of* Ψ.
(ii) *The dimension of* $\text{Conv}(\Psi)$ *is equal to the rank of the Weyl group, that is, the number of fundamental roots.*

Proof The Weyl group acts transitively on Ψ by Corollary 6.6.4, and it acts on the ambient space of Ψ by orthogonal transformations by Theorem 4.2.3 (ii). It follows that all elements of Ψ are at the same distance from the origin, and the conclusion of (i) follows from Lemma 8.4.6.

Since the vertices of $\text{Conv}(\Psi)$ are the orbit of a single point under the action of the Weyl group acting by reflections, the dimension of $\text{Conv}(\Psi)$ is bounded by the dimension of the ambient space in which W acts by reflections, and the latter is equal to the rank of W. Conversely, it follows from Proposition 8.2.21 that any fundamental root of W is the difference of two weights in Ψ, and this establishes the reverse inclusion. □

One can also see from the explicit descriptions of the weights of minuscule representations given in Section 8.1 that all the weights in a minuscule representation are equidistant from the origin.

Definition 8.4.8 The convex hull of the weights of a minuscule representation $L(\mathfrak{g}, \omega)$ is known as the *weight polytope*, $P(\mathfrak{g}, \omega)$, of $L(\mathfrak{g}, \omega)$. More generally, for any simple Lie algebra $\mathfrak{g}$ with Weyl group W, and any weight ω, we define $P(\mathfrak{g}, \omega)$ to be the convex hull of the set $W.\omega$.

Note that since the Weyl group acts by orthogonal transformations on the set of vertices, this induces an action on the polytope itself.

The weight polytope has an interesting relationship with root systems of types E_6 and E_7.

Example 8.4.9 Example 8.1.7 gives explicit coordinates for the weights of the module $L(E_7, \omega_6)$. By Theorem 8.2.22, there are four W-orbits of weights, corresponding to the four possible Euclidean distances between weights. In this case, the four distances are 0, $\sqrt{32}$, $\sqrt{64}$ and $\sqrt{96}$; the last three correspond to pairs of ideals of the heap that can be trivialized to antichains of sizes 1, 2 and 3. There is only one root length, corresponding to differences of weights at distance $\sqrt{32}$. Using this, we can give an explicit description of the root system in type E_7.

If λ_1 and λ_2 are two of the 56 weights at mutual distance $\sqrt{32}$, an exhaustive check shows that one of two things can happen. Either we have (a) $\lambda_1 = \pm v_{i,j}$ and

$\lambda_2 = \pm v_{i,k}$ for $j \neq k$ and the same choice of signs, or (b) $\lambda_1 = v_{i,j}$ and $\lambda_2 = -v_{k,l}$ (or vice versa), where $\{i, j, k, l\}$ has cardinality 4. In case (a), the difference $\lambda_1 - \lambda_2$ is of the form $4(\varepsilon_i - \varepsilon_j)$, where $0 \leq i, j \leq 7$ and $i \neq j$; this gives 56 distinct roots. In case (b), the difference $\lambda_1 - \lambda_2$ is of the form

$$\sum_{i=0}^{7} \pm 2\varepsilon_i,$$

where there are four $+$ and four $-$ signs in the sum; this gives 70 distinct roots. It follows that there are 126 roots in the root system of type E_7.

By the data in Appendix B, we see that the highest root in type E_7 is

$$2\alpha_1 + 3\alpha_2 + 4\alpha_3 + 3\alpha_4 + 2\alpha_5 + \alpha_6 + 2\alpha_7,$$

which in the above coordinates is $-4\varepsilon_1 + 4\varepsilon_2$.

The root system of type E_7 could also be constructed using the Weyl group action and the coordinates of the fundamental roots given in 8.1.7.

Lemma 8.4.10 *If λ_1 and λ_2 are two of the 56 weights of $L(E_7, \omega_6)$ whose difference is a root, then the line segment between λ_1 and λ_2 lies entirely in the boundary of the weight polytope.*

Proof The weight polytope P has dimension 7 by Proposition 8.4.7 (ii). By inspection, the 56 weights lie in the hyperplane $\sum_{i=0}^{7} x_i = 0$ of $\mathbb{R}^8$, which must therefore be the affine hull of P. It follows that, for any $0 \leq i \leq 7$, not all the points in any open neighbourhood of a point $x \in P$ have the same ε_i-coordinate.

Example 8.4.9 shows that the distance between the two weights is $\sqrt{32}$. Theorem 8.2.22 then shows that by acting by a suitable Weyl group element, we may assume that

$$\lambda_1 = v_{0,1} = (3, 3, -1, -1, -1, -1, -1, -1)$$

and

$$\lambda_2 = v_{0,2} = (3, -1, 3, -1, -1, -1, -1, -1).$$

Let x be a point on the line segment between λ_1 and λ_2, so that

$$x = (3, 3 - 4t, -1 + 4t, -1, -1, -1, -1, -1)$$

for some $0 \leq t \leq 1$. Since the ε_0-coordinate of x is 3, which is the largest allowable coordinate of a point of the weight polytope P, every neighbourhood of x must include points with ε_0-coordinate strictly greater than 3, some of which lie in the hyperplane $\sum_{i=0}^{7} x_i = 0$. The conclusion follows. □

Exercise 8.4.11

(i) Use Exercise 8.1.11 to show that one can obtain a root system of type E_6 by taking those roots in type E_7 (in the notation of Example 8.4.9) that are orthogonal to the vector $v_{0,7}$.

(ii) Show that a root satisfying the hypotheses of (i) has one of the following three forms:

(a) $4(\varepsilon_i - \varepsilon_j)$ where $1 \leq i, j \leq 6$ and $i \neq j$;
(b) $\sum_{i=0}^{7} \pm 2\varepsilon_i$, where there are four $+$ signs and four $-$ signs, and furthermore, ε_0 and ε_7 appear with signs opposite to each other; or
(c) $\pm 4(\varepsilon_0 - \varepsilon_7)$.

(iii) Show that the highest root is $-4\varepsilon_0 + 4\varepsilon_7$.
(iv) Deduce from (ii) that the number of roots in a root system of type E_6 is $30 + 40 + 2 = 72$.

Exercise 8.4.12 Let P be the weight polytope of one of the minuscule representations of type E_6, considered as a subset of $\mathbb{R}^8$ under the identifications of Exercise 8.1.11.

(i) Show that the affine hull of P is the intersection of the hyperplane $\sum_{i=0}^{7} x_i = 0$ and a certain hyperplane of the form

$$3(x_0 + x_7) - (x_1 + x_2 + \cdots + x_6) = \pm 8,$$

where the choice of sign depends on the representation being considered.
(ii) Show that for any scalars $x_0, \ldots, x_5$, there exists a unique point in P of the form

$$(x_0, x_1, \ldots, x_5, t_1, t_2),$$

for some $t_1, t_2 \in \mathbb{R}$.
(iii) State and prove an analogue of Lemma 8.4.10 for the minuscule representations in type E_6.

Exercise 8.4.13 Let P be the weight polytope of one of the spin representations in type D_n, with coordinates described as in Exercise 8.1.9.

(i) Show that the affine hull of P is all of $\mathbb{R}^n$.
(ii) Give another proof of the construction of the root system of type D_n given in Exercise 5.3.27, this time by using the weight polytope and adapting the argument of Example 8.4.9.
(iii) Show that the highest root in type D_n in these coordinates is $4\varepsilon_1 + 4\varepsilon_2$.
(iv) State and prove an analogue of Lemma 8.4.10 for the spin representations in type D_n.

Exercise 8.4.14 Let P be the weight polytope of the natural representation in type D_n, with coordinates given in Example 8.1.6. Repeat Exercise 8.4.13 for this case.

Exercise 8.4.15 Let P be the weight polytope of the minuscule representation in type C_n, with coordinates described as in Exercise 8.1.9.

(i) Show that the affine hull of P is all of $\mathbb{R}^n$.
(ii) Give another proof of the construction of the root system of type C_n given in Exercise 5.3.24, this time by using the weight polytope and adapting the argument of Example 8.4.9.
(iii) Recall from Exercise 5.3.23 that the highest root in type C_n is

$$2(\alpha_1 + \alpha_2 + \cdots + \alpha_{l-1}) + \alpha_l$$

and that the highest short root is

$$\alpha_1 + 2(\alpha_2 + \cdots + \alpha_{l-2} + \alpha_{l-1}) + \alpha_l.$$

Show that, in these coordinates, the highest root is $8\varepsilon_1$ and the highest short root is $4\varepsilon_1 + 4\varepsilon_2$.

(iv) Explain why the analogue of Lemma 8.4.10 fails for the minuscule representation in type C_n, even though the underlying polytope is the same as the one in Exercise 8.4.14, where the property holds.

Exercise 8.4.16 Let P be the weight polytope of one of the spin representations in type B_n, with coordinates described as in Example 8.1.5.

(i) Show that the affine hull of P is all of $\mathbb{R}^n$.

(ii) Give another proof of the construction of the root system of type B_n given in Exercise 5.3.31, this time by using the weight polytope and adapting the argument of Example 8.4.9.

(iii) Observe from Appendix B that the highest root in type B_n is

$$\alpha_1 + 2(\alpha_2 + \alpha_3 + \cdots + \alpha_l)$$

and recall from Exercise 6.4.9 that the highest short root is

$$\alpha_1 + \alpha_2 + \cdots + \alpha_l.$$

Show that, in these coordinates, the highest root is $4\varepsilon_1 + 4\varepsilon_2$ and the highest short root is $4\varepsilon_1$.

(iv) Show that the analogue of Lemma 8.4.10 fails for the spin representation in type B_2.

(v) State and prove an analogue of Lemma 8.4.10 for the spin representation in type B_n if $n > 2$.

Exercise 8.4.17 Let P be the weight polytope of one of the minuscule representations in type A_n, with coordinates described as in Exercise 8.1.8.

(i) Show that the affine hull of P is a hyperplane in $\mathbb{R}^n$ of the form

$$x_1 + x_2 + \cdots + x_n = c$$

for some suitable constant c.

(ii) Give another proof of the construction of the root system of type A_n given in Exercise 5.3.20, this time by using the weight polytope and adapting the argument of Example 8.4.9.

(iii) Show that in these coordinates the highest root is $4\varepsilon_1 - 4\varepsilon_n$.

(iv) Show that the analogue of Lemma 8.4.10 fails for the natural representation in type A_1.

(v) State and prove an analogue of Lemma 8.4.10 for the minuscule representations in type A_n if $n > 1$.

Exercise 8.4.18 Show that, with the usual coordinates, scalar multiplication by -1 is an automorphism of the polytope $P(A_5, \omega_3)$, although no element of the Weyl group effects this automorphism. Deduce that the Weyl group of a weight polytope of a minuscule representation is not always the full group of automorphisms of the polytope.

Exercise 8.4.19 Let $P = P(B_n, \omega_n)$ be an n-dimensional hypercube of side D, acted on by the Weyl group $W = W(B_n)$. Use the Trivialization Theorem and Exercise 8.4.16 to show the following.

(i) If x_1 and x_2 are vertices of P at distance $\sqrt{k}D$ and k is even, then there there is a sequence of $k/2$ commuting reflections in W exchanging x_1 and x_2. Furthermore, all the reflections in the sequence must correspond to long roots.

(ii) If x_1 and x_2 are vertices of P at distance $\sqrt{k}D$ and k is odd, then there there is a sequence of $\lfloor k/2 \rfloor + 1$ commuting reflections in W exchanging x_1 and x_2. Furthermore, all but one of the reflections in the sequence must correspond to long roots.

(iii) The bounds in (i) and (ii) above are sharp: it is not possible to exchange the two vertices with a shorter sequence of commuting reflections.

8.5 Faces of weight polytopes

By Theorem 8.4.2, we can think of a polytope Π_n as a bounded intersection of finitely many closed half spaces. The part of Π_n that lies in one of the hyperplanes is called a *facet*, and each facet is an $(n-1)$-dimensional polytope. Iterating this construction gives rise to a set of k-dimensional polytopes Π_k (called k-*faces*) for each $0 \leq k \leq n$. A *flag* in a polytope P is a sequence of faces

$$\emptyset = P_{-1} \subset P_0 \subset P_1 \subset P_2 \subset \cdots \subset P_n = P$$

such that P_i is an i-face of P. The polytope P is said to be *regular* if the group $\text{Aut}(P)$ acts transitively on the flags of P. The k-*skeleton* of P is defined to be the union of the sets Π_i for all $i \leq k$; that is, the union of all the faces of dimension at most k.

Exercise 8.5.1 Show that a polytope is the convex hull of its facets, and deduce that a polytope is the convex hull of the set Π_0.

Exercise 8.5.2 Show that the facets of a polytope P lie in the (topological) boundary of P.

Exercise 8.5.3 Show that the (topological) boundary of a polytope P consists precisely of the union of its facets. Deduce that if a point p of P lies in no facets, then p is an interior point.

Definition 8.5.4 Let W be a finite Weyl group with generating set S, and recall the definition of the chamber C from Section 4.3. For each subset $I \subseteq S$, we define the subset C_I of the closure $\overline{C}$ to be

$$C_I := \{v \in V_\Gamma : (\alpha_i, v) = 0 \text{ for } i \in I \text{ and } (\alpha_i, v) > 0 \text{ for } i \in S \backslash I\}.$$

In particular, we have $C = C_\emptyset$.

Exercise 8.5.5 Show that the sets C_I, as I ranges over all subsets of S, partition the closure $\overline{C}$ of the fundamental chamber.

Definition 8.5.6 Let $\chi \in \overline{C}$. By Exercise 8.5.5, we have $\chi \in C_I$ for a unique $I \subseteq S$. Let $M = M_\chi = S \backslash I$, that is,

$$M_\chi = S \backslash \{s_i : s_i(\chi) = \chi\}.$$

A subset $K \subseteq S$ is said to be *M-connected* if every $k \in K$ can be connected to some $m \in M$ by a path within K.

Theorem 8.5.7 (Casselman; Satake; Borel–Tits) *Maintain the above notation, let $\chi \in \overline{C}$, and let $P = \Pi(W.\chi)$ be the convex hull of the finite set $W.\chi$.*

(i) *There is a bijection between the M_χ-connected subsets of S and the W-orbits of the faces of the convex hull $W.\chi$ given by associating the M_χ-connected subset I with the convex hull of the set $W_I.\chi$.*
(ii) *The bijection of (i) associates subsets of cardinality k with faces of dimension k.*

Proof Part (i) is proved in Casselman [12, theorem 3.1] and the remark immediately following it.

Part (ii) follows from (i) and the observation that the affine hull of $W_I.\chi$ has dimension $|I|$. □

Proposition 8.5.8 *Let $P = P(\mathfrak{g}, \omega_p)$ be the weight polytope of a minuscule representation of a simple Lie algebra $\mathfrak{g}$, and assume that* $\dim(P) \geq 2$. *Then the polytope P satisfies the hypotheses of Theorem 8.5.7, and the Weyl group W of $\mathfrak{g}$ acts transitively on the set of faces of $P(\mathfrak{g}, \omega)$ of dimensions* 0, 1 *and (except in type A_l)* 2.

Proof The assertion for dimension 0 is a restatement of the familiar fact that the Weyl group acts transitively on the weights of a minuscule representation.

It follows that the polytope $P = P(\mathfrak{g}, \omega_p)$ satisfies the hypotheses of Theorem 8.5.7. In this case, the set M_{ω_p} consists of the single vertex p, and the M_{ω_p}-connected subsets are the subsets of Γ are the connected subsets of Γ that contain p. Clearly, there is only one such set of cardinality 1, which proves the assertion for dimension 1. Since there is only one such set of cardinality 1, Theorem 8.5.7 (ii) shows that there is only one orbit of faces of dimension 1.

If $\mathfrak{g}$ is not of type A_l, every minuscule weight is an endpoint of the corresponding finite type Dynkin diagram Γ. It follows that the vertex p above is an endpoint of Γ. This implies that there is a unique vertex q of Γ that is adjacent to p, and that $\{p, q\}$ is the unique M_{ω_p}-connected subset of size 2. The proof is completed by another application of Theorem 8.5.7. □

Example 8.5.9 Consider the weight polytope $P = P(B_l, \omega_l)$. As in Example 8.1.5, the 2^l weights can be given by coordinates $(\pm 2, \pm 2, \ldots, \pm 2)$. The highest weight is $(2, 2, \ldots, 2)$, and (as always) this is the unique weight χ in the closure $\overline{C}$ of the fundamental chamber. The only fundamental reflection that does not fix χ is s_l, so the M_χ-connected subsets of the Dynkin diagram B_l are the l sets of the form $M_{\chi,k} = \{i : k \leq i \leq l\}$ for some fixed $1 \leq k \leq l$. In particular, the parabolic subgroup obtained by omitting the generator s_1 is the stabilizer of one of the facets of the hypercube.

Theorem 8.5.7 (i) shows that a representative of the W-orbit of faces corresponding to $M_{\chi,k}$ is $\Pi(W_I.\chi)$, where $I = \{s_k, s_{k+1}, \ldots, s_l\}$ and $|I| = l - k + 1$. The vertices $W_I.\chi$ are the $2^{|I|}$ vertices all of whose leftmost $l - |I|$ coordinates are equal to 2, and whose rightmost $|I|$ coordinates take all possible values in the set $\{\pm 2\}$. We recognize these vectors as the coordinates of the vertices of an $|I|$-dimensional hypercube. This proves the expected result that there is one W-orbit of faces of dimension j for each $1 \leq j \leq l$, and that the j-dimensional faces are isometric copies of the j-dimensional hypercube.

It follows from the previous paragraph by inducting on l that the l-dimensional hypercube is a regular polytope. This is easy checked for the base case $l = 0$, so suppose $l > 0$ and suppose that

$$\emptyset \subset P_0 \subset P_1 \subset P_2 \subset \cdots \subset P_{l-1} \subset P$$

and

$$\emptyset \subset Q_0 \subset Q_1 \subset Q_2 \subset \cdots \subset Q_{l-1} \subset P$$

are two flags in the hypercube. Since W acts transitively on the facets, we may assume without loss of generality that $P_{l-1} = Q_{l-1}$. By applying another element of W to both flags, we may further assume that $P_{l-1} = Q_{l-1}$ is the W-orbit representative of the $(l-1)$-dimensional faces given by Theorem 8.5.7. The proof is then completed by the inductive hypothesis.

Exercise 8.5.10 The polytopes $P = P(A_l, \omega_k)$ are known as *hypersimplices* or *Johnson polytopes*.

(i) Prove that P is a regular polytope if $k = 1$ and $k = l$. Prove also that the polytopes $P(A_l, \omega_1)$ and $P(A_l, \omega_l)$ are simplices.
(ii) Prove that if $k \notin \{1, l\}$, then the Weyl group has two orbits on two-dimensional faces, and that P is not regular.

Exercise 8.5.11 The polytopes $P = P(C_l, \omega_1)$ are known as *hyperoctahedra*. Prove that P is a regular polytope, all of whose faces are hyperoctahedra of lower dimension.

Exercise 8.5.12 Show that the polytopes $P = P(D_l, \omega_1)$ are l-dimensional hyperoctahedra, which were proved to be regular in Exercise 8.5.11. Show that P has one $W(D_l)$-orbit of faces in all dimensions except dimension $l - 1$, in which case there are two orbits of faces.

Example 8.5.13 Consider the weight polytope $P = P(D_l, \omega_l)$; this is known as the *half cube* or *demihypercube*. As in Exercise 8.1.9, the 2^{l-1} weights can be given by coordinates $(\pm 2, \pm 2, \ldots, \pm 2)$, in which the entry -2 appears an even number of times. The highest weight is $(2, 2, \ldots, 2)$, and this is the unique weight χ in the closure $\overline{C}$ of the fundamental chamber. The only fundamental reflection that does not fix χ is s_l.

It is convenient to divide the M_χ-connected subsets of the Dynkin diagram D_l into three types: (a) the singleton set $\{l\}$, (b) the sets containing $l - 2$ and $l - 1$, and (c) the sets containing $l - 2$ but not $l - 1$. The M_χ-connected subsets in (b) are of the form

$$M_{\chi,k} = \{i : k \leq i \leq l - 2\} \cup \{l - 1, l\}$$

for some $1 \leq k \leq l-2$, and the M_χ-connected subsets in (c) are of the form

$$M_{\chi,k} = \{i : k \leq i \leq l-2\} \cup \{l\}.$$

The unique singleton set in (a) corresponds to the unique $W = W(D_l)$ orbit of 1-faces. By considering the set $W_I.\chi$ as in Example 8.5.9, we recognize the subsets I appearing in (b) as isometric copies of $|I|$-dimensional half cubes, and we recognize the subsets I' appearing in (c) as isometric copies of $|I'|$-dimensional simplices (see Exercise 8.5.10 (i)).

The number of vertices in a k-dimensional half cube is 2^{k-1}, and the number of vertices in a k-dimensional simplex is $k+1$. Note that $2^{k-1} = k+1$ only occurs if $k = 3$. It follows that if $l \geq 5$, then the half cube is not a regular polytope, because the four-dimensional faces form two distinct isometry classes.

Note also that there are two W-orbits of three-dimensional faces of P, even though both are isometric to three-dimensional simplices. The simplices in one orbit extend to four-dimensional simplices, and those of the other orbit do not.

Exercise 8.5.14 Show that the four-dimensional half cube, $P(D_4, \omega_4)$, is isometric to a four-dimensional hyperoctahedron, and is therefore a regular polytope. (Hint: consider the relationship between $P(D_4, \omega_4)$ and $P(D_4, \omega_1)$.)

Exercise 8.5.15 State and prove a version of Example 8.5.13 for the weight polytope $P = P(D_l, \omega_{l-1})$, and show that P is isometric to the l-dimensional half cube.

Example 8.5.16 The weight polytopes $P(E_6, \omega_1)$ and $P(E_6, \omega_5)$ are isometric. We follow Coxeter's notation and denote this polytope by 2_{21}; let P be such a polytope. Arguing as in Example 8.5.13, we see that there is one $W = W(E_6)$-orbit of k-faces of P for $k \leq 3$, and that these faces are isometric to simplices. There are two orbits of 4-faces, both isometric to simplices. There are also two orbits of 5-faces, one consisting of 5-simplices and the other consisting of five-dimensional hyperoctahedra. As in Example 8.5.13, it is clear that P is not a regular polytope.

Since E_6 is simply laced, there is only one possible length for roots by Theorem 4.2.3 (iii). By Proposition 4.3.18 (iv), the stabilizer of the highest root, θ, is generated by the subgroup W_I with $I = \{s_1, s_2, s_3, s_4, s_5\}$. The set I corresponds to the vertices of the affine Dynkin diagram $E_6^{(1)}$ that are not adjacent to vertex 0. Using Theorem 8.5.7, we recognize W_I as the stabilizer of one of the members of the orbit of five-dimensional simplex shaped faces. It follows that the action of W on the 5-simplex shaped faces is equivalent to the action of W on the roots. Recall from Exercise 8.4.11 that there are 72 such roots; it follows that that there are 72 five-simplex shaped faces, and that the Weyl group W has order

$$72|W_I| = 72 \times 6! = 51840.$$

Exercise 8.5.17 The weight polytope $P = P(E_7, \omega_6)$ is called 3_{21} in Coxeter's notation. Show that:

(i) there is one $W = W(E_7)$-orbit of k-faces of P for $k \leq 4$, and that these faces are isometric to simplices;
(ii) there are two orbits of 5-faces, both isometric to simplices; and

(iii) there are two orbits of 6-faces, one consisting of 6-simplices and the other consisting of six-dimensional hyperoctahedra.
(iv) Show that P is not a regular polytope.
(v) Show that the action of W on the six-dimensional hyperoctahedral faces is equivalent to the action of W on the roots of E_7. Deduce that P has 126 six-dimensional hyperoctahedral faces.

Exercise 8.5.18 Consider the 126 hyperoctahedral faces of 3_{21}, described in Exercise 8.5.17.

(i) Show that each of the 126 faces described in (v) is perpendicular to the (positive or negative) root α that it corresponds to.
(ii) Show that the 126 faces fall into 63 parallel pairs, where the pairing is induced by multiplication by -1.
(iii) Show that the 24 vertices of a parallel pair of such faces are all moved by the action of the reflection s_α.

Exercise 8.5.19 Use Proposition 4.3.18 (iv) to show that the polytope $P(E_8, \omega_1)$ is isometric to the convex hull of the 240 roots in type E_8.

Exercise 8.5.20 The polytope of Exercise 8.5.19 is called 4_{21} in Coxeter's notation. (It is not the weight polytope of a minuscule representation.) Show that

(i) there is one $W = W(E_8)$-orbit of k-faces of P for $k \leq 5$, and that these faces are isometric to simplices;
(ii) there are two orbits of 6-faces, both isometric to simplices; and
(iii) there are two orbits of 7-faces, one consisting of 7-simplices and the other consisting of 7-dimensional hyperoctahedra.
(iv) Show that P is not a regular polytope.

Exercise 8.5.21 Consider the group $W = W(E_8)$ acting on the set R consisting of its 240 roots.

(i) Use Proposition 7.1.1 to show that there are at least five W-orbits on $R \times R$.
(ii) Use Theorem 8.5.7 to show that all ordered pairs of roots at a mutual angle of $\pi/3$ are W-conjugate.
(iii) Use Proposition 4.3.18 to show that all pairs of orthogonal roots are W-conjugate.
(iv) Show that the Weyl group $W(E_8)$ acts as a rank 5 permutation group on R, and that two ordered pairs of roots are W-conjugate if and only if the angles between each pair are equal.

Remark 8.5.22 We will meet purely combinatorial analogues of Example 8.5.16 and Exercise 8.5.17 later; see Exercise 10.2.2 and Proposition 9.2.8 respectively.

8.6 Graphs from minuscule representations

Let $L(\mathfrak{g}, \omega)$ be a minuscule representation with Weyl group W, and let $d_0, d_1, \ldots, d_{N-1}$ be as in Proposition 8.2.20, so that N is the rank of W acting on the weights $\Lambda(\mathfrak{g}, \omega)$.

Let $D(\mathfrak{g}, \omega)$ be the set of ordered pairs of weights, and for each $\delta \in D(\mathfrak{g}, \omega)$, let

$$\overrightarrow{\delta} = \overrightarrow{(\lambda_i, \lambda_j)} = \lambda_j - \lambda_i.$$

Note that the set $D(\mathfrak{g}, \omega)$ is closed under negation.

Let

$$0 = d_0 < d_1 < \cdots < d_{N-1}$$

be the possible Euclidean distances between weights, as in Proposition 8.2.20. For each $1 \leq i \leq N-1$, define the graph $G_i(\mathfrak{g}, \omega)$ to have vertices indexed by the weights of $L(\mathfrak{g}, \omega)$, and with an edge between weights λ and μ if $|\lambda - \mu| = d_i$.

Since the Weyl group acts transitively on the weights by orthogonal transformations, it follows that the graphs $G_i(\mathfrak{g}, \omega)$ are all *regular*, in other words, each vertex is incident to a fixed number, $v_i(\mathfrak{g}, \omega)$, of edges.

For $\alpha \in D(\mathfrak{g}, \omega)$, let $m_\alpha(\mathfrak{g}, \omega)$ be the cardinality of the set

$$\{\delta \in D(\mathfrak{g}, \omega) : \overrightarrow{\delta} = \alpha\}.$$

By Theorem 8.2.22 (ii), the quantity $m_\alpha(\mathfrak{g}, \omega)$ depends only on the magnitude $d_i = |\overrightarrow{\delta}|$, so we write $m_i(\mathfrak{g}, \omega) = m_\alpha(\mathfrak{g}, \omega)$.

Let $R_i(\mathfrak{g}, \omega)$ be the cardinality of the set

$$\{\alpha \in D(\mathfrak{g}, \omega) : |\alpha| = d_i\}.$$

Theorem 8.6.1 *With the above notation, we have*

$$m_i(\mathfrak{g}, \omega) \times R_i(\mathfrak{g}, \omega) = \dim(L(\mathfrak{g}, \omega)) \times v_i(\mathfrak{g}, \omega).$$

Proof Define the set

$$S_i = \{(\lambda, \alpha) : \lambda \in L(\mathfrak{g}, \omega),\ \alpha \in D(\mathfrak{g}, \omega),\ \lambda + \alpha \in L(\mathfrak{g}, \omega),\ |\alpha| = d_i\}.$$

The result will follow by enumerating the size of S_i in two different ways.

The number of elements of S_i with a fixed λ is $v_i(\mathfrak{g}, \omega)$, and there are $\dim(L(\mathfrak{g}, \omega))$ choices for λ. On the other hand, the number of elements of S_i with a fixed α is $m_i(\mathfrak{g}, \omega)$, and there are $R_i(\mathfrak{g}, \omega)$ choices for α. □

Example 8.6.2 Maintain the notation of Exercise 8.1.8. The graphs $G_1(A_{n-1}, \omega_k)$ are known as the *Johnson graphs*, $J(n, k)$. Let F be a principal subheap corresponding to the minuscule representation $L(A_{n-1}, \omega_k)$.

In this case, the coordinates of a weight have entries in the set $\{\pm 2\}$, and the weights at distance d_1 from a given one are obtained by choosing one of the k entries equal to $+2$ and one of the $n-k$ entries equal to -2, and exchanging them. We then have

$$\dim(L(A_{n-1}, \omega_k)) = \binom{n}{k},$$

and $v_1(A_{n-1}, \omega_k) = k(n-k)$.

It follows from Exercise 5.3.20 that there are $n(n-1)$ roots in type A_{n-1}. Since there is only one root length in type A, it follows from the Trivialization Theorem that the relative content of a pair of ideals (I, I') of F is a root if and only if the pair is trivializable to a singleton heap. In turn, this means that $R_1(A_{n-1}, \omega_k) = n(n-1)$. The same argument shows that $m_1(A_{n-1}, \omega_k)$ is equal to the number of weights moved by a reflection. In this case, a reflection acts as a transposition exchanging an entry of $+2$ with an entry of -2, and fixing the remaining $n-2$ entries, $k-1$ of which are equal to +2. It follows that

$$m_1(A_{n-1}, \omega_k) = \binom{n-2}{k-1}.$$

Theorem 8.6.1 then gives a combinatorial proof of the identity

$$\binom{n-2}{k-1} \times n(n-1) = \binom{n}{k} \times k(n-k).$$

The typical application of Theorem 8.6.1 is to calculate one of the four quantities involved when the other three are known.

Exercise 8.6.3 Consider the spin representation $L(D_l, \omega_l)$. The graph $G_1(D_l, \omega_l)$ is called the *half cube graph*. Verify that $R_1(D_l, \omega_l) = 2l(l-1)$, $\dim(D_l, \omega_l) = 2^{l-1}$ and

$$v_1(D_l, \omega_l) = \binom{l}{2}.$$

Deduce that $m_1(D_l, \omega_l) = 2^{n-3}$.

Example 8.6.4 Consider the 27-dimensional representation $L(E_6, \omega_5)$. The graph $G_1(E_6, \omega_5)$ is called the *Schläfli graph*. Recall from Exercise 8.4.11 that there are 72 roots. The argument of Example 8.6.2 shows that $R_1(E_6, \omega_5) = 72$.

By inspection of the corresponding principal subheap, we see that there are six pairs of ideals (I, I') with $\chi(I, I') = \alpha_p$ for any vertex p of the Dynkin diagram. It follows that each reflection exchanges six pairs of weights, and that $m_i(E_6, \omega_5) = 6$. (This fact will be the foundation for the study of Schläfli double sixes, which we will come to in Section 10.2.) Since $6 \times 72 = 27 \times 16$, it follows from Theorem 8.6.1 that the Schläfli graph is regular of degree 16, that is, $v_1(E_6, \omega_5) = 16$.

We can also prove this directly: upon restriction to a $\mathfrak{g}(D_5)$ subalgebra as in Proposition 8.2.9 (iv), the representation $L(E_6, \omega_5)$ decomposes into a direct sum of a 16-dimension spin representation, a ten-dimensional natural representation and a one-dimensional trivial representation. The submodule corresponding to the distance d_1 is the spin representation.

Exercise 8.6.5 Use Theorem 8.2.22 to show that the graph $G_2(E_6, \omega_5)$ is regular of degree 10; that is, $v_2(E_6, \omega_5) = 10$. Show that there are 135 unordered pairs $\{\lambda_1, \lambda_2\}$ of weights at mutual distance $\sqrt{2}D$ in the polytope 2_{21}, where D is the minimal nonzero distance between weights. Show also that $W(E_6)$ acts transitively on these 135 unordered pairs.

Exercise 8.6.6 The graph $G_1(E_7, \omega_6)$ is known as the *Gosset graph*. State and prove a version of Example 8.6.4 for the representation $L(E_7, \omega_6)$, proving that we have $m_1(E_7, \omega_6) = 12$, $R_1(E_7, \omega_6) = 126$, $\dim(E_7, \omega_6) = 56$ and $v_1(E_7, \omega_6) = 27$.

Exercise 8.6.7

(i) State and prove a version of Exercise 8.6.3 for the representation $L(D_l, \omega_{l-1})$.
(ii) State and prove a version of Example 8.6.4 for the representation $L(E_6, \omega_1)$.

(iii) State and prove a version of Example 8.6.4 for the representation $L(D_l, \omega_1)$, proving that we have $m_1(D_l, \omega_1) = 2$, $R_1(D_l, \omega_1) = 2l(l-1)$, $\dim(D_l, \omega_1) = 2l$ and $v_1(D_l, \omega_1) = 2(l-1)$.

Exercise 8.6.8 Recall from Exercise 5.3.27 (and Exercise 8.6.7) that there are $2l(l-1)$ roots in type D_l. Let $W = W(D_l)$, and let $P = P(D_l, \omega_1)$ be the weight polytope corresponding to the natural representation, and let D be the minimum nonzero distance between two vertices of P. Recall that the polytope P is an l-dimensional hyperoctahedron by Exercise 8.5.12. Use the results of Exercise 5.3.29 and the other exercises of Section 5.3 to prove the following.

(i) Two vertices v_1 and v_2 of P lie at distance $\sqrt{2}D$ if and only if $v_1 - v_2 = \pm\alpha \pm \beta$ is a sum of a strongly orthogonal pair of positive roots.
(ii) The correspondence in (i) gives a 2 to $l-1$ correspondence between (a) unordered positive strongly orthogonal pairs of roots and (b) the l long diagonals of P.
(iii) For each long diagonal $v_1 - v_2$ as in (ii), there are precisely $l-1$ unordered strongly orthogonal pairs of positive roots $\{\alpha, \beta\}$ such that $v_1 - v_2 = \pm\alpha \pm \beta$.
(iv) Use the coordinates of Exercise 5.3.29 (ii) and Example 8.1.6 to describe the $l-1$ pairs in (iii) for any long diagonal $\pm 8\varepsilon_i$, where $1 \leq i \leq n$.

8.7 Notes and references

1 The definition of minuscule system given in Definition 8.1.1 differs slightly from the one given in Green [30]. In [30], a minuscule system is defined to satisfy condition (iii), together with the condition that "$\lambda + \alpha \in \Psi$ if and only if $c = -1$, and $\lambda - \alpha \in \Psi$ if and only if $c = +1$".

2 The results about orbitals and suborbits (Proposiion 8.2.1) are well known; our treatment follows Cameron [9, section 9].

3 Coxeter [17] uses the notation α_n, β_n and γ_n to refer to the n-dimensional simplex, hypercube and hyperoctahedron; these are the polytopes $P(A_n, \omega_1)$, $P(B_n, \omega_n)$ and $P(C_n, \omega_1)$ in our notation. The book [17] also uses the terms "measure polytope" and "cross polytope" to refer to the hypercube and hyperoctahedron, respectively, and uses the term "cell" to mean what we call a "facet".

Coxeter denotes the n-dimensional half cube $P(D_n, \omega_n)$ by $\mathrm{h}\gamma_n$. Coxeter also uses the following notation for polytopes with $W(E_n)$ symmetry.

Coxeter's notation	Our notation
2_{21}	$P(E_6, \omega_1)$, $P(E_6, \omega_5)$
1_{22}	$P(E_6, \omega_6)$
3_{21}	$P(E_7, \omega_6)$
2_{31}	$P(E_7, \omega_1)$
1_{32}	$P(E_7, \omega_7)$
4_{21}	$P(E_8, \omega_1)$
2_{41}	$P(E_8, \omega_7)$
1_{42}	$P(E_8, \omega_8)$

The notation p_{qr} is intended to suggest a tree-shaped graph with three branches emerging from a vertex, of lengths p, q and r; in the above three examples, this gives the Dynkin diagrams of types E_6, E_7 and E_8.

4 Parker and Röhrle [59] prove a version of the branching rule for minuscule representations given in Proposition 8.2.9 (iv). They work in the context of universal Chevalley groups over algebraically closed fields, and their strategy is to examine single cosets appearing in certain double cosets of the Weyl group.

5 Kumar [41, proposition 1.5] proves another result about double cosets. If μ is a minuscule weight, then the number of irreducible submodules in the tensor product $L(\lambda) \otimes L(\mu)$ is the number of double cosets $W_\lambda \backslash W / W_\mu$, where $W(\lambda)$ and $W(\mu)$ are the stabilizers of the highest weights λ and μ, respectively.

6 Wenzl [89] considers the n-fold tensor products of minuscule representations of types E_6 and E_7 and describes their *centralizer algebras*; that is, the endomorphisms of the tensor product that commute with the Lie algebra action.

7 It would be interesting to have a conceptual proof of Proposition 8.2.20.

8 Theorem 8.2.22 is new to the best of our knowledge.

9 Theorem 8.3.10 is probably well-known but we are not aware of an explicit statement of the theorem in the literature. For example, Carter [11, section 13.8] constructs the minuscule representations in types E_6 and E_7 by working directly with the roots R, rather than the set of weights.

10 The fundamental result Theorem 8.4.2 appears as Ziegler [92, theorem 1.1], where it is called the "main theorem for polytopes".

11 There are notions of polytope automorphism that are more general than the one in Definition 8.4.3. A *combinatorial automorphism* of a polytope is a permutation of the vertex set that maps faces to faces. An *affine automorphism* (respectively, *Euclidean automorphism*) is a permutation of the vertex set induced by an affine (respectively, Euclidean) automorphism. It is immediate that every Euclidean automorphism is affine, and that every affine automorphism is combinatorial. The automorphisms of Definition 8.4.3 are Euclidean automorphisms.

12 According to Casselman [12], Theorem 8.5.7 is implicit in work of Satake and Borel & Tits ([68, lemma 5] and [7, section 12.16]). Casselman's proof is essentially the same as the earlier proofs.

13 Theorem 8.6.1 is new (although not difficult given the definitions).

14 The half cube graphs $G_1(D_l, \omega_l)$ and the Johnson graphs $G_1(A_{n-1}, \omega_k)$ play a role in the theory of matroids; see Chepoi [13, section 2].

9

The 28 bitangents

In Chapters 9 and 10, we will study the rich combinatorics associated with the two 27-dimensional minuscule representations in type E_6, and the 56-dimensional minuscule representation in type E_7.

The numbers 12, 27, 28 and 56 figure prominently in many of these combinatorial structures, for reasons that we now outline. The 56 weights of $L(E_7, \omega_7)$ fall into 28 positive–negative pairs; these pairs correspond in algebraic geometry to the 28 bitangents to a plane quartic curve. It is not necessary for our purposes to understand the underlying geometric terminology, but for completeness, we remark that a nonsingular plane quartic C is a nonhyperelliptic genus 3 curve embedded by its canonical linear system. Each of the 28 bitangents is tangent to C at two points, which may coincide.

If any one of the 28 bitangents is removed, the remaining bitangents are in natural correspondence with the 27 lines on a cubic surface, and also with the weights of either of the minuscule representations $L(E_6, \omega_1)$ or $L(E_6, \omega_5)$.

Each reflection in $W(E_7)$ moves 24 of the 56 opposite weights in 12 opposite pairs. These pairs correspond to distinguished subsets of 12 bitangents among the 28. These subsets are the Steiner complexes, which are the main object of study in Chapter 9.

We start by studying the Gosset graph $G_1(E_7, \omega_6)$ in Section 9.1. The vertices of the graph may be identified with the 56 lines on the del Pezzo surface of degree 2, which is discussed in Section 9.2. The 28 bitangents, which we introduce in Section 9.3, may be identified with opposite pairs of vertices in the Gosset graph. We review the Hesse–Cayley notation for bitangents in Section 9.4.

In Section 9.5, we define Steiner complexes as certain subsets of size 12 of the 28 bitangents, and investigate their combinatorics. Finally, in Section 9.6, we show how the Steiner complexes naturally give rise to a symplectic bilinear form.

These ideas are further developed in Chapter 10, which is largely concerned with substructures of the 28 bitangents. For example, the Steiner complexes not containing a given bitangent can be identified with subsets of the 27 lines of size 12, known as Schläfli double sixes.

9.1 The Gosset graph

Recall from Section 6.5 that the weights of the unique minuscule representation $L(E_7, \omega_6)$ in type E_7 may be conveniently labelled by symbols $(i, j)^{\pm}$, where

$0 \leq i < j \leq 7$. We have another way to label the weights by symbols $\pm v_{i,j}$, where $0 \leq i < j \leq 7$, as in Example 8.1.7. Fortunately, these two notations are naturally compatible.

Proposition 9.1.1 *The bijection between the two parametrizations of the weights of the minuscule module in type E_7 is given by*

$$(i, j)^+ \leftrightarrow v_{i,j}$$

and

$$(i, j)^- \leftrightarrow -v_{i,j}.$$

Proof If $0 \leq j < i \leq 7$, we define $(j, i)^{\pm}$ to mean $(i, j)^{\pm}$ for convenience during the following argument.

By Example 8.1.7, the lowest weight vector is $v_{0,7}$, and the highest is $-v_{0,7}$. These correspond to the symbols $(0, 7)^+$ and $(0, 7)^-$ in the other notation, since each symbol corresponds to an orbit of ideals in the full heap $\mathrm{FH}(E_7^{(1)}(6))$ under the action of the infinite cyclic group of automorphisms.

We next observe that each of the generators s_p with $1 \leq p \leq 6$ acts as the permutation $(p, p+1)$ in the sense that

$$s_p(v_{i,j}) = v_{s_p(i),s_p(j)}$$

and

$$s_p((i, j)^{\pm}) = (s_p(i), s_p(j))^{\pm}.$$

The highest root θ in type E_7 is given by $-4\varepsilon_1 + 4\varepsilon_2$ in the standard coordinates, by Example 8.4.9. Proposition 5.4.14 shows that the action of s_0 and s_θ on the weights is the same. It follows that s_0 acts by the permutation $(0, 1)$, in the sense that

$$s_\theta(v_{i,j}) = v_{s_0(i),s_0(j)}.$$

On the other hand, it follows from the constructions that the generator s_0 acts as the permutation $(0, 1)$ in the sense that

$$s_0((i, j)^{\pm}) = (s_0(i), s_0(j))^{\pm}.$$

Let $H \cong S_8$ be the subgroup of $W(E_7^{(1)})$ generated by $s_0, s_1, \ldots, s_6$. Since every weight is in an H-orbit containing either the highest or the lowest weight vector, the bijection between the two descriptions of the weights is indeed as claimed. □

We know from Example 8.4.9 that, given any two weights of $L(E_7, \omega_6)$, there are four possible relative positions between the two weights with respect to the action of W, according as the Euclidean distance between them is 0, $\sqrt{32}$, $\sqrt{64}$ or $\sqrt{96}$. It will be useful for later purposes to introduce the following terminology for these four relationships, although the full meaning of the words will not be apparent until later.

Definition 9.1.2 Let λ_1 and λ_2 be weights of $L(E_7, \omega_6)$, regarded as points in Euclidean space as in Example 8.1.7. We say that the pair of weights λ_1 and λ_2 is *equal* (respectively, *skew*, *incident*, *opposite*) if the Euclidean distance between the weights is 0 (respectively, $\sqrt{32}$, $\sqrt{64}$, $\sqrt{96}$.)

It is sometimes helpful to have the following characterization of the relationships of skewness, incidence and oppositeness, without reference to Euclidean distance.

Lemma 9.1.3 *Let $(i, j)^+$ be a weight of $L(E_7, \omega_6)$.*

(i) *The weight $(k, l)^\pm$ is skew to $(i, j)^+$ if and only if either*
   (a) *the sign is positive and $|\{i, j\} \cap \{k, l\}| = 1$ or*
   (b) *the sign is negative and $|\{i, j\} \cap \{k, l\}| = 2$.*
(ii) *The weight $(k, l)^\pm$ is incident to $(i, j)^+$ if and only if either*
   (a) *the sign is positive and $|\{i, j\} \cap \{k, l\}| = 2$ or*
   (b) *the sign is negative and $|\{i, j\} \cap \{k, l\}| = 1$.*
(iii) *The weight $(k, l)^\pm$ is opposite to $(i, j)^+$ if and only if it is equal to $(i, j)^-$.*

We have an analogous result for $(i, j)^-$ by reversing the roles of $+$ and $-$ above.

Proof This follows from the definitions and Proposition 9.1.1 using a case by case check. □

The Gosset graph was defined in Exercise 8.6.6 as the graph $G_1(E_7, \omega_6)$. In other words, it is the graph with 56 vertices in bijection with the 56 weights of the minuscule module in type E_7, and with an edge between two vertices if and only if the corresponding weights are skew.

Thanks to Lemma 9.1.3, we may identify the vertices of the Gosset graph with the edges of two disjoint copies of the complete graph K_8 on eight vertices. Recall from Example 8.4.9 that 56 of the roots in type E_7 may be identified with ordered pairs from the set $\{0, 1, \ldots, 7\}$, and the other 70 are in bijection with the size 4 subsets of an eight-element set.

Proposition 9.1.4 *Let α be a root of the Lie algebra of type E_7, and maintain the notation of Example 8.4.9.*

(i) *If we have $\alpha = 4(\varepsilon_i - \varepsilon_j)$, then the reflection $s_\alpha \in W(E_7)$ acts as the transposition (i, j).*
(ii) *Suppose that there is a disjoint union*

$$\{0, 1, \ldots, 7\} = K_0 \dot{\cup} K_1$$

for which $|K_0| = |K_1| = 4$ and

$$\alpha = \left(\sum_{i \in K_0} 2\varepsilon_i\right) - \left(\sum_{j \in K_1} 2\varepsilon_j\right).$$

Then we have

$$s_\alpha((i, j)^\pm) = \begin{cases} (k, l)^\mp & \text{if } \{i, j, k, l\} \in \{K_0, K_1\}, \\ (i, j)^\pm & \text{otherwise.} \end{cases}$$

Proof This follows from the formula for the reflection s_α and the identifications of Proposition 9.1.1. Part (i) follows from the same reasoning that shows that the fundamental reflection s_i (where $1 \le i \le 6$) acts by permuting the coordinates at positions i and $i + 1$.

Part (ii) also follows less obviously from the same reasoning. The key observation here is that the vector

$$\alpha = \left(\sum_{i \in K_0} 2\varepsilon_i\right) - \left(\sum_{j \in K_1} 2\varepsilon_j\right)$$

is orthogonal to $v_{a,b}$ unless a and b come from the same K_l (with $l \in \{0, 1\}$). One then checks that if $a, b \in K_l$ for the same l, then s_α sends $v_{a,b}$ to $-v_{c,d}$, where $K_l = \{a, b, c, d\}$. □

Remark 9.1.5 The operations s_α appearing in Proposition 9.1.4 (ii) are known as *bifid transformations.*

Example 9.1.6 Let

$$\alpha = 2(\varepsilon_0 + \varepsilon_1 + \varepsilon_4 + \varepsilon_6 - \varepsilon_2 - \varepsilon_3 - \varepsilon_5 - \varepsilon_7)$$

and, correspondingly, let $K_0 = \{0, 1, 4, 6\}$ and $K_1 = \{2, 3, 5, 7\}$. Then we have

$$s_\alpha((3,4)^+) = (3,4)^+,$$
$$s_\alpha((1,6)^-) = (0,4)^+,$$
$$s_\alpha((2,3)^+) = (5,7)^-.$$

9.2 Del Pezzo surfaces

Du Val [21] showed how n points in general position in the complex projective plane define a discrete set of curves that are in bijection with the vertices of a certain polytope in n-dimensional space, in such a way that the intersection properties of the curves are mirrored by the metrical properties of the vertices of the polytope.

A *del Pezzo surface* is a certain type of nonsingular surface appearing in algebraic geometry. Over an algebraically closed field, every del Pezzo surface is either a product of two projective lines, or the blow up (in the sense of algebraic geometry) of $9 - d$ points in general position in the projective plane, with $1 \le d \le 9$. Here, "general position" means that no three points are collinear, no six lie on a conic, and no eight lie on a cubic having a node at one of them. These conditions characterize del Pezzo surfaces, so that du Val's construction produces del Pezzo surfaces if $n \le 9$. In modern terminology, du Val proved the following.

Theorem 9.2.1 (du Val) *Let $n \le 8$, and consider the polytope P that corresponds in the above sense to lines (i.e., curves of self-intersection -1) on the blow up of n points in general position on the complex projective plane. Let D be the minimal nonzero distance between two vertices of P.*

(i) *The polytope P is Coxeter's pure Archimedean polytope $(n-4)_{21}$.*
(ii) *If v_1 and v_2 are vertices of P, then we have*

$$|v_1 - v_2|^2 = (c+1)D,$$

where c is the intersection number of the curves corresponding to v_1 and v_2.

Proof Both parts are proved in [21, section 3]. □

The important cases of Theorem 9.2.1 for minuscule representations are the cases $n = 6$ and $n = 7$. The corresponding polytopes are called 2_{21} and 3_{21} in Coxeter's notation. The weights of either $L(E_6, \omega_1)$ or $L(E_6, \omega_5)$ form the vertices of an isometric copy of 2_{21}. The weights of $L(E_7, \omega_6)$ form the vertices of an isometric copy of 3_{21}, and the coordinates given in Example 8.1.7 are precisely the same as those used by du Val [21, section 7 (ii)].

Definition 9.2.2 For $3 \leq k \leq 8$, define the Dynkin diagram Γ_k as follows:

k	Γ_k
3	$A_2 \cup A_1$
4	A_4
5	D_5
6	E_6
7	E_7
8	E_8

Theorem 9.2.3 *Let $4 \leq k \leq 8$. Let Σ be a del Pezzo surface obtained by blowing up k points in the complex projective plane, no three of which are collinear, no six of which lie on a conic, and no eight of which lie on a cubic having a node at one of them.*

(i) *The set of lines of Σ is in bijection with the set of left cosets $W(Q_k)/W(Q_{k-1})$.*
(ii) *The group $W(\Gamma_k)$ acts transitively on the set of lines of Σ and the action respects the matrix of intersection numbers.*

Proof This is proved in Dolgachev [20, corollary 8.2.12]. □

Theorem 9.2.1 shows that two lines on a del Pezzo surface are skew if and only if the corresponding vertices on the polytope are at mutual distance D. For later purposes, it will be helpful to understand larger configurations of mutually skew lines. We will call two weights of $L(E_7, \omega_6)$ *skew* if their corresponding lines are skew, meaning that they have mutual intersection number 0.

Lemma 9.2.4 *Consider the minuscule representation $L(E_7, \omega_6)$, and let $W = W(E_7)$. Let F be the corresponding parabolic subheap, label the elements of the vertex chain $\varepsilon^{-1}(6) \cap F$ as*

$$F(6, 1) < F(6, 2) < F(6, 3)$$

and number the other vertex chains similarly.

(i) *The stabilizer of ω_6 is $W_1 = W_{\{s_1,s_2,s_3,s_4,s_5,s_7\}}$.*
(ii) *If ω is a weight that is skew to ω_6, then W_1 acting on the orbit $W_1.\omega$ is isomorphic to $W(E_6)$ acting on the weights of $L(E_6, \omega_5)$.*
(iii) *There are 27 weights that are skew to ω_6. There is a unique highest weight among these 27, which is $\omega_5 - \omega_6$.*

Proof Part (i) follows from Theorem 4.3.10 (iv). The orbits of W_1 acting on the weights are described in Proposition 8.2.9. By Theorem 9.2.1, the orbit corresponding

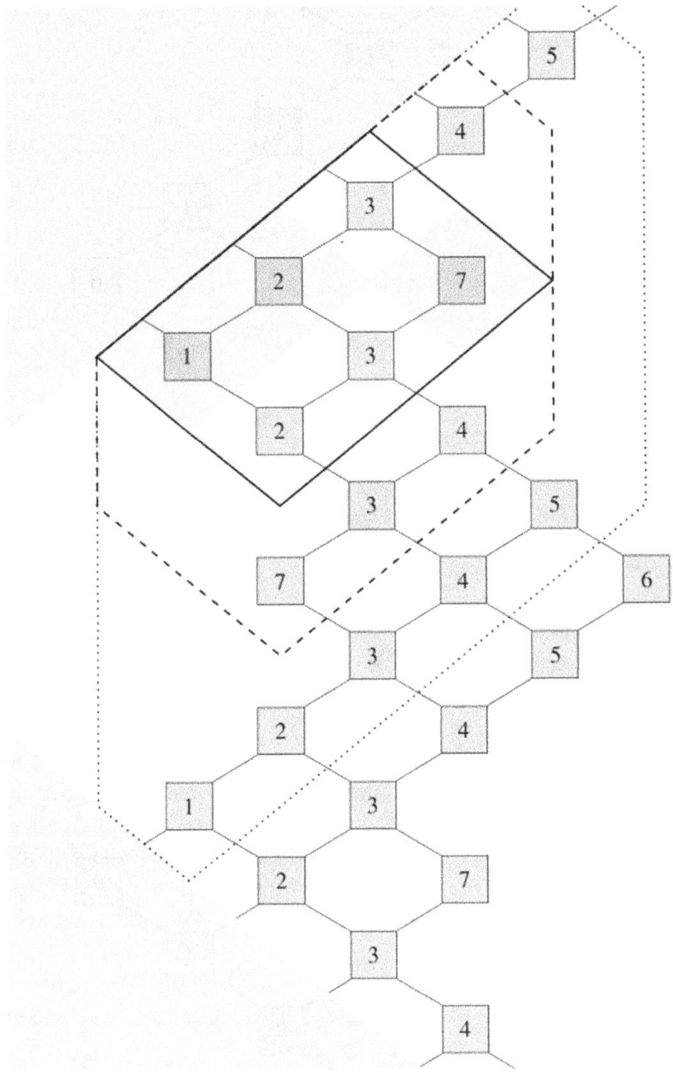

Figure 9.1 The subheaps corresponding to Lemmas 9.2.4 (dotted line), 9.2.5 (dashed line), 9.2.6 (solid line) and 9.2.7 (shaded)

to the weights skew to ω_6 corresponds to the weights at minimal nonzero distance from ω_6, which is $S_{6,3}$ in the notation of Proposition 8.2.9.

Recall that $S_{6,3}$ is the subheap given by

$$\{x \in F : x \not\geq F(6,3) \text{ and } x \not\leq F(6,2)\}$$

and shown enclosed by the dotted line in Figure 9.1.

After discarding unnecessary labels, $S_{6,3}$ is isomorphic in **Heap** to the principal subheap F' of the full heap $\mathrm{FH}(E_6^{(1)}(5))$. The assertions of (ii) now follow from properties of F'.

The first assertion of (iii) follows because F' has 27 ideals. The second assertion follows because the highest of the 27 weights corresponds to the ideal

$$I = \{x \in F : x \not\geq F(6, 3)\},$$

which has weight $\omega_5 - \omega_6$, because $I \in J_6^+(F)$, $I \in J_5^-(F)$, and $I \in J_k^0(F)$ for $k \neq 5, 6$. □

Lemma 9.2.5 *Maintain the notation of Lemma 9.2.4.*

(i) *Every ordered pair of skew weights (λ_1, λ_2) is W-conjugate to the pair*

$$O_2 = (\omega_6, \omega_5 - \omega_6).$$

(ii) *The point stabilizer W_2 in W of O_2 is the parabolic subgroup $W_{\{s_1,s_2,s_3,s_4,s_7\}}$.*
(iii) *If ω is a weight that is skew to both members of O_2, then W_2 acting on the orbit $W_2.\omega$ is isomorphic to $W(D_5)$ acting on the weights of $L(D_5, \omega_4)$.*
(iv) *There are 16 weights that are skew to both elements of O_2. There is a unique highest weight among these 16, which is $\omega_4 - \omega_5$.*

Proof To prove (i), we may assume that $\lambda_1 = \omega_6$, because W acts transitively on the weights of a minuscule representation. By applying a suitable element of W_1, we can then arrange by Lemma 9.2.4 for $\lambda_2 = \omega_5 - \omega_6$, which completes the proof.

The stabilizer W_2 is the stabilizer of $W(D_5)$ acting as in Lemma 9.2.4 (iii), and part (ii) follows by applying Theorem 4.3.10 (iv).

If F is the principal subheap for $L(E_7, \omega_6)$ with vertex chains numbered as in Lemma 9.2.4, let F'' be the subheap whose highest element is $F(4, 5)$ (the highest occurrence of 4), and whose lowest element is $F(7, 2)$ (the second highest occurrence of 7 in F); the subheap F'' is shown enclosed by the dashed line in Figure 9.1.

Lemma 9.2.4 reduces the problem in (iii) to that of finding the orbit of weights in the spin representation of Lemma 9.2.4 (iii) that are skew to the highest weight of $L(E_6, \omega_5)$. By applying the argument of Lemma 9.2.4 (ii) to the subheap F'', the proof of (iii) follows.

The first assertion of (iv) follows because F'' has 16 ideals. The highest weight of F'' corresponds to the ideal $\langle F(4, 5)\rangle$ of F, which has weight $\omega_4 - \omega_5$, and the second assertion follows. □

Lemma 9.2.6 *Maintain the notation of Lemma 9.2.4.*

(i) *Every ordered triple of mutually skew weights $(\lambda_1, \lambda_2, \lambda_3)$ is W-conjugate to the triple*

$$O_3 = (\omega_6, \omega_5 - \omega_6, \omega_4 - \omega_5).$$

(ii) *The point stabilizer W_3 in W of O_3 is the parabolic subgroup $W_{\{s_1,s_2,s_3,s_7\}}$.*
(iii) *If ω is a weight that is skew to all three members of O_3, then W_3 acting on the orbit $W_3.\omega$ is isomorphic to S_4 in its natural action on pairs of objects, corresponding to the minuscule representation $L(A_4, \omega_3)$.*
(iv) *There are ten weights that are skew to all three elements of O_3. There is a unique highest weight among these ten, which is $\omega_3 - \omega_4$.*

Proof This is proved in a similar way to Lemma 9.2.5, but using Lemma 9.2.5 itself instead of Lemma 9.2.4.

In this case, the subheap with ten ideals (shown enclosed by the solid line in Figure 9.1) has maximal element $F(3, 6)$ and minimal element $F(2, 3)$. □

Lemma 9.2.7 *Maintain the notation of Lemma 9.2.4.*

(i) *Every ordered quadruple of mutually skew weights* $(\lambda_1, \lambda_2, \lambda_3, \lambda_4)$ *is* W-*conjugate to the quadruple*

$$O_4 = (\omega_6, \omega_5 - \omega_6, \omega_4 - \omega_5, \omega_3 - \omega_4).$$

(ii) *The point stabilizer* W_4 *in* W *of* O_4 *is the parabolic subgroup* $W_{\{s_1,s_2,s_7\}}$.

(iii) *There exist weights that are skew to all four members of* O_4. *If* ω *is such a weight, then* W_4 *acting on the orbit* $W_4.\omega$ *is isomorphic to* $S_3 \times S_2$ *in its natural action on pairs* (a, b), *where* $a \in \{1, 2, 3\}$ *and* $b \in \{4, 5\}$.

(iv) *There are six weights that are skew to all four elements of* O_4.

Proof This is proved in a similar way to Lemma 9.2.6, but using Lemma 9.2.6 itself instead of Lemma 9.2.5.

In this case, the subheap with six ideals (shown in the shaded area of Figure 9.1) has three elements α, β, γ, with $\varepsilon(\alpha) = 1$, $\varepsilon(\beta) = 2$, $\varepsilon(\gamma) = 7$, and $\beta > \alpha$. □

Proposition 9.2.8 *Consider the minuscule representation* $L(E_7, \omega_6)$, *and let* $W = W(E_7)$.

(i) *There is one* W-*orbit of* 7-*tuples of skew weights, and there are no* k-*tuples of skew weights with* $k > 7$. *The set stabilizer of such a* 7-*tuple is the full symmetric group* S_7.

(ii) *There are two* W-*orbits of* 6-*tuples of skew weights; the elements of one orbit extend to skew* 7-*tuples and have trivial stabilizer, and the elements of the other orbit do not extend and have stabilizer of order* 2.

(iii) *There is one* W-*orbit of* 5-*tuples of skew weights, and the stabilizer of one is of order* 2.

(iv) *Each* 5-*tuple extends uniquely to a nonextendable skew* 6-*tuple.*

(v) *Each* 5-*tuple extends in two ways to an extendable skew* 6-*tuple.*

(vi) *Each* 5-*tuple extends uniquely to a skew* 7-*tuple.*

Proof The six ideals of the subheap $\{\alpha, \beta, \gamma\}$ in the proof of Lemma 9.2.7 are

$$\emptyset, \quad \{\alpha\}, \quad \{\alpha, \beta\}, \quad \{\gamma\}, \quad \{\alpha, \gamma\}, \quad \{\alpha, \beta, \gamma\}.$$

These are labelled by the pairs

$$(1, 4), \quad (2, 4), \quad (3, 4), \quad (1, 5), \quad (2, 5), \quad (3, 5)$$

respectively. With this labelling, two weights are skew if their corresponding ideals have labels differing in one coordinate.

By Lemma 9.2.7, $O_4 \cup \{(3, 4)\}$ is a 5-tuple of skew weights, and any such 5-tuple is conjugate to it. The stabilizer of this 5-tuple is generated by the transposition of $S_3 \times S_2$ that exchanges 1 and 2. This completes the proof of (iii).

Suppose we extend the above 5-tuple to a skew 6-tuple by adding a weight differing from $(3, 4)$ in the second coordinate; the only way to achieve this is to add $(3, 5)$. It is not possible to add any of the six elements listed to this collection to form a skew 7-tuple, which proves (iv). The point stabilizer of the 6-tuple is the transposition of $S_3 \times S_2$ exchanging 1 and 2.

Alternatively, we may extend the 5-tuple by adding an element differing in the first coordinate, either $(1, 4)$ or $(2, 4)$. Without loss of generality, suppose that we add $(1, 4)$. The stabilizer of this 6-tuple is trivial, and the 6-tuple extends uniquely to a skew 7-tuple by adding $(2, 4)$. This proves the assertions of (ii), (v) and (vi), so it remains to prove (i).

Suppose that $(\lambda_1, \dots, \lambda_7)$ and $(\mu_1, \dots, \mu_7)$ are two skew 7-tuples. We can find $w \in W$ to move the extendable 6-tuple $(\lambda_1, \dots, \lambda_6)$ to $(\mu_1, \dots, \mu_6)$ by (ii), and by (vi) (applied to any 5-subset of the 6-tuple) w must move the first 7-tuple to the second, in some order. Part (i) now follows by taking the second 7-tuple to be a permutation of the first 7-tuple. □

Lemma 9.2.9 *Maintain the notation of Lemma 9.2.4.*

(i) *If ω is a weight that is incident to ω_6 (in the sense of Definition 9.1.2), then W_1 acting on the orbit $W_1.\omega$ is isomorphic to $W(E_6)$ acting on the weights of $L(E_6, \omega_1)$.*
(ii) *There are 27 weights that are incident to ω_6. There is a unique highest weight among these 27, which is $\omega_1 - \omega_6$.*

Proof The proof follows the same lines as the proof of Lemma 9.2.4, using the subheap $S_{6,2}$ instead of $S_{6,3}$. □

Lemma 9.2.10 *Maintain the notation of Lemma 9.2.4.*

(i) *Every ordered pair of incident weights (λ_1, λ_2) is W-conjugate to the pair*

$$O'_2 = (\omega_6, \omega_1 - \omega_6).$$

(ii) *The point stabilizer W'_2 in W of O'_2 is the parabolic subgroup $W_{\{s_2,s_3,s_4,s_5,s_7\}}$.*
(iii) *If ω is a weight that is incident to both members of O'_2, then W'_2 acting on the orbit $W'_2.\omega$ is isomorphic to $W(D_5)$ acting on the weights of $L(D_5, \omega_5)$.*
(iv) *There are ten weights that are incident to both elements of O'_2. There is a unique highest weight among these ten, which is $\omega_5 - \omega_1 - \omega_6$.*

Proof The proof follows the same lines as that of Lemma 9.2.5, but with Lemma 9.2.9 playing the role of 9.2.4. The role of the subheap F'' is played by the subheap with maximal element $F(5, 2)$ and minimal element $F(5, 1)$. □

Lemma 9.2.11 *Maintain the notation of Lemma 9.2.4.*

(i) *Every ordered triple of incident weights $(\lambda_1, \lambda_2, \lambda_3)$ is W-conjugate to the pair*

$$O'_3 = (\omega_6, \omega_1 - \omega_6, \omega_5 - \omega_1 - \omega_6).$$

(ii) *The point stabilizer* W_3' *in* W *of* O_3' *is the parabolic subgroup* $W_{\{s_2, s_3, s_4, s_7\}}$.
(iii) *The only weight incident to all three weights of* O_3' *is* $\omega_6 - \omega_5$.

Proof This is proved in a similar way to Lemma 9.2.10, but using Lemma 9.2.10 itself instead of Lemma 9.2.9.

The relevant subheap to consider in this case has maximal element $F(4, 2)$ and minimal element $F(4, 1)$. The weight mentioned in part (iii) corresponds to the ideal $\langle F(6, 1)\rangle$. □

Proposition 9.2.12 *Consider the minuscule representation* $L(E_7, \omega_6)$*, and let* $W = W(E_7)$.

(i) *There is one* W*-orbit of* 4*-tuples of incident weights, and there are no* k*-tuples of incident weights with* $k > 4$.
(ii) *Denote the set (respectively, point) stabilizer of a* 4*-tuple of incident weights by* H_1 *(respectively,* H_2*). Then* $H_2 \cong W(D_4)$ *is conjugate to the subgroup* W_3' *of Lemma 9.2.11 (ii), and* $H_1/H_2 \cong S_4$.
(iii) *There is one* W*-orbit of triples of incident weights. Each triple of incident weights extends uniquely to a* 4*-tuple of incident weights.*
(iv) *There is one* W*-orbit of pairs of incident weights. Each pair of incident weights can be extended in ten different ways to a triple of incident weights, and in five different ways to a* 4*-tuple of incident weights.*

Proof Part (i) follows from Lemma 9.2.11, as does the first assertion of (ii). The second assertion of (ii) is proved using the argument in the last paragraph of Proposition 9.2.8, together with Lemma 9.2.11 (i).

The first assertion of (iii) follows from Lemma 9.2.10 (i), and the second assertion follows from Lemma 9.2.11.

The first assertion of (iv) follows from Lemma 9.2.10 (i). Since triples of incident weights exist, this implies that every pair of incident weights extends to at least one triple. Lemma 9.2.10 (iv) shows that each pair extends to ten different triples, and Lemma 9.2.11 shows that each triple extends to a unique 4-tuple. For a given incident pair, there are two possible intermediate triples that extend the pair to a given quadruple containing it, and it follows from this that each incident pair is contained in $10/2 = 5$ incident 4-tuples. This completes the proof of (iv). □

Exercise 9.2.13 Let $n > 4$. Recall from Example 8.5.13 that the n-dimensional half cube $P(D_n, \omega_n)$ has one $W(D_n)$-orbit of simplex shaped faces, except in dimension 3, where there are two orbits. Furthermore, these two orbits may be distinguished by the fact that tetrahedra in one orbit extend to 4-simplex shaped faces, and tetrahedra in the other orbit do not. Use the proof techniques of Section 9.2 to prove the following.

(i) If $k \neq 1, 3$ then the pointwise stabilizer of a k-simplex shaped face is isomorphic to the symmetric group S_{n-k-1}, and is conjugate to the parabolic subgroup $\langle s_1, s_2, \ldots, s_{n-k-2}\rangle$.
(ii) The pointwise stabilizer of a 1-simplex shaped face is isomorphic to $S_{n-2} \times S_2$, and is conjugate to the parabolic subgroup generated by $S \backslash \{s_{n-2}, s_n\}$.
(iii) The formulae of (i) apply when $k = 3$ in the case of an extendable tetrahedron.

(iv) The pointwise stabilizer of a nonextendable tetrahedron (i.e., 3-simplex shaped face) is isomorphic to the symmetric group S_{n-3}, and is conjugate to the parabolic subgroup

$$\langle s_1, s_2, \ldots, s_{n-4} \rangle.$$

Exercise 9.2.14 Maintain the notation of Exercise 9.2.13, and use the techniques of that exercise to prove the following.

(i) If $k \neq 1, 3$ then the setwise stabilizer of a k-simplex shaped face is isomorphic to $S_{n-k-1} \times S_{k+1}$, and is conjugate to the parabolic subgroup generated by $S \backslash \{s_{n-k-1}, s_{n-1}\}$.
(ii) The setwise stabilizer of a 1-simplex shaped face is isomorphic to $S_{n-2} \times S_2 \times S_2$, and is conjugate to the maximal parabolic subgroup generated by $S \backslash \{s_{n-2}\}$.
(iii) The formulae of (i) apply in the case $k = 3$ in the case of an extendable tetrahedron.
(iv) The setwise stabilizer of a nonextendable tetrahedron is isomorphic to $S_{n-3} \times S_4$, and is conjugate to the maximal parabolic subgroup generated by $S \backslash \{s_{n-3}\}$.

Exercise 9.2.15 Use the results of Exercise 9.2.13 to prove the following, still under the assumption that $n > 4$.

(i) The point stabilizer of a nonextendable 3-simplex face of $P(D_n, \omega_n)$ is identical to the point stabilizer of any of its 2-simplex shaped faces, and any 2-simplex shaped face is contained in a unique nonextendable 3-simplex shaped face.
(ii) Any 2-simplex shaped face is contained in precisely $n - 3$ extendable 3-simplex shaped faces.

9.3 Bitangents

Let V be an eight-dimensional vector space with basis $\{e_0, e_1, \ldots, e_7\}$ over the field $\mathbb{F}_2$ of two elements. Let V' be the kernel of the linear map from V to $\mathbb{F}_2$ that sends $\sum_{i=0}^{7} \lambda_i e_i$ to $\sum_{i=0}^{7} \lambda_i$; note that the vectors in V' are those whose support has even cardinality. Let V'' be the one-dimensional subspace of V (or V') spanned by the all-ones vector $\sum_{i=0}^{7} e_i$, and define $\overline{V}$ (respectively, $\overline{V'}$) to be the vector space V/V'' (respectively, V'/V'') of dimension 7 (respectively, 6).

Let Q be the set of 28 vectors in V each of whose support has size 2, and let $\overline{Q}$ be the image of Q in $\overline{V}$. (Note that $\overline{Q}$ also has size 28.) In algebraic geometry, the elements of $\overline{Q}$ are in natural correspondence with the *28 bitangents to a plane quartic curve*.

It can be shown that the six-dimensional vector space $\overline{V'}$ supports a natural action of the symplectic group Sp(6, 2), and that this group acts doubly transitively on the set $\overline{Q'}$. From this point of view, any pair of elements of $\overline{Q'}$ is essentially interchangeable with any other. However, the triples of $\overline{Q'}$ turn out to be of two different types: "azygetic" (or "asyzygetic"), meaning that the sum of the three elements in the triple is equal to a fourth element of $\overline{Q'}$; or "syzygetic", if this is not the case.

Let R be the set of 63 nonzero elements of $\overline{V'}$. Given any element $r \in R$, it turns out that there are six possible pairs $\{q_1, q_2\}$ of elements in $\overline{Q'}$ for which $q_1 + q_2 = r$. The (disjoint) union of these six pairs is known as a Steiner complex.

It may not immediately be clear why one might want to study these structures outside the context of algebraic geometry. However, it turns out that the underlying combinatorial framework of the 28 bitangents emerges naturally from a study of the full heap $\mathrm{FH}(E_7^{(1)}(6))$, as we shall show.

By Theorem 9.2.3, the 56 weights of $L(E_7, \omega_6)$ are in bijection with the 56 lines on a del Pezzo surface Σ of degree 2. It follows from Proposition 4.3.17 (ii) that the longest element w_0 of the Weyl group $W(E_7)$ acts as -1 on the weights. By Dolgachev [20, example 8.2.3], the action of w_0 on Σ is the so-called *Geiser transformation*, and the 28 conjugate pairs of lines under this transformation are in natural correspondence with the 28 bitangents to the plane quartic curve. For our purposes, we will think of a bitangent as a pair of opposite weights in $L(E_7, \omega_6)$.

As we shall see, there is a lot of interesting combinatorial algebraic geometry associated with the 28 bitangents to a plane quartic curve, and it can be accessed via the combinatorics of the associated minuscule representation.

An interesting feature is that the action of $W(E_7)$ on the 28 bitangents is doubly transitive. This result is easier to understand in a more general context, as we now show.

Lemma 9.3.1 *Suppose that V_F is a minuscule representation of a simple Lie algebra $\mathfrak{g}$, and that the associated Weyl group W acts as a rank 4 permutation group on the weights. Then one of the following possibilities holds:*

(i) *$\mathfrak{g}$ has type A_l for $l \geq 5$ and $V_F \cong L(A_l, \omega_3)$;*
(ii) *$\mathfrak{g}$ has type A_l for $l \geq 5$ and $V_F \cong L(A_l, \omega_{l-2})$;*
(iii) *$\mathfrak{g}$ has type B_3 and V_F is the spin representation;*
(iv) *$\mathfrak{g}$ has type D_6 or D_7 and V_F is one of the two spin representations in each case; or*
(v) *$\mathfrak{g}$ has type E_7 and $V_F \cong L(E_7, \omega_6)$.*

Proof This follows from Theorem 8.2.22 (i) by solving for $N = 4$.

Suppose we are in type A_l and that $\min(k+1, l+2-k) = 4$. If $k + 1 \leq l + 2 - k$, then we must have $k = 3$ and $l \geq 5$, which is part (i). The other possibility is that $l + 2 - k \leq k + 1$, which implies that $k \geq 3$ and $l = k + 2$, which is part (ii).

Parts (iii), (iv) and (v) are immediate from the formulae given. □

Lemma 9.3.2 *Suppose that V_F is a minuscule representation of a simple Lie algebra $\mathfrak{g}$, and that the associated Weyl group W acts as a rank 4 permutation group on the weights. Suppose also that the weights of V_F are closed under negation. Then one of the following possibilities holds:*

(i) *$\mathfrak{g}$ has type A_5 and $V_F \cong L(A_5, \omega_3)$;*
(ii) *$\mathfrak{g}$ has type B_3 and V_F is the spin representation;*
(iii) *$\mathfrak{g}$ has type D_6 and V_F is one of the two spin representations in each case; or*
(iv) *$\mathfrak{g}$ has type E_7 and $V_F \cong L(E_7, \omega_6)$.*

Proof By assumption, the negative of the highest weight must be one of the weights in each case. The lowest weights corresponding to the various highest weights are listed in Proposition 6.6.8.

In type A, the lowest weight corresponding to highest weight ω_3 (respectively, ω_{l-2}) is $-\omega_{l-2}$ (respectively, $-\omega_3$). For the hypotheses to be satisfied, we need $l - 2 = 3$; in other words, l is 5 and the highest weight is ω_3.

A similar check shows that the situations of Lemma 9.3.1 (iii), (iv) and (v) all satisfy the required hypotheses, except when $\mathfrak{g}$ has type D_7. In the latter case, a highest weight of ω_6 (respectively, ω_7) corresponds to a lowest weight of $-\omega_7$ (respectively, $-\omega_6$). □

Lemma 9.3.3 *Let V_F be a minuscule representation with associated Weyl group W. Suppose that W acts as a rank 4 permutation group on the weights of V_F, and that the weights are closed under negation. Let D be the minimal nonzero Euclidean distance between weights. Let $\{\lambda, -\lambda\}$ be a pair of opposite weights, and let μ be a weight not equal to either of the pair. Then either μ is distance D from λ and $\sqrt{2}D$ from $-\lambda$, or vice versa.*

Proof Since W acts transitively on the weights by orthogonal transformations, all the weights are equidistant from the origin. This means that if $\{\lambda, -\lambda\}$ is an arbitrary pair of opposite weights and μ is a weight not equal to either of the pair, then the points $\{\lambda, -\lambda, \mu\}$ form a right angled triangle with the right angle at the vertex μ.

If D is the minimal nonzero distance between weights, the other possible nonzero distances are $\sqrt{2}D$ and $\sqrt{3}D$; this follows by an argument like that of Example 8.4.9 for the simply laced types A_5, D_6 and E_7, and by an ad hoc check for type B_3. In particular, we have $|\lambda - (-\lambda)| = \sqrt{3}D$. Pythagoras' theorem then shows that

$$|\lambda - \mu|^2 + |\mu + \lambda|^2 = \sqrt{3}D,$$

which can only happen if μ is distance D from one of the pair $\{\lambda, -\lambda\}$, and distance $\sqrt{2}D$ from the other one. □

Lemma 9.3.4 *Maintain the hypotheses of Lemma 9.3.2. If $w \in W$ fixes each pair $\{\lambda, -\lambda\}$ of opposite weights, then w acts as the same scalar ± 1 on all the weights. Moreover, if w is not the identity, then W does not have type A_5, $w = w_0$ and w acts as the scalar -1.*

Proof Theorem 8.2.22 shows that there are four possible Euclidean distances between weights, corresponding to the four orbits. It follows that there are four possible values for the angle between two weights, and a calculation in shows that these angles are

$$0, \ \cos^{-1}\left(\frac{1}{3}\right), \ -\cos^{-1}\left(\frac{1}{3}\right), \ \pi;$$

let us denote these angles by $0, \theta, -\theta, \pi$ for short.

Choose a weight λ, and define $w(\lambda) = c\lambda$ for some $c \in \{\pm 1\}$. Let $\mu \notin \{\pm\lambda\}$ be any other weight. By Lemma 9.3.3, either λ and μ are at angle θ, or λ and $-\mu$ are at angle θ (but not both). Without loss of generality, suppose that λ and μ are at angle θ.

Since w acts by orthogonal transformations, the angle between λ and μ is the same as the angle between $w.\lambda$ and $w.\mu$. Since $w.\mu = \pm\mu$, it follows by the above paragraph that w acts by the same scalar c on all of the weights $\{\pm\lambda, \pm\mu\}$. This shows that w acts as the same scalar ± 1 on all the weights, proving the first assertion.

The second assertion follows from Lemma 9.3.2 and Proposition 4.3.17. □

Proposition 9.3.5 *Let V_F be a minuscule representation with associated Weyl group W. Suppose that W acts as a rank 4 permutation group on the weights of V_F, and that the weights are closed under negation. Then W acts doubly transitively on the set of pairs of opposite weights $\{\lambda, -\lambda\}$.*

Proof Let $\{\lambda_1, -\lambda_1\}$, $\{\lambda_2, -\lambda_2\}$, $\{\mu_1, -\mu_1\}$ and $\{\mu_2, -\mu_2\}$ be pairs of opposite weights, chosen so that $\lambda_1 \neq \pm\lambda_2$ and $\mu_1 \neq \pm\mu_2$. We need to prove the existence of some $w \in W$ such that w sends the pair $\{\lambda_1, -\lambda_1\}$ to $\{\mu_1, -\mu_1\}$ and $\{\lambda_2, -\lambda_2\}$ to $\{\mu_2, -\mu_2\}$. By Lemma 9.3.3, we may assume without loss of generality that μ_1 is distance D from λ_1 and distance $\sqrt{2}D$ from $-\lambda_1$, and also that μ_2 is distance D from λ_2 and distance $\sqrt{2}D$ from $-\lambda_2$. Theorem 8.2.22 (ii) now proves the existence of the required element w. □

Corollary 9.3.6 *The Weyl group $W(E_7)$ acts doubly transitively on the 28 bitangents.*

Proof This follows by combining Lemma 9.3.2 (iv) with Proposition 9.3.5. □

Corollary 9.3.6 shows that any pair of bitangents is interchangeable with any other. However there are different types of triples of bitangents, as we will now see.

Definition 9.3.7 Let D be the minimal nonzero distance between two weights in $L(E_7, \omega_6)$. Let $l = \{\lambda_1, \lambda_2\}$, $m = \{\mu_1, \mu_2\}$ and $n = \{\nu_1, \nu_2\}$ be three distinct bitangents (i.e., pairs of opposite weights). We say that the triple $\{l, m, n\}$ is *syzygetic* if there exists a triple $\{\lambda_i, \mu_j, \nu_k\}$ at mutual distance D, where $i, j, k \in \{1, 2\}$. We say that the triple $\{l, m, n\}$ is *azygetic* if there exists a triple $\{\lambda_i, \mu_j, \nu_k\}$ at mutual distance $\sqrt{2}D$, where $i, j, k \in \{1, 2\}$.

Lemma 9.3.8 *A triple of bitangents cannot be both syzygetic and azygetic.*

Proof Suppose that $\{l, m, n\}$ is a syzygetic triple of bitangents, and write $l = \{\lambda_1, \lambda_2\}$, $m = \{\mu_1, \mu_2\}$ and $n = \{\nu_1, \nu_2\}$. Without loss of generality, assume that $\{\lambda_1, \mu_1, \nu_1\}$ is a triple of weights at mutual distance D.

Suppose that $\{\lambda_1, \mu_j, \nu_k\}$ is a triple of weights at mutual distance $\sqrt{2}D$. We must have $j = 2$, which in turn forces $k = 1$, but λ_1 and ν_1 are at mutual distance D by Lemma 9.3.3.

A similar argument shows that λ_2 cannot be involved in a triple of weights at mutual distance $\sqrt{2}D$, and this completes the proof. □

Lemma 9.3.9 *A triple of bitangents is either syzygetic or azygetic.*

Proof Suppose that $\{l, m, n\}$ is a triple of bitangents, but not a syzygetic one, and write $l = \{\lambda_1, \lambda_2\}$, $m = \{\mu_1, \mu_2\}$ and $n = \{\nu_1, \nu_2\}$. By re-indexing if necessary, we may assume that λ_1 is at distance D from μ_1 by Lemma 9.3.3.

Since neither $\{\lambda_1, \mu_1, \nu_1\}$ nor $\{\lambda_1, \mu_1, \nu_2\}$ is a triple at mutual distance D by hypothesis, Lemma 9.3.3 shows that ν_1 is at distance D from one of the pair $\{\lambda_1, \mu_1\}$, and at distance $\sqrt{2}D$ from the other. By re-indexing the ν_i if necessary, we may assume that ν_1 is at distance D from λ_1. More applications of Lemma 9.3.3 now show that $\{\lambda_1, \mu_2, \nu_2\}$ is a triple of weights at mutual distance $\sqrt{2}D$. This shows that $\{l, m, n\}$ is azygetic, and completes the proof. □

The previous two lemmas show that, from the point of view of the group action, there are precisely two types of triples of bitangents: syzygetic and azygetic.

Corollary 9.3.10 *Any pair of bitangents extends to an azygetic triple in exactly ten ways, and extends to a syzygetic triple in* 16 *ways.*

Proof By Corollary 9.3.6, it is enough to prove the statement for a fixed pair of bitangents. Lemma 9.3.9 and the fact that the number of bitangents is $2 + 10 + 16$ show that it is enough to prove the statement about azygetic triples. Finally, we observe that the bitangents forming an azygetic triple with $\{[0, 1], [1, 7]\}$ are those of the form $[0, i]$ or $[i, 7]$ with $2 \leq i \leq 6$. □

Definition 9.3.11 Let $k \geq 3$. We call a k-tuple of bitangents *syzygetic* (respectively, *azygetic*) if any of its 3-element subsets is syzygetic (respectively, azygetic).

Proposition 9.3.12 *Maintain the above notation. Let* $k \geq 3$, *and let* $L = \{l_1, l_2, \ldots, l_k\}$ *be a* k*-tuple of bitangents, where* $l_i = \{\lambda_{i,1}, \lambda_{i,2}\}$.

(i) *If* L *is syzygetic, then it is possible to re-index the weights so that the set* $\{\lambda_{1,a}, \lambda_{2,a}, \ldots, \lambda_{k,a}\}$ *are at mutual distance* D, *simultaneously for* $a = 1$ *and* $a = 2$.
(ii) *If* L *is azygetic, then it is possible to re-index the weights so that the set* $\{\lambda_{1,a}, \lambda_{2,a}, \ldots, \lambda_{k,a}\}$ *are at mutual distance* $\sqrt{2}D$, *simultaneously for* $a = 1$ *and* $a = 2$.

Proof The proof is by induction on k, and the base case, $k = 3$, follows from the definitions, so suppose that $k \geq 4$.

We deal with the proof of (i), because the proof of (ii) is similar. We therefore suppose that $\{l_1, l_2, \ldots, l_k\}$ is a syzygetic k-tuple, where $l_i = \{\lambda_{i,1}, \lambda_{i,2}\}$ for all i. By induction, we may assume that the $(k-1)$-tuple $\{\lambda_{1,a}, \lambda_{2,a}, \ldots \lambda_{k-1,a}\}$ is at mutual distance D, simultaneously for $a = 1$ and $a = 2$.

By re-indexing l_k if necessary, we may arrange for $\lambda_{k,1}$ to be at distance D from $\lambda_{1,1}$. Suppose for a contradiction that $\lambda_{k,1}$ is at distance $\sqrt{2}D$ from $\lambda_{2,1}$. Now $\{l_1, l_2, l_k\}$ is syzygetic, so there is a triple $\{\lambda_{1,p}, \lambda_{2,q}, \lambda_{k,r}\}$ of weights at mutual distance D. By replacing p, q, r by $3 - p, 3 - q, 3 - r$ respectively, we may assume that $p = 1$. The assumption on l_k then forces $r = 1$. We must have $q = 1$ for $\lambda_{2,q}$ to be at distance D from $\lambda_{1,p}$, but we must have $q = 2$ for $\lambda_{2,q}$ to be at distance D from $\lambda_{k,r}$, and this is the required contradiction. It follows that $\lambda_{k,1}$ is at distance D from each of the $\lambda_{i,1}$ for $i < k$. By Lemma 9.3.3, the same is then true for the elements $\lambda_{i,2}$ for $i \leq k$, and this completes the proof. □

Proposition 9.3.13 *Let* $W = W(E_7)$, *and let* D *be the minimal nonzero distance between two weights of* $L(E_7, \omega_6)$.

(i) *If* $k \geq 3$, *there is a* 2:1 *correspondence from the set of* k*-tuples of weights of* $L(E_7, \omega_6)$ *at mutual distance* D *to the set of* k*-tuples of syzygetic bitangents.*

(ii) *If* $k \geq 3$, *there is a* 2:1 *correspondence from the set of* k*-tuples of weights of* $L(E_7, \omega_6)$ *at mutual distance* $\sqrt{2}D$ *to the set of* k*-tuples of azygetic bitangents.*
(iii) *The maximum cardinality of a set of syzygetic bitangents is* 7, *and all such subsets are conjugate under the action of* W.
(iv) *The maximum cardinality of a set of azygetic bitangents is* 4, *and all such subsets are conjugate under the action of* W.

Proof Consider a set of k syzygetic bitangents. By Proposition 9.3.12, these must be of the form

$$\{\lambda_1, \mu_1\}, \{\lambda_2, \mu_2\}, \ldots \{\lambda_k, \mu_k\},$$

where $\{\lambda_1, \ldots, \lambda_k\}$ is a set of weights at mutual distance D. The same is true of the weights μ_i, since $\mu_i = -\lambda_i$ for all i. By Lemma 9.3.3, these two are the only k-tuples of weights with this property among the $2k$ weights listed. This proves (i), and (ii) follows by a similar argument.

Part (iii) follows from (i) and Proposition 9.2.8 (i). Similarly, (iv) follows from (ii) and Proposition 9.2.12 (i). □

Definition 9.3.14 The 7-tuples of bitangents appearing in Proposition 9.3.13 (iii) are known as *Aronhold sets*.

Exercise 9.3.15 Show that the set stabilizer of an Aronhold set in $W(E_7)$ is isomorphic to $S_7 \times \mathbb{Z}_2$.

Exercise 9.3.16 Show that the point stabilizer of a set of four azygetic bitangents in $W(E_7)$ is isomorphic to $W(D_4) \times \mathbb{Z}_2$.

Exercise 9.3.17 Let λ_1 and λ_2 be two distinct weights of $L(E_7, \omega_6)$.

(i) Suppose that λ_1 and λ_2 are two skew weights of $L(E_7, \omega_6)$. Show that the pair $\{\lambda_1, \lambda_2\}$ is the unique pair of weights whose sum is $\lambda_1 + \lambda_2$.
(ii) Suppose that λ_1 and λ_2 are two incident weights of $L(E_7, \omega_6)$. Show that there are precisely five other pairs of weights $\{\mu_1, \mu_2\}$ for which $\lambda_1 + \lambda_2 = \mu_1 + \mu_2$. Use Lemma 9.3.3 to show that, in each case, the set

$$\{\lambda_1, \lambda_2, -\mu_1, -\mu_2\}$$

is a mutually incident quadruple of weights. (Compare with Proposition 9.2.12 (iv).)

9.4 Hesse–Cayley notation

The following useful notation for the 28 bitangents is often called the *Hesse–Cayley notation*.

Definition 9.4.1 Let $0 \leq i < j \leq 7$. We define the bitangent $[i, j] = [j, i]$ to be the one associated by the above construction with the pair of opposite weights $\pm v_{i,j}$, or $(i, j)^{\pm}$, in $L(E_7, \omega_6)$.

Figure 9.2 The five types of triples of bitangents, (A), (B), (C), (D) and (E).

Definition 9.4.2 Let T be a triple of bitangents. We say that T is:

(i) of type (A) if $T = \{[i, j], [i, k], [i, l]\}$ and i, j, k, l are distinct;
(ii) of type (B) if $T = \{[i, j], [j, k], [k, i]\}$ and i, j, k are distinct;
(iii) of type (C) if $T = \{[i, j], [i, k], [l, m]\}$ and i, j, k, l, m are distinct;
(iv) of type (D) if $T = \{[i, j], [k, l], [m, n]\}$ and i, j, k, l, m, n are distinct;
(v) of type (E) if $T = \{[i, j], [j, k], [k, l]\}$ and i, j, k, l are distinct.

The five types of triples of bitangents are shown in Figure 9.2. A specific example of a triple of bitangents corresponds to a labelling of the vertices in each of the five pictures by letters i, j, k, and so on. Recall from Section 9.1 that the weights of $L(E_7, \omega_6)$ may be identified with the edges of two copies of K_8; say, a copy with solid edges and a copy with dashed edges. The solid and dashed edges in the figures then show how to select a weight from each pair of weights in a bitangent so as to form a set at mutual distance D in types (A), (B) and (C), and a set at mutual distance $\sqrt{2}D$ in types (D) and (E), as in Proposition 9.3.12. For example, the triple of weights $\{(0, 1)^+, (0, 2)^+, (3, 4)^-\}$ is at mutual distance $\sqrt{2}D$ and corresponds to a triple of bitangents of type (C), where in this case $+$ corresponds to solid and $-$ to dashed.

Lemma 9.4.3 *Each triple of bitangents falls into precisely one of the types described in Definition 9.4.2.*

Proof From the definitions, type (A) (respectively, (B), (C), (D), (E)) involves four (respectively, three, five, six, four) distinct indices. Type (A) has an index that occurs in each of the three bitangents, and type (E) does not. It follows that the five types are mutually exclusive.

To prove the types are exhaustive, first note that if the bitangents involve six distinct indices, we are in type (D). If not, then two of the three bitangents must be of the form $\{[i, j], [i, k]\}$. There are four further cases, according as the third bitangent (a) shares none of the three indices shown (type (C)); (b) shares only i (type (A)); (c) shares exactly one of j or k (type (E)) or (d) shares both of j and k (type (B)). This completes the proof. □

Lemma 9.4.4 *Maintain the notation of Definition 9.4.2. A triple of bitangents of type (A), (B) or (C) is syzygetic, whereas a triple of bitangents of type (D) or (E) is azygetic.*

Proof The proof proceeds by considering triples of weights in $L(E_7, \omega_6)$.

The triple $\{(1, 2)^+, (1, 3)^+, (1, 4)^+\}$ is a set of weights at mutual distance D corresponding to a triple of syzygetic bitangents of type (A), proving the assertion in this case. Type (B) is proved similarly, by considering the triple $\{(1, 2)^+, (1, 3)^+, (2, 3)^+\}$, and type (C) is proved similarly by considering the triple $\{(1, 2)^+, (1, 3)^+, (4, 5)^-\}$.

The triple $\{(1,2)^+, (3,4)^+, (5,6)^+\}$ is a set of weights at mutual distance $\sqrt{2}D$ corresponding to a triple of azygetic bitangents of type (D). Type (E) is proved similarly by considering the triple $\{(1,2)^+, (2,3)^-, (3,4)^+\}$. □

Proposition 9.4.5 *There are two orbits of $W(E_7)$ on unordered triples of bitangents: the syzygetic triples and the azygetic triples.*

Proof Since $W = W(E_7)$ acts on the weights of $L(E_7, \omega_6)$ by orthogonal transformations, it is clear that no syzygetic triple can be W-conjugate to an azygetic triple.

Recall from the proof of Proposition 9.1.1 that W has a subgroup isomorphic to S_8 that acts naturally on the eight indices underlying the Hesse–Cayley notation. It follows that the unordered triples of type (A) are all in the same W-orbit, and the same is true for types (B), (C), (D) and (E).

Let us call a triple of weights syzygetic (respectively, azygetic) if it corresponds to a syzygetic (respectively, azygetic) triple of bitangents when the negatives of the weights are added.

The bifid transformation s_7 corresponding to the partition $\{0,1,2,3\} \dot\cup \{4,5,6,7\}$ moves the triple $\{(0,4)^+, (0,5)^+, (0,1)^+\}$ of type (A) to the triple $\{(0,4)^+, (0,5)^+, (2,3)^-\}$ of type (C). The same bifid transformation moves the triple $\{(0,4)^+, (4,1)^+, (0,1)^+\}$ of type (B) to the triple $\{(0,4)^+, (4,1)^+, (2,3)^-\}$ of type (C). It now follows from Lemma 9.4.4 that all the syzygetic triples are W-conjugate.

The same bifid transformation as above moves the triple $\{(0,4)^+, (4,5)^-, (5,1)^+\}$ of type (E) to the triple $\{(0,4)^+, (6,7)^+, (5,1)^+\}$ of type (D). It now follows from Lemma 9.4.4 that all the azygetic triples are W-conjugate. □

Definition 9.4.6 We call a subgraph of the complete graph K_n

(i) a *fan* (or r-*fan*) if it consists of r edges all sharing a common vertex,
(ii) a *triangle* if it is of the form $\{[a,b],[b,c],[c,a]\}$ for a, b, c distinct, and
(iii) a *square* if it is of the form $\{[a,b],[b,c],[c,d],[d,a]\}$ for a, b, c, d distinct.

We call a subset of bitangents a *fan*, *triangle* or *square* if the corresponding subgraph of K_8 has the same property.

Lemma 9.4.7 *If G is the subgraph of K_8 corresponding to a syzygetic set of bitangents, then any connected component of G is either a fan or a triangle.*

Proof Let G_0 be a connected component of G. If G_0 has one or two edges, then G_0 is a fan, so assume that G_0 has at least three edges.

Let G_1 be one of the connected subgraphs of G_0 with exactly three edges. By Lemma 9.4.3, G_1 is of type (A) or (B). Suppose G_1 is of type (B). There is no way to add a new edge to a syzygetic triple of type (B) in such a way that the new edge shares a vertex with the triangle of type (B) without creating a subgraph of type (E), which would contradict Lemma 9.4.4. It follows that $G_1 = G_0$ and that G_0 is a triangle, as required.

We may therefore assume that G_1 is of type (A). In this case, the only way to add a new edge to G_1 that shares a vertex with the 3-fan G_1 without creating a subgraph

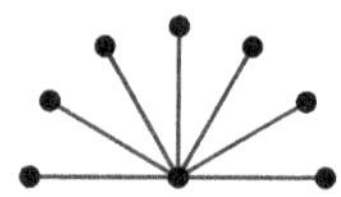

Figure 9.3 An Aronhold set as described in Proposition 9.4.8 (i)

Figure 9.4 An Aronhold set as described in Proposition 9.4.8 (ii).

of type (E) is to extend it to a 4-fan. This procedure may be iterated to show that G_0 itself is a fan, and this completes the proof. □

Proposition 9.4.8 *Any Aronhold set is either a 7-fan or the disjoint union of a 4-fan with a triangle. More precisely, any Aronhold set is either:*

(i) *of the form* $\{[i, j] : 0 \leq j \leq 7,\ j \neq i\}$ *for some fixed* i *with* $0 \leq i \leq 7$*; that is, the result of acting a permutation of* S_8 *on the indices of the set*

$$\{[0, 1],\ [0, 2],\ [0, 3],\ [0, 4],\ [0, 5],\ [0, 6],\ [0, 7]\},$$

or

(ii) *the result of acting a permutation of* S_8 *on the indices of the set*

$$\{[0, 1],\ [0, 2],\ [0, 3],\ [0, 4],\ [5, 6],\ [6, 7],\ [5, 7]\}.$$

Proof Each element of the Aronhold set determines an edge in the complete graph K_8, and we can classify the sets by the corresponding graphs with seven edges.

Let G be such a graph. By Lemma 9.4.7, every connected component of G is a fan or a triangle. However, by Lemma 9.4.4, G has no subgraph of type (D), which means that G has at most two connected components. By Lemma 9.4.7, these components are either fans or triangles. If G were a triangle or a disjoint union of two triangles, this would not give the required seven edges, so at least one component must be a fan. If there were two fans of sizes k and l, this would use a total of $k + l + 2 = 9$ vertices, which is impossible.

We conclude that either G is a fan, or G is the union of a triangle with a fan (necessarily a 4-fan), which completes the proof. □

The Aronhold sets in Proposition 9.4.8 are shown in Figures 9.3 and 9.4. A specific element of these sets corresponds to a labelling of the eight vertices in the diagram by the numbers $0, 1, \ldots, 7$.

Proposition 9.4.9 *Any maximal set of azygetic bitangents is either a square or a collection of four edges partitioning the vertices. More precisely, any maximal set of azygetic bitangents is either:*

Figure 9.5 The two types of maximal sets of azygetic bitangents, as in Proposition 9.4.9 (i) (left) and (ii) (right).

(i) *the result of acting a permutation of S_8 on the indices of the set*

$$\{[0, 1],\ [1, 2],\ [2, 3],\ [3, 4]\},$$

as shown on the left hand side of Figure 9.5, or

(ii) *the result of acting a permutation of S_8 on the indices of the set*

$$\{[0, 1],\ [2, 3],\ [4, 5],\ [6, 7]\},$$

as shown on the right hand side of Figure 9.5.

Proof The only way to extend an azygetic triple of type (E) to an azygetic 4-tuple without creating a subset of type (A), (B) or (C) is by creating a configuration of the type described in (i).

The only way to extend an azygetic triple of type (D) to an azygetic 4-tuple without creating a subset of type (C) is by creating a configuration of the type described in (ii). □

Corollary 9.4.10 *There are* 288 *Aronhold sets, and* 315 4*-tuples of azygetic bitangents.*

Proof In Proposition 9.4.8 there are eight Aronhold sets of type (i). To construct an Aronhold set in part (ii), we need to choose three vertices to form a triangle, and then one point to form the base of the 4-fan. This is a total of

$$\binom{8}{3} \times 5 = 280$$

choices, and this proves the first assertion.

In Proposition 9.4.9 (i), there are

$$\binom{8}{4} = 70$$

ways to select the four vertices of a square, and three ways to form a square from the four vertices, for a total of 210 choices. In part (ii), the number of ways to partition eight points into four pairs is

$$7 \times 5 \times 3 \times 1 = 105.$$

This proves the second assertion. □

Corollary 9.4.11 *Any azygetic triple of bitangents can be uniquely completed to an azygetic 4-tuple.*

Proof It follows from the definitions that any azygetic triple of type (D) may be completed uniquely to a collection of four edges partitioning the vertices, and that any azygetic triple of type (E) may be completed uniquely to a square. This results in an azygetic 4-tuple in each case by Proposition 9.4.9. Lemma 9.4.3 then shows that all the cases have been considered. □

Lemma 9.4.12 *There are precisely* 315 *8-tuples of weights of* $L(E_7, \omega_6)$ *that form a cube of side* D*, where* D *is the minimal nonzero distance between weights. Each such cube consists of the eight weights in an azygetic quadruple of bitangents.*

Proof We will establish a bijection between the cubes described and the azygetic quadruples of bitangents, and then the conclusion will follow from Corollary 9.4.10.

The set T_8 of eight weights in an azygetic quadruple of bitangents contain a 4-element subset T_4 at mutual distance $\sqrt{2}D$ by Proposition 9.3.12 (ii). Since two opposite weights are at mutual distance $\sqrt{3}D$, it follows that we have $T_8 \backslash T_4 = -T_4$, and that the eight weights form an isometric copy of the vertex set of a cube of side D.

Conversely, if T_8 is an 8-tuple of weights that is an isometric copy of a cube of side D, we can use the fact that the cube is a bipartite graph to find a unique decomposition $T_8 = T \dot{\cup} T'$, where each of T and T' is a set of weights at mutual distance $\sqrt{2}D$. We further observe that $T = -T'$, which means that the eight weights form four bitangents. The properties of the set T and the definition of azygetic show that these four bitangents are azygetic.

The above two constructions are inverse to each other, and this completes the proof. □

Exercise 9.4.13 Use Corollary 9.4.10 and the result of Exercise 9.3.15 to show that the order of $W(E_7)$ is

$$7! \times 2 \times 288 = 2903\,040.$$

Exercise 9.4.14 Use Proposition 9.2.12 (ii), Corollary 9.4.10 and the result of Exercise 9.3.16 to show that the order of $W(E_7)$ is

$$|W(D_4)| \times 4! \times 2 \times 315.$$

Deduce from Exercise 9.4.13 that $|W(D_4)| = 192$, confirming the result of Exercise 3.3.15.

Exercise 9.4.15 Consider the 315 cubes of side D among the weights of $L(E_7, \omega_6)$ as described in Lemma 9.4.12. Show that precisely 45 of these 315 cubes include any given pair $\{\lambda, -\lambda\}$ of opposite weights.

Exercise 9.4.16

(i) Show that the action of S_8 on the 378 pairs of bitangents has two orbits, of sizes 168 and 210.

(ii) Show that the action of S_8 on the azygetic triples of bitangents has two orbits, of sizes 420 and 840.

(iii) Show that the action of S_8 on the syzygetic triples of bitangents has three orbits, of sizes 56, 1680 and 280.

(iv) Show that the action of S_8 on the syzygetic quadruples of bitangents has four orbits, of sizes 560, 280, 1680 and 2520.

Exercise 9.4.17 Prove the assertion in the introduction to Section 9.3: a triple of bitangents is azygetic if and only if the sum of the corresponding three elements of $\overline{Q'}$ is equal to a fourth element of $\overline{Q'}$.

9.5 Steiner complexes

The following result gives an easy way to distinguish the two orbits of triples of orthogonal roots in type E_7, mentioned in Proposition 4.4.5.

Proposition 9.5.1 *Let $T = \{\alpha, \beta, \gamma\}$ be a triple of orthogonal roots of type E_7. Then T is W-conjugate to a strongly orthogonal triple of roots and only if $\alpha + \beta + \gamma = 2\lambda$ for some weight λ of $L(E_7, \omega_6)$.*

Proof The proof of Proposition 4.4.5 constructs an example of a strongly orthogonal triple. This consists of the highest root in type E_7,

$$2\alpha_1 + 3\alpha_2 + 4\alpha_3 + 3\alpha_4 + 2\alpha_5 + \alpha_6 + 2\alpha_7,$$

the highest root in the D_6 subsystem,

$$\alpha_2 + 2\alpha_3 + 2\alpha_4 + 2\alpha_5 + \alpha_6 + \alpha_7,$$

and the root α_6. The sum of these is

$$2\alpha_1 + 4\alpha_2 + 6\alpha_3 + 5\alpha_4 + 4\alpha_5 + 3\alpha_6 + 3\alpha_7.$$

In our coordinates for $L(E_7, \omega_6)$, a calculation shows that this is equal to

$$-6\varepsilon_0 + 2\varepsilon_1 + 2\varepsilon_2 + 2\varepsilon_3 + 2\varepsilon_4 + 2\varepsilon_5 + 2\varepsilon_6 - 6\varepsilon_7,$$

which is equal to $-2v_{0,7}$ as required. Since the weights of $L(E_7, \omega_6)$ form a single W-orbit, Proposition 4.4.5 shows that every orthogonal triple of roots in the orbit of a strongly orthogonal triple must sum to twice a weight.

To complete the proof, it suffices to show that at least one orthogonal triple of roots does not sum to twice a weight. The sum $\alpha_1 + \alpha_3 + \alpha_5$ of the triple $\{\alpha_1, \alpha_3, \alpha_5\}$ of orthogonal fundamental roots sums in our coordinates to

$$4(\varepsilon_1 + \varepsilon_3 + \varepsilon_5) - 4(\varepsilon_2 + \varepsilon_4 + \varepsilon_6),$$

which by inspection is not equal to twice a weight. □

Exercise 9.5.2 If α and β are orthogonal positive roots in type E_7, let $\alpha \times \beta$ be the unique positive root for which $\{\alpha, \beta, \alpha \times \beta\}$ is a strongly orthogonal triple. Let us call a positive root γ in type E_7 "type (a)" if it is one of the 28 roots listed in case (a) of Example 8.4.9, and "type (b)" if it is one of the 35 roots listed in case (b) of Example 8.4.9.

(i) Use Proposition 9.5.1 to show that it is not possible for three roots of type (a) to form a strongly orthogonal triple.

(ii) Show that it is also not possible for one root of type (a) and two roots of type (b) to form a strongly orthogonal triple.
(iii) Deduce that $\alpha \times \beta$ is of type (a) if exactly one of α and β is of type (b), and that $\alpha \times \beta$ is of type (b) otherwise.

A simple example of a strongly orthogonal triple is the following.

Corollary 9.5.3 *In type E_7, the set $\{\alpha_4, \alpha_6, \alpha_7\}$ forms a strongly orthogonal triple of fundamental roots.*

Proof The sum $\alpha_4 + \alpha_6 + \alpha_7$ is equal in coordinates to

$$-2\varepsilon_0 - 2\varepsilon_1 - 2\varepsilon_2 - 2\varepsilon_3 + 6\varepsilon_4 - 2\varepsilon_5 + 6\varepsilon_6 - 2\varepsilon_7,$$

which by inspection is $2v_{4,6}$. The proof now follows from Proposition 9.5.1. □

Exercise 9.5.4 Show that the triple of fundamental roots $\{\alpha_4, \alpha_6, \alpha_7\}$ in type E_7 is the unique triple satisfying the condition of Corollary 9.5.3.

Lemma 9.5.5 *If $\{\lambda, -\lambda\}$ is a pair of opposite weights of $L(E_7, \omega_6)$, and $\{\alpha, \beta, \gamma\}$ is a triple of mutually orthogonal positive roots for which $s_\alpha s_\beta s_\gamma(\lambda) = -\lambda$, then $\{\alpha, \beta, \gamma\}$ is W-conjugate to a strongly orthogonal triple of roots.*

Proof By Theorem 5.4.13, the action of a reflection s_α on a weight will either fix it, or translate it by $\pm\alpha$. Now E_7 is a simply laced system, and Example 8.4.9 shows that $|(\lambda - (-\lambda))| = \sqrt{3}D$. It follows that no individual reflection α, β and γ can act trivially, which means that we have

$$\lambda - (-\lambda) = \pm\alpha \pm \beta \pm \gamma$$

for suitable (independent) choices of signs. The conclusion follows from Proposition 9.5.1. □

Exercise 9.5.6 Use Exercise 9.4.15 and Lemma 9.5.5 to show that, given an opposite pair of weights $\{\lambda, -\lambda\}$ of $L(E_7, \omega_6)$, there are precisely 45 triples of mutually commuting reflections $\{s_\alpha, s_\beta, s_\gamma\}$ such that $s_\alpha s_\beta s_\gamma(\lambda) = -\lambda$.

Definition 9.5.7 The *Steiner complex* associated with a reflection $s_\alpha \in W(E_7)$ is the subset of the 28 bitangents that are moved by the action of s_α. Two bitangents in the Steiner complex that are exchanged by the action of s_α are said to be *opposite*.

Theorem 9.5.8

(i) *Each Steiner complex contains* 12 *bitangents; that is, six pairs of opposite bitangents.*
(ii) *There are* 63 *distinct Steiner complexes, forming a single orbit under the action of $W(E_7)$.*
(iii) *We obtain* 28 *Steiner complexes by acting permutations of S_8 on the set*

$$\{[0,2],[0,3],[0,4],[0,5],[0,6],[0,7],[1,2],[1,3],[1,4],[1,5],[1,6],[1,7]\},$$

as shown in Figure 9.6.

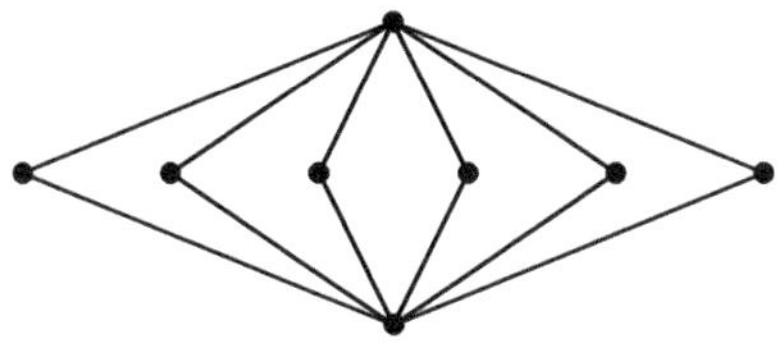

Figure 9.6 A Steiner complex as described in Theorem 9.5.8 (iii)

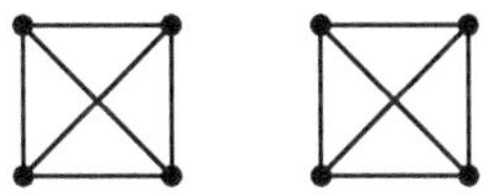

Figure 9.7 A Steiner complex as described in Theorem 9.5.8 (iv)

(iv) *We obtain the other* 35 *Steiner complexes by acting permutations of* S_8 *on the set*

$$\{[0,1],[0,2],[0,3],[1,2],[1,3],[2,3],[4,5],[4,6],[4,7],[5,6],[5,7],[6,7]\},$$

as shown in Figure 9.7.

(v) *If* $\{\alpha, \beta\}$ *is one of the six pairs of opposite bitangents in a Steiner complex* Σ, *then the other ten elements of* Σ *are precisely those bitangents* γ *for which* $\{\alpha, \beta, \gamma\}$ *is an azygetic triple.*

(vi) *Any pair of bitangents* $\{\alpha, \beta\}$ *taken together with the ten bitangents forming an azygetic triple with them form a Steiner complex in which the bitangents in the original pair are opposite to each other.*

(vii) *Each bitangent* ω *lies in* 27 *Steiner complexes, and the stabilizer of* ω *acts transitively on these* 27 *Steiner complexes.*

Proof As in Example 8.6.2, there is only one W-orbit of roots (where $W = W(E_7)$), and a pair of ideals (I, I') of the principal subheap F associated with $L(E_7, \omega_6)$ is relatively trivializable to a singleton if and only if $\chi(I, I')$ is a root.

Exercise 8.6.6 shows that $m_1(E_7, \omega_6) = 12$, and combined with the above paragraph and Theorem 5.4.13, we see that each reflection s_α moves $12 \times 2 = 24$ weights. Since the action of W on the weights is linear, and the weights are closed under negation by Example 8.1.7, these 24 weights form 12 opposite pairs, and this gives the required set of 12 bitangents, proving (i).

The reflections in type E_7 are described explicitly in Proposition 9.1.4. The 28 reflections in part (i) of that result act as transpositions of S_8. A typical transposition is the one exchanging symbols 0 and 1. This will move precisely the 12 bitangents listed in (iii). There are $\binom{8}{2} = 28$ transpositions in S_8; these will correspond to bitangents obtained by acting permutations of S_8 on the 12 bitangents listed, proving (iii).

The 35 reflections in Proposition 9.1.4 (ii) correspond to the 35 partitions of eight elements into two parts of size 4. One of the corresponding reflections is the generators s_7, which acts as a bifid transformation, as described in Example 8.1.7. This description shows that s_7 moves the 12 bitangents listed in (iv). The other reflections of this type correspond to the other partitions of eight elements into two parts of size 4, which are obtained by acting permutations of S_8 on the 12 bitangents listed, proving (iv).

To prove (ii), we recall from Example 8.4.9 that E_7 has 126 roots, and these correspond to $126/2 = 63$ reflections, so the number of Steiner complexes is at most 63. However, the Steiner complexes constructed in (iii) and (iv) are distinct, so there are exactly $28 + 35 = 63$ Steiner complexes. Since type E_7 is simply laced, there is one orbit of roots and all the reflections are conjugate under the action of $W(E_7)$, completing the proof of (ii).

To prove (v), it suffices by Corollary 9.3.6 to consider a fixed pair of bitangents. Since the set of 12 bitangents listed in the proof of Corollary 9.3.10 contains all ten bitangents azygetic to the given pair, and agrees with the Steiner complex listed in (iv), the result follows.

By Corollary 9.3.6, any pair of bitangents is W-conjugate to some fixed opposite pair occurring in a Steiner complex. Part (vi) now follows from (v).

To prove (vii), we recall that $W = W(E_7)$ acts transitively on the Steiner complexes by (ii), and W acts transitively on the bitangents by Corollary 9.3.6. It follows that every bitangent lies in the same number, t, of Steiner complexes. By counting the pairs (ω, Σ) for which Σ is a Steiner complex and $\omega \in \Sigma$, we find that $t = 63 \times 12/28 = 27$, proving the first assertion of (vii). The second assertion of (vii) follows by Corollary 9.3.6. □

Exercise 9.5.9 Show that the 24 vertices in an opposite pair of half cube shaped faces of 3_{21}, as described in Exercise 8.5.18, are precisely the vertices involved in the 12 bitangents of a Steiner complex, and that this gives a bijective correspondence between the 63 opposite pairs of faces and the 63 Steiner complexes.

Exercise 9.5.10 The function from pairs of bitangents to Steiner complexes in Theorem 9.5.8 (vi) induces a map f from pairs of bitangents to positive roots in type E_7. Show that f is 6 to 1. (This gives a combinatorial proof of the fact that $\binom{28}{2} = 63 \times 6$.)

Lemma 9.5.11 *Let α and β be distinct positive roots of E_7 such that $\gamma = \alpha + \beta$ is a root, and let Σ_α, Σ_β and Σ_γ be the corresponding Steiner complexes.*

(i) *We have $\Sigma_\gamma = \Sigma_\alpha \,\Delta\, \Sigma_\beta$, where Δ denotes symmetric difference.*
(ii) *We have $|\Sigma_\alpha \cap \Sigma_\beta| = 6$.*

Proof We argue using the Lie algebra, $\mathfrak{g}(E_7)$. Let λ be a weight of $L(E_7, \omega_6)$. If α' is any positive root, we have

$$H_{\alpha'}(\lambda) = c\lambda$$

for some integer c, where c is odd (± 1) if $s_{\alpha'}$ moves λ, and c is even (zero) if $s_{\alpha'}$ fixes λ.

By Lemma 7.2.7, we have $H_\alpha + H_\beta = H_\gamma$. The argument of the previous paragraph now shows that s_γ fixes a weight λ if and only if either (a) both s_α and s_β fix λ, or (b) both s_α and s_β move λ. Part (i) follows by considering the action of $W(E_7)$ on pairs $\{\pm\lambda\}$ of weights.

Part (ii) now follows from (i) and Theorem 9.5.8 (i). □

Lemma 9.5.12 *Suppose that α and β are positive orthogonal roots of E_7, and Σ_α and Σ_β are the corresponding Steiner complexes.*

(i) *The intersection $\Sigma_\alpha \cap \Sigma_\beta$ consists of a 4-tuple of azygetic bitangents. These four bitangents consist of two opposite pairs in Σ_α, and the same is true for Σ_β.*
(ii) *If γ is the unique positive root for which $\{\alpha, \beta, \gamma\}$ is a strongly orthogonal triple, then we have $|\Sigma_\alpha \cup \Sigma_\beta \cup \Sigma_\gamma| = 28$.*

Proof By Proposition 4.4.5 (i), there is only one orbit of orthogonal pairs of roots, so without loss of generality we may take α and β to be the fundamental roots α_4 and α_6, respectively.

Recall the explicit coordinates for the weights of $L(E_7, \omega_6)$ given in Example 8.1.7. If a weight λ is moved by both s_4 and s_6, then it follows that we have $\lambda = \pm v_{i,j}$, where $i \in \{4, 5\}$ and $j \in \{6, 7\}$. It follows that precisely eight weights are moved by both s_4 and s_6, and that precisely $8/2 = 4$ bitangents are moved, in pairs, by each of s_4 and s_6. Furthermore, the quadruple

$$-v_{5,7},\quad -v_{4,6},\quad v_{5,6},\quad v_{4,7}$$

is at mutual distance $\sqrt{2}D$. The assertions of (i) now follow.

By Corollary 9.5.3, we may assume that $\gamma = s_7$. If a weight μ is fixed by both s_4 and s_6, then μ must have coordinates of the form

$$\mu = (c_1, c_2, c_3, c_4, d, d, e, e).$$

It is not possible for a weight of this form to have an occurrence of 3 in the first four coordinates and another in the last four coordinates, and the same is true for occurrences of -3. It follows that μ is moved by s_7. We deduce that any weight is moved by either s_4, s_6 or s_7, and (ii) follows. □

Proposition 9.5.13 *Let α and β be orthogonal positive roots of E_7, and let γ be the unique positive root such that $\{\alpha, \beta, \gamma\}$ is a strongly orthogonal triple, as in Proposition 4.4.5. Let Σ_α, Σ_β and Σ_γ be the corresponding Steiner complexes.*

(i) *We have*

$$|\Sigma_\alpha \cap \Sigma_\beta| = |\Sigma_\alpha \cap \Sigma_\gamma| = |\Sigma_\beta \cap \Sigma_\gamma| = |\Sigma_\alpha \cap \Sigma_\beta \cap \Sigma_\gamma| = 4$$

and $|\Sigma_\alpha \cup \Sigma_\beta \cup \Sigma_\gamma| = 28$.
(ii) *We have $\Sigma_\gamma = \Sigma \backslash (\Sigma_\alpha \ \Delta\ \Sigma_\beta)$, where Σ is the set of all 28 bitangents and Δ denotes symmetric difference.*
(iii) *The three partitions of the 4-element set $T = \Sigma_\alpha \cap \Sigma_\beta \cap \Sigma_\gamma$ into pairs split the set T into opposite pairs of bitangents in Σ_α, Σ_β and Σ_γ, respectively.*

Proof Lemma 9.5.12 shows that

$$|\Sigma_\alpha \cap \Sigma_\beta| = |\Sigma_\alpha \cap \Sigma_\gamma| = |\Sigma_\beta \cap \Sigma_\gamma| = 4,$$

and that $|\Sigma_\alpha \cup \Sigma_\beta \cup \Sigma_\gamma| = 28$. Since each Steiner set is of size 12 by Theorem 9.5.8 and $8 + 8 + 8 + 4 = 28$, it must be the case that $|\Sigma_\alpha \cap \Sigma_\beta \cap \Sigma_\gamma| = 4$. Assertions (i) and (ii) follow from these observations. Part (iii) follows from (ii) and Lemma 9.5.12 (i). □

Lemma 9.5.14 *Let $T = \{b_1, b_2, b_3, b_4\}$ be a set of four bitangents of the form $\Sigma_\alpha \cap \Sigma_\beta \cap \Sigma_\gamma$ for some strongly orthogonal triple $\{\alpha, \beta, \gamma\}$ of roots of type E_7. There is a unique Steiner complex T_1 (respectively, T_2, T_3) such that b_1 and b_4 (respectively, b_2 and b_4, b_3 and b_4) is a pair of opposite bitangents in the Steiner complex. Furthermore, we have $\{T_1, T_2, T_3\} = \{\Sigma_\alpha, \Sigma_\beta, \Sigma_\gamma\}$.*

Proof The four bitangents can be split into pairs in three distinct ways. The result now follows from Lemma 9.5.12 (i), and Theorem 9.5.8 (v) and (vi). □

Proposition 9.5.15 *Let Σ_α and Σ_β be distinct Steiner complexes corresponding to positive roots α and β.*

(i) *The set $\Sigma_\alpha \cap \Sigma_\beta$ has size 4 or 6.*
(ii) *If $|\Sigma_\alpha \cap \Sigma_\beta| = 4$, then the roots α and β are orthogonal.*
(iii) *If $|\Sigma_\alpha \cap \Sigma_\beta| = 6$, then $\gamma = \alpha \pm \beta$ is also a root for some choice of sign, and the roots $\{\pm\alpha, \pm\beta, \pm\gamma\}$ form a type A_2 subsystem.*
(iv) *If $|\Sigma_\alpha \cap \Sigma_\beta| = 6$, then no pair of the bitangents in the set $\Sigma_\alpha \cap \Sigma_\beta$ is opposite in the sense of Definition 9.5.7.*

Proof There are three possibilities for α and β, corresponding to parts (ii), (iii) and (iv) of Proposition 7.1.1.

If Proposition 7.1.1 (ii) applies, $\gamma = \alpha - \beta$ is a root, and by exchanging the roles of α and β if necessary, we may assume that γ is a positive root. Lemma 9.5.11 applied to the roots γ and β then shows that $|\Sigma_\alpha \cap \Sigma_\beta| = 6$.

If Proposition 7.1.1 (iv) applies, $\gamma = \alpha + \beta$ is a root. Lemma 9.5.11 applied to the roots α and β then shows that $|\Sigma_\alpha \cap \Sigma_\beta| = 6$.

Finally, if Proposition 7.1.1 (iii) applies, Lemma 9.5.12 shows that $|\Sigma_\alpha \cap \Sigma_\beta| = 4$.

The assertions of (i), (ii) and (iii) follow from the above observations.

Proposition 7.1.8 shows that if ω and ω' are opposite bitangents in Σ_α, meaning that they are exchanged by the action of s_α, then it is not possible for s_β to move both of ω and ω', and part (iv) follows from this. □

Lemma 9.5.16 *Let $S = \{b_1, b_2, b_3, b_4\}$ be a 4-tuple of azygetic bitangents, and let Σ_α be a Steiner complex. If Σ_α contains three elements of S, then it must contain the fourth.*

Proof Up to permutations in S_8, every Steiner complex looks like one of the two listed in Theorem 9.5.8 (iii) and (iv), so it suffices to consider these two.

Consider the Steiner system

$$\Sigma = \{[0, 2], [0, 3], [0, 4], [0, 5], [0, 6], [0, 7], [1, 2], [1, 3], [1, 4], [1, 5], [1, 6], [1, 7]\}.$$

There are no azygetic triples of type (E) contained in Σ. The azygetic triples of type (D) contained in Σ are either of the form $\{[0, a], [a, 1], [1, b]\}$ or of the form $\{[1, a], [a, 0], [0, b]\}$. The first triple is completed by adding $[b, 0]$ and the second by adding $[b, 1]$; in both cases, the bitangent added is already in the Steiner complex.

The other case to consider involves the Steiner complex

$$\Sigma' = \{[0, 1], [0, 2], [0, 3], [1, 2], [1, 3], [2, 3], [4, 5], [4, 6], [4, 7], [5, 6], [5, 7], [6, 7]\}.$$

Any azygetic 4-tuple contained in Σ' corresponds to a refinement of the partition $\{0, 1, 2, 3\} \cup \{4, 5, 6, 7\}$ into four sets of size 2. A routine check then shows that any azygetic triple of type (E) contained in Σ' may be completed to an azygetic 4-tuple by adding a bitangent already present in Σ'. On the other hand, any azygetic triple of type (D) contained in Σ' consists of a set $\{[a, b], [b, c], [c, d]\}$ where

$$\{a, b, c, d\} \in \{\{0, 1, 2, 3\} \cup \{4, 5, 6, 7\}\}.$$

This triple is completed by the bitangent $[d, a]$, which is then also in Σ', completing the proof. □

Lemma 9.5.17 *If Σ_α and Σ_β are Steiner complexes, and $\Sigma = \Sigma_\alpha \cap \Sigma_\beta$ has size 4, then Σ is an azygetic 4-tuple of bitangents. Conversely, any azygetic 4-tuple of bitangents is equal to the intersection of two Steiner complexes $\Sigma_\alpha \cap \Sigma_\beta$.*

Proof For the first assertion, we may apply Proposition 9.5.15 (ii) and assume as in the proof of Lemma 9.5.12 that $\alpha = \alpha_4$ and $\beta = \alpha_6$. The four bitangents moved by α_4 and α_6 are $[4, 6]$, $[4, 7]$, $[5, 6]$ and $[5, 7]$. This forms a square, and is azygetic by Proposition 9.4.9.

The second assertion follows from the first and Proposition 9.3.13 (iv). □

Lemma 9.5.18 *Let α_1, α_2 and α_3 be a strongly orthogonal triple of roots in type E_7. Let β be a fourth positive root. Then β is orthogonal to either one or three roots in the triple $\{\alpha_1, \alpha_2, \alpha_3\}$.*

Proof The definition of strongly orthogonal triple means that β cannot be orthogonal to precisely two of the elements in the triple, so it suffices to prove that β must be orthogonal to at least one of the elements.

Let $\Sigma_{\alpha_1}, \Sigma_{\alpha_2}, \Sigma_{\alpha_3}$ and Σ_β be the Steiner complexes corresponding to $\alpha_1, \alpha_2, \alpha_3$ and β. By Proposition 9.5.13, there is a partition of the 28 bitangents into sets A_1, A_2, A_3, C of sizes 8, 8, 8 and 4 respectively, such that $\Sigma_{\alpha_i} = A_i \cup C$. By Proposition 9.5.15, each set $B \cap \Sigma_{\alpha_i}$ has size 4 or 6, and to prove the result, it remains to show that $|B \cap \Sigma_{\alpha_i}| = 4$ for some i.

Suppose towards a contradiction that $|B \cap \Sigma_{\alpha_i}| = 6$ for all i. Since B has size 12 by Theorem 9.5.8 (i), we cannot have $|B \cap C| \in \{0, 1, 2, 4\}$, which forces $|B \cap C| = 3$. By Lemma 9.5.17, C is an azygetic 4-tuple. Lemma 9.5.16 then shows that $|B \cap C|$ cannot have size 3, which completes the proof. □

Curiously, there does not seem to be an obvious proof of Lemma 9.5.18 using the basic properties of root systems.

Exercise 9.5.19 Adapt the argument of Proposition 9.5.1 (using Proposition 4.4.3 in the place of Proposition 4.4.5) to show that a pair of orthogonal roots $\{\alpha, \beta\}$ in type D_l (where $l > 4$) is $W(D_l)$-conjugate to a strongly orthogonal pair if and only if $\alpha + \beta = 2\lambda$, where λ is a weight of $L(D_l, \omega_1)$.

Exercise 9.5.20 Let $n > 4$. Recall from Example 8.5.13 that the n-dimensional half cube, $P(D_n, \omega_n)$ has two $W = W(D_n)$-orbits of tetrahedral (3-simplex shaped) faces: those that extend to 4-simplex shaped faces, and those that do not. Let T be a 3-simplex shaped face of $P(D_n, \omega_n)$, and let α and β be vectors associated with opposite (i.e., skew) edges in the boundary of T. Use Exercise 9.5.19 to prove the following.

(i) Show that if T is a nonextendable face of $P(D_n, \omega_n)$, then α and β are orthogonal roots that form a strongly orthogonal pair.
(ii) Show that if T is an extendable face of $P(D_n, \omega_n)$, then α and β are orthogonal roots that do not form a strongly orthogonal pair.

Exercise 9.5.21 Let T_1 be the tetrahedron (in $\mathbb{R}^4$) with vertices

$$\{(2, 0, 0, 0),\ (0, 2, 0, 0),\ (0, 0, 2, 0),\ (0, 0, 0, 2)\},$$

let T_2 be the tetrahedron with vertices

$$\{(-2, 2, 2, 2),\ (2, -2, 2, 2),\ (2, 2, -2, 2),\ (2, 2, 2, -2)\},$$

and let T_3 be the tetrahedron with vertices

$$\{(2, 2, 2, 2),\ (2, -2, -2, 2),\ (2, -2, 2, -2),\ (2, 2, -2, -2)\}.$$

(i) Show that T_2 is one of the 3-simplex shaped faces of the half cube $P(D_4, \omega_3)$.
(ii) Show that T_3 is one of the 3-simplex shaped faces of the half cube $P(D_4, \omega_4)$.
(iii) Now interpret the coordinates of T_1, T_2 and T_3 as coordinates in projective space. Show that if $\{i, j, k\} = \{1, 2, 3\}$ and P is any of the four vertices of T_i, then T_j and T_k are in perspective from P. (In other words, the projective line through P and any vertex of T_j passes through a unique vertex of T_k.)

Exercise 9.5.22 Use Corollary 9.3.6 to show that the set of Steiner complexes form a (28, 12, 11)-design. (This means that the Steiner complexes are a collection of subsets of the 28 bitangents, each of size 12, such that any pair of bitangents is contained in precisely 11 Steiner complexes.)

Exercise 9.5.23 Prove the assertions about R in the introduction to Section 9.3, as follows.

(i) Show that for each element $r \in R$, there are six possible pairs $\{q_1, q_2\}$ of elements in $\overline{Q'}$ for which $q_1 + q_2 = r$.
(ii) Show that the six pairs of elements in (i) are pairwise disjoint.
(iii) Show that the union of the six pairs gives a Steiner complex, and that any Steiner complex arises in this way.

9.6 Symplectic structure

The Steiner complexes give rise to a natural symplectic bilinear form, which we now describe.

Definition 9.6.1 Let S be a set, and let $D(S)$ be the set of pairs $\{T, \overline{T}\}$ as T ranges over the subsets of S and $\overline{T} = S \backslash T$. We define the operation $+_\Delta$ on T by

$$\{T, \overline{T}\} +_\Delta \{U, \overline{U}\} = \{T \Delta U, \overline{T \Delta U}\},$$

where Δ denotes symmetric difference. (This is well-defined, because of the elementary fact that

$$T \,\Delta\, \overline{U} = \overline{T} \,\Delta\, U = \overline{T \,\Delta\, U}.)$$

Lemma 9.6.2 *The operation $+_\Delta$ of Definition 9.6.1 makes $D(S)$ into an abelian group under $+_\Delta$. The identity element is $\{S, \emptyset\}$.*

Proof Commutativity follows from Definition 9.6.1, and associativity is inherited from the associativity of the “exclusive or” operation $*$, where

$$T * U := T \,\Delta\, \overline{U}.$$

The claim about the identity element follows by a routine check, and every element is self-inverse. □

Definition 9.6.3 Let V_Σ be the set of size 64 consisting of the 63 Steiner complexes (each of size 12) together with the set Σ_0 of all 28 bitangents.

Lemma 9.6.4 *Let $S = \Sigma_0$. The set*

$$\{\{T, \overline{T}\} : T \in V_\Sigma\}$$

is a subgroup of $(S, +_\Delta)$.

Proof Let Σ_α and Σ_β be distinct Steiner complexes. If $\Sigma_\alpha \cap \Sigma_\beta$ has size 4, then $\overline{\Sigma_\alpha \,\Delta\, \Sigma_\beta}$ is a Steiner complex by Propositions 9.5.15 and 9.5.13.

By Proposition 9.5.15, the only other possibility is that $\Sigma_\alpha \cap \Sigma_\beta$ has size 6. In this case, Lemma 9.5.11 shows that $\Sigma_\alpha \,\Delta\, \Sigma_\beta$ is a Steiner complex.

The proof follows from these two observations. □

Definition 9.6.5 If $T, U \in V_\Sigma$, then we define $T +_\Sigma U$ to be the unique element of the pair

$$\{T, \overline{T}\} +_\Delta \{U, \overline{U}\}$$

that lies in V_Σ.

Proposition 9.6.6 *Let V_Σ be as above, and let Σ_α and Σ_β be distinct Steiner complexes associated with the positive roots α and β.*

(i) *The set V_Σ is an abelian group under the operation $+_\Sigma$.*
(ii) *The set V_Σ is a six-dimensional vector space over the field with two elements, $\mathbb{F}_2$.*
(iii) *If α and β are not orthogonal, then $\Sigma_\alpha +_\Sigma \Sigma_\beta = \Sigma_\gamma$, where $\gamma = \pm\alpha \pm \beta$ and $\{\pm\alpha, \pm\beta, \pm\gamma\}$ is a root system of type A_2.*
(iv) *If α and β are orthogonal, then $\Sigma_\alpha +_\Sigma \Sigma_\beta = \Sigma_\gamma$, where $\{\alpha, \beta, \gamma\}$ is a strongly orthogonal triple.*

Proof Part (i) follows from the definitions and Lemma 9.6.4.

Since every element of V_Σ has order 2, a familiar fact from group theory shows that V_Σ is abelian. It then follows from the classification of finite abelian groups that

V_Σ is elementary abelian, which shows that V_Σ is an $\mathbb{F}_2$-vector space. Part (ii) then follows from the fact (see Example 8.4.9) that there are $2^6 - 1 = 63$ positive roots in type E_7.

Part (iii) follows from the definitions, using Lemma 9.5.11 (ii) and Proposition 9.5.15 (iii).

Part (iv) follows from the definitions and Proposition 9.5.13 (ii). □

Proposition 9.6.7 *There is a symplectic $\mathbb{F}_2$-bilinear form, $B(,)$ on the vector space V_Σ, where*

$$B(T, U) = \begin{cases} 0 & \text{if } T = \Sigma_0 \text{ or } U = \Sigma_0, \\ 0 & \text{if } T = U, \\ 1 & \text{if } |T \cap U| = 6, \\ 0 & \text{if } |T \cap U| = 4. \end{cases}$$

Proof The conditions in the definition are exhaustive by Proposition 9.5.15. We first show that B is bilinear. For this, it suffices to prove that it is always the case that

$$B(T, U_1 +_\Sigma U_2) = B(T, U_1) + B(T, U_2).$$

This follows quickly from the definitions in the cases $T = \Sigma_0$, $U_1 = \Sigma_0$, $U_2 = \Sigma_0$ or $U_1 = U_2$, so we assume that $U_1 = \Sigma_\alpha$ and $U_2 = \Sigma_\beta$ are distinct Steiner complexes, and that $T = \Sigma_\delta$ is also a Steiner complex.

To prove (iii) and (iv), we recall the standard bilinear form on the root space defined in Proposition 7.1.1 for which $(\alpha', \alpha') = 2$ for all roots α'. With these conventions, we have

$$B(\Sigma_{\beta_1}, \Sigma_{\beta_2}) = (\beta_1, \beta_2) \bmod 2$$

for all (not necessarily distinct) roots β_1 and β_2.

Suppose now that $|U_1 \cap U_2| = 6$. By Proposition 9.6.6 (iii), we have $U_1 +_\Sigma U_2 = \Sigma_\gamma$, where γ is the unique element of the form $\pm\alpha \pm \beta$ that is a positive root. The bilinearity of B in this case follows from the observation that $\gamma - \alpha - \beta$ is a linear combination of fundamental roots with even integer coefficients.

Finally, suppose that $|U_1 \cap U_2| = 4$. In this case, Proposition 9.6.6 (iv) shows that we have $U_1 +_\Sigma U_2 = \Sigma_\gamma$, where $\{\alpha, \beta, \gamma\}$ is a strongly orthogonal triple. Lemma 9.5.18 shows that δ is orthogonal to either one or three elements of $\{\alpha, \beta, \gamma\}$, which in turn means that either zero or two terms in the list $B(T, U_1 +_\Sigma U_2)$, $B(T, U_1)$, $B(T, U_2)$ are nonzero. It follows that $B(T, U_1 +_\Sigma U_2) = B(T, U_1) + B(T, U_2)$, which shows that B is bilinear in this case.

Recall from Definition 1.2.1 that a bilinear form is called symplectic if it is skew-symmetric and nondegenerate. Skew-symmetry holds automatically over $\mathbb{F}_2$. To prove nondegeneracy, it must be shown that for every nonzero element $T \in V_\Sigma$, there exists $U \in V_\Sigma$ with $B(T, U) = 0$. By the analysis in the case $|U_1 \cap U_2| = 6$, we see that this is equivalent to the assertion that any positive root in type E_7 has another positive root that is not orthogonal to it. The latter claim follows from the observations that there is only one orbit of roots, and not all pairs of roots are orthogonal. □

9.7 Notes and references

1 More details about del Pezzo surfaces may be found in Hartshorne's book [33, section V.4]. The introduction of Serganova and Skorobogatov [69] summarizes some of Manin's insights from [53] concerning the relationship between the hierarchy in Definition 9.2.2, root systems, and del Pezzo surfaces.

2 Further combinatorial details of bitangents, Hesse–Cayley notation and Steiner complexes may be found in Dickson's contribution to the classic text Miller, Blichfeldt and Dickson [58, section XIX].

3 Manivel's paper [54] contains further discussion about the relationship between Lie algebras, the 28 bitangents and Steiner sets.

4 Our approach to bitangents largely follows Dolgachev's book [20, chapter 6]. Dolgachev uses the term "azygetic", but the synonymous term "asyzygetic" is also found in the literature (for example, in [58]).

5 It also follows from the theory in [20, section 5.4] that any pair of Steiner complexes intersects in six or in four elements. Dolgachev calls such a pair "azygetic" in the first case, and "syzygetic" in the second. The notion of Steiner complexes discussed here is the special case $g = 3$ (where g stands for "genus") of the theory in [20, section 5.4].

6 It is well-known (see, for example, Collins [15, chapter 1]) that the group $W(E_7)$ is isomorphic to $\mathrm{Sp}(6, 2) \times \mathbb{Z}_2$. Since $\mathrm{Sp}(6, 2)$ is a simple group, the $\mathbb{Z}_2$ factor is the central subgroup of $W(E_7)$ generated by the longest element. Since the longest element fixes each bitangent, this gives an action of $\mathrm{Sp}(6, 2)$ on the 28 bitangents, and another explanation for the existence of the symplectic structure in Section 9.6. The symplectic form on Steiner complexes emerges naturally in the algebraic geometry approach [20, section 5.4].

7 A set of three tetrahedra in projective space that satisfies the conditions of Exercise 9.5.21 (iii) is called a *triad of desmic tetrahedra*.

10

Exceptional structures

It follows from Example 8.2.11 and Exercise 8.2.12 that there is a tower of vector spaces

$$L(A_5, \omega_3) \subset L(D_6, \omega_6) \subset L(E_7, \omega_6)$$

of dimensions 20, 32 and 56. This inclusion respects the decomposition of each space into weight spaces. This corresponds to a chain

$$W(A_5)/W(A_2 \cup A_2) \subset W(D_6)/W(A_5) \subset W(E_7)/W(E_6)$$

of containments of cosets of parabolic subgroups. (In the case of $W(D_6)$, we may use either of the two parabolic subgroups of type A_5.) This chapter is about the surprisingly rich combinatorial structure of this chain of inclusions.

Section 10.1 introduces the famous configuration of 27 lines on a cubic surface. The symmetry group of this configuration is well-known to be the group $W(E_6)$. We shall see that the incidence properties of these lines turn out to be governed by the combinatorics of the minuscule module $L(E_6, \omega_5)$. We develop these ideas further in Section 10.2, which investigates the properties of Schläfli double sixes; these are particular collections of 12 lines on the cubic surface.

Section 10.3 introduces the theory of 2-graphs. Just as a graph may be regarded as a collection of distinguished pairs (i.e., edges) in a vertex set, a 2-graph may be regarded as a collection of distinguished triples in a vertex set. We shall see that there is an interesting 2-graph on the set of 28 bitangents in which the distinguished triples are the azygetic triples. This 2-graph has interesting subgraphs on 16 and 10 points. In each of these three cases, the elements of the 2-graph may be identified with opposite pairs of weights in one of the minuscule modules $L(E_7, \omega_6)$, $L(D_6, \omega_6)$ or $L(A_5, \omega_3)$.

Section 10.4 introduces generalized quadrangles, which are a particular type of incidence structure. It turns out that there is a canonical way to remove one of the vertices in one of the 2-graphs constructed above to create an incidence structure on the remaining vertices, and this construction results in three generalized quadrangles, called GQ(2,4), GQ(2,2) and GQ(2,1). The first of these, GQ(2,4), has a close direct relationship with the 27 lines on a cubic surface.

Section 10.5 sketches some of the connections between minuscule representations and invariant forms.

10.1 The 27 lines on a cubic surface

A cubic surface is an algebraic surface in three-dimensional projective space defined by a single polynomial that is homogeneous of degree 3. The *Cayley–Salmon theorem* in algebraic geometry states that a smooth cubic surface over an algebraically closed field contains 27 straight lines.

A good example of a cubic surface is the *Fermat cubic*, which is given by the equation

$$w^3 + x^3 + y^3 + z^3 = 0.$$

If $a, b \in \mathbb{C}$ are cube roots of -1, then the points with projective coordinates

$$(w : aw : y : by)$$

form a projective line, and this line is entirely contained in the projective surface. The same is true for all nine choices of a and b, and all three decompositions of the four coordinates into pairs. This gives an explicit description of the 27 lines in this case.

A well-known result in algebraic geometry is that the del Pezzo surface Σ resulting from blowing up six points in general position in the projective complex plane is in fact a cubic surface. The cubic surfaces below are assumed to arise as del Pezzo surfaces in this way. We can then understand the combinatorics of the 27 lines in terms of Weyl groups and minuscule representations, using the results of Section 9.2.

More precisely, Theorem 9.2.3 shows that the lines of Σ are in bijection with the left cosets $W(E_6)/W(D_5)$. By Proposition 8.1.3 (iii), we may identify these with the weights of either $L(E_6, \omega_1)$ or $L(E_6, \omega_5)$, and we will usually work with the latter. By Proposition 8.2.20 (i), there are three possible distances between weights of $L(E_6, \omega_5)$, namely 0, D and $\sqrt{2}D$ for some D. By Theorem 9.2.1, the weight polytope is Coxeter's polytope 2_{21}, and two weights are at distance D (respectively, $\sqrt{2}D$) if and only if the corresponding lines are skew (respectively, incident).

By construction, the Weyl group $W(E_6)$ acts on the 27 lines and respects their incidence relations. Recall that the weights of $L(E_6, \omega_5)$ are defined explicitly in Exercise 8.1.11 as the set

$$\Psi_6 = \{-v_{i,7} : 1 \leq i \leq 6\} \cup \{v_{i,j} : 1 \leq i < j \leq 6\} \cup \{-v_{0,i} : 1 \leq i \leq 6\}.$$

Definition 10.1.1 We denote the line corresponding to the weight $-v_{0,i}$ (respectively, $v_{i,j}$, $-v_{i,7}$) of $L(E_6, \omega_5)$ by E_i (respectively, $F_{i,j}$, G_i).

The 27 lines are also naturally identified with the 27 bitangents other than $[0, 7]$. The line E_i (respectively, $F_{i,j}$, G_i) corresponds to the bitangent $[0, i]$ (respectively, $[i, j]$, $[i, 7]$).

Exercise 10.1.2 Show that two of the 27 lines are skew if and only if they correspond to a pair of weights of $L(E_6, \omega_5)$ whose difference is a root.

Exercise 10.1.3 Use Theorem 9.2.1 to show that the incidence relations between the 27 lines are as follows:

(i) the lines E_i are mutually skew;
(ii) the lines G_i are mutually skew;

(iii) the line E_i is skew to G_j if $i = j$, but incident otherwise;
(iv) the line F_{ij} is incident to E_k and to G_k if $k \in \{i, j\}$, but is skew to E_k and to G_k otherwise;
(v) the line F_{ij} is incident to F_{kl} if $\{i, j\}$ and $\{k, l\}$ are disjoint, and is skew to F_{kl} otherwise.

Exercise 10.1.4 Identify the 27 lines with the weights of $L(E_6, \omega_5)$ as above, and let F be the principal subheap of the full heap $\varepsilon : \text{FH}(E_6^{(1)}(5)) \to E_6^{(1)}$ shown in Figure 6.12. Let ω and ω' be weights corresponding to the distinct lines L and L', and let I and I' be the corresponding ideals of F. Define a relation $\leq_s$ on the 27 lines by the condition that $L \leq_s L'$ if and only if (a) L and L' are skew and (b) $\omega' \geq \omega$ in the usual order on weights (in other words, $\omega' - \omega$ is a positive root).

(i) Show that the reflexive transitive extension of $\leq_s$ gives a partial order, $\leq$, on the 27 lines.
(ii) Show that $L \leq L'$ if and only if $I \subseteq I'$. Deduce that $\leq$ is not a total order.
(iii) Show that, with respect to the partial order of (i), G_1 is the unique minimal element and E_6 is the unique maximal element. Show also that we have

$$G_1 < G_2 < G_3 < G_4 < G_5 < G_6$$

and

$$E_1 < E_2 < E_3 < E_4 < E_5 < E_6.$$

(iv) Let $1 \leq i < j \leq 6$ and $1 \leq k < l \leq 6$. Show that if F_{ij} and F_{kl} are comparable under $\leq$, then we have $F_{ij} < F_{kl}$ if and only if $(i, j) < (k, l)$ in the lexicographic order.
(v) Show that performing the above construction starting with the full heap $\varepsilon : \text{FH}(E_6^{(1)}(1)) \to E_6^{(1)}$ results in a partial order that is opposite to $\leq$.

Exercise 10.1.5 Let (W, S) be the Coxeter system of type E_6, let $I = S \backslash \{s_6\}$, and let W_I be the corresponding parabolic subgroup. Let $\leq$ be the order on the 27 lines constructed in Exercise 10.1.4.

(i) Use Exercise 5.4.8 to show that we have $w \in W^I$ if and only if we have

$$w(E_1) < w(E_2) < w(E_3) < w(E_4) < w(E_5) < w(E_6).$$

(ii) Show that a statement analogous to (i) also holds for the lines G_i.

Lemma 10.1.6 *Each of the 27 lines is incident with 10 others and skew to 16 others.*

Proof By Example 8.6.4, the Schläfli graph $v_1(E_6, \omega_5)$ has 27 vertices and is regular of degree 16. By Theorem 9.2.1, two lines are skew if and only if the corresponding weights are at distance D, which happens if and only if they are adjacent in the graph. □

Lemma 10.1.7 *Let λ and μ be distinct weights of $L(E_6, \omega_5)$, and let L and M be the corresponding bitangents.*

(i) *The triple $\{L, M, [0, 7]\}$ is syzygetic if and only if $|\lambda - \mu| = D$.*
(ii) *The triple $\{L, M, [0, 7]\}$ is azygetic if and only if $|\lambda - \mu| = \sqrt{2}D$.*

Proof As in Exercise 8.1.11, the weights of $L(E_6, \omega_5)$ may be embedded isometrically in those of $L(E_7, \omega_6)$ in such a way that the weights of $L(E_6, \omega_5)$ are all at the same distance from $+v_{0,7}$ and all at the same distance from $-v_{0,7}$.

Assume that we have such an embedding, and suppose that $|\lambda - \mu| = D$. By Lemma 9.3.3, there exists $\nu \in \{\pm v_{0,7}\}$ such that λ, μ, ν are at mutual distance D, from which (i) follows. A similar argument proves (ii). □

Lemma 10.1.8 *Among the 27 lines on a cubic surface, the maximum size of a subset of mutually skew lines is 6. All 6-tuples of mutually skew lines are conjugate under the action of $W(E_6)$.*

Proof Observe from Exercise 8.1.11 that the set Ψ_6 consists precisely of the 27 weights that are at distance D from the weight $-v_{0,7}$, which is the highest weight of $L(E_7, \omega_6)$.

Any 6-tuple of mutually skew lines corresponds to a collection of weights in Ψ_6 at mutual distance D. By adding $-v_{0,7}$ as a seventh point, we obtain a collection of weights in $L(E_7, \omega_6)$ at mutual distance D. By Proposition 9.2.8 (i), all such 7-tuples of weights are conjugate under $W(E_7)$, and furthermore, they are conjugate by an element of $W(E_7)$ fixing $-v_{0,7}$. The result then follows from the fact that the stabilizer of $-v_{0,7}$ is precisely the subgroup $W(E_6)$ of $W(E_7)$ obtained by omitting the generator labelled 6. □

Lemma 10.1.9 *Among the 27 lines, there are 72 6-subsets of mutually skew lines.*

Proof By Corollary 9.4.10, there are 288 Aronhold sets, and by Proposition 9.3.13, there are 576 7-sets of mutually skew weights in $L(E_7, \omega_6)$, all of which are conjugate under the action of $W(E_7)$ by Proposition 9.2.8 (i).

As in the proof of Lemma 10.1.8, there is a bijection between 7-tuples of skew weights in $L(E_7, \omega_6)$ containing the highest weight vector and 6-tuples of skew weights in $L(E_6, \omega_5)$, so we enumerate the former set.

The set Ω of all (A, λ), where A one of the 576 7-sets mentioned above, and $\lambda \in A$ clearly has cardinality 576×7. Since $W(E_7)$ acts transitively on the weights of $L(E_7, \omega_6)$, each weight λ lies in the same number, N, of sets A. Since there are 56 choices for λ, we have

$$N = \frac{576 \times 7}{56} = 72,$$

and this completes the proof. □

Example 10.1.10 Recall that the principal subheap F of $L(E_6, \omega_5)$ is shown in Figure 6.12. Let us label the elements of the vertex chain $\varepsilon^{-1}(p)$ as $F(p, 1) < F(p, 2) < \cdots$ as usual. Notice that the ideals of F in the chain

$$\emptyset \subset \langle F(1, 1)\rangle \subset \langle F(2, 1)\rangle \subset \langle F(3, 1)\rangle \subset \langle F(4, 1)\rangle \subset \langle F(5, 1)\rangle$$

each differ from the adjacent ideals in the chain by a singleton. It follows that the corresponding set C of six weights are at mutual distance D. Observe that the set

$$\{\lambda_1 - \lambda_2 : \lambda_1, \lambda_2 \in C,\ \lambda_1 \neq \lambda_2\}$$

is a root system of type A_5 with fundamental weights $\alpha_1, \alpha_2, \alpha_3, \alpha_4, \alpha_5$. By Theorem 4.3.10 (iv), these 30 roots are precisely those that are orthogonal to the highest root

$$\theta_l = \alpha_1 + 2\alpha_2 + 3\alpha_3 + 2\alpha_4 + \alpha_5 + 2\alpha_6.$$

By inspection of the heap F, Theorem 5.4.13 shows that the reflection s_{θ_l} sends the ideals

$$\emptyset,\ \langle F(1,1)\rangle,\ \langle F(2,1)\rangle,\ \langle F(3,1)\rangle,\ \langle F(4,1)\rangle,\ \langle F(5,1)\rangle$$

respectively to the ideals

$$\langle F(6,2)\rangle, \langle F(6,2), F(1,2)\rangle, \langle F(6,2), F(2,3)\rangle, \langle F(3,4)\rangle, \langle F(4,3)\rangle, \langle F(5,2)\rangle = F.$$

Let us denote the weights corresponding to the ideals in the first list by $\lambda_1, \dots, \lambda_6$ (in order), and the weights arising from the second list by $\mu_1, \dots, \mu_6$ (in order). The remarks about orthgonal roots above show that $|\lambda_i - \mu_j|$ is equal to D if $i = j$, and $\sqrt{2}D$ otherwise.

Definition 10.1.11 A 12-subset $L_1, \dots, L_6, M_1, \dots, M_6$ of the 27 lines is called a *(Schläfli) double six* if:

(i) the L_i are mutually skew;
(ii) the M_i are mutually skew;
(iii) L_i is skew to M_j if $i = j$, and incident otherwise.

We will use the notation

$$\begin{pmatrix} L_1 & L_2 & L_3 & L_4 & L_5 & L_6 \\ M_1 & M_2 & M_3 & M_4 & M_5 & M_6 \end{pmatrix}$$

to denote the Schläfli double six above. (The order of the columns of the matrix, and the order of the rows of the matrix, is not important, so there are $2 \times 6!$ ways to denote the same double six using this notation.) We denote the corresponding matrix of weights by

$$\begin{pmatrix} \lambda_1 & \lambda_2 & \lambda_3 & \lambda_4 & \lambda_5 & \lambda_6 \\ \mu_1 & \mu_2 & \mu_3 & \mu_4 & \mu_5 & \mu_6 \end{pmatrix}.$$

Example 10.1.12 In the notation of Definition 10.1.1, the double six of Example 10.1.10 is given by

$$\begin{pmatrix} E_1 & E_2 & E_3 & E_4 & E_5 & E_6 \\ G_1 & G_2 & G_3 & G_4 & G_5 & G_6 \end{pmatrix}.$$

In this case, the line E_6 (respectively, G_1) corresponds to the highest (respectively, lowest) weight of $L(E_6, \omega_5)$, which in the notation of Exercise 8.1.11 is $-v_{0,6}$ (respectively, $-v_{1,7}$). Notice that the lines appearing the top half of the double six form a single orbit under the parabolic subgroup $\langle s_1, s_2, s_3, s_4, s_5\rangle$, and that the same is true of the lines appearing in the bottom half.

10.2 Combinatorics of double sixes

We next describe some of the combinatorial properties of Schläfli double sixes. It turns out that these are closely related to those of Steiner complexes.

Theorem 10.2.1 *Let*

$$\begin{pmatrix} L_1 & L_2 & L_3 & L_4 & L_5 & L_6 \\ M_1 & M_2 & M_3 & M_4 & M_5 & M_6 \end{pmatrix}$$

be a double six with corresponding weights

$$\begin{pmatrix} \lambda_1 & \lambda_2 & \lambda_3 & \lambda_4 & \lambda_5 & \lambda_6 \\ \mu_1 & \mu_2 & \mu_3 & \mu_4 & \mu_5 & \mu_6 \end{pmatrix}.$$

(i) *Every 6-set of mutually skew lines is contained in a double six.*
(ii) *There are 36 double sixes, and they form a single orbit under the action of $W(E_6)$.*
(iii) *The vectors $\{\pm(\lambda_1 - \mu_1)\}$ are roots in type E_6 under the usual identifications. This induces a bijection between the Schläfli double sixes and the positive roots in type E_6. We denote the double six corresponding to the root α by Σ_α.*
(iv) *The double six Σ_α is equal to the Steiner complex Σ_α, identifying the root system of type E_6 as a subset of the root system of type E_7, and identifying the 27 lines with bitangents as in Definition 10.1.1.*
(v) *The 30 nonzero vectors of the form $\{\pm(\lambda_i - \lambda_j)\}$ are precisely those vectors orthogonal to the positive root α corresponding to the double six. These 30 vectors form a root system of type A_5.*

Proof An example of a double six appears in Definition 10.1.11, namely Σ_{θ_l}. It corresponds to the highest root in type E_6 and contains a 6-tuple of skew lines.

Part (i) then follows from Lemma 10.1.8, which shows that the 6-tuples of skew lines form a single orbit under $W(E_6)$.

By Exercise 8.4.11, there are 36 positive roots in type E_6. It follows that by acting $W(E_6)$ on the double six constructed in Definition 10.1.11, there must be at least 36 double sixes. On the other hand, part (i) and Lemma 10.1.8 show that there are at most $72/2 = 36$ double sixes, and this proves (ii) and (iii). (Note that the root corresponding to a Schläfli double six is only defined up to sign, because the two rows of the matrix may be interchanged.)

The construction of Σ_{θ_l} shows that the 12 lines in the double six are all moved by s_{θ_l}. Part (ii) then shows that the same is true for an arbitrary double six, with respect to a suitable reflection. Identifying the lines with bitangents as in Definition 10.1.1, we see that these 12 lines in the double six Σ_α are the 12 bitangents moved by s_α; the latter is equal to the Steiner complex Σ_α by definition, proving (iv).

Part (v) holds in the case $\alpha = \theta_l$ by Example 10.1.10, and the general case follows from (ii). □

Exercise 10.2.2 Use Lemma 10.1.7, Theorem 9.2.1 and Proposition 9.2.8 to prove the following about the action of $W = W(E_6)$ on the 27 lines.

(i) There are two W-orbits of 5-tuples of skew lines; the elements of one orbit extend to skew 6-tuples, and the orbits of the other do not.
(ii) There is one W-orbit of 4-tuples of skew lines.
(iii) Each skew 4-tuple extends uniquely to a nonextendable skew 5-tuple.
(iv) Each skew 4-tuple extends in two ways to an extendable skew 5-tuple.

Exercise 10.2.3 Use Theorem 8.5.7, Theorem 10.2.1 and Example 10.1.12 to show that each skew 6-tuple of lines appearing in a double six corresponds to the vertices in a five-dimensional simplex shaped face of the polytope 2_{21}. Show furthermore that a pair of 6-tuples lies in the same double six if and only if the corresponding five-dimensional faces are parallel.

Exercise 10.2.4 Show that the k-tuples of skew lines of Exercise 10.2.2 correspond to faces of the polytope 2_{21} as follows.

(i) A skew 4-tuple of lines corresponds to a 3-simplex shaped face of 2_{21}.
(ii) An extendable skew 5-tuple of lines corresponds via Theorem 8.5.7 to to a 4-simplex shaped face of 2_{21} that is contained both in a 5-simplex shaped face and in a five-dimensional hyperoctahedral face. (This corresponds to the parabolic subgroup $\langle s_2, s_3, s_4, s_5 \rangle$.)
(iii) A nonextendable skew 5-tuple of lines corresponds via Theorem 8.5.7 to a 4-simplex shaped face of 2_{21} that is contained only in five-dimensional hyperoctahedral faces. (This corresponds to the parabolic subgroup $\langle s_3, s_4, s_5, s_6 \rangle$.)

Theorem 10.2.1 shows that 36 of the Steiner complexes can be understood in terms of double sixes. Theorem 9.5.8 (ii) and (vii) show that there are $27 = 63 - 36$ other Steiner complexes, and that all 27 contain the bitangent $[0, 7]$. These 27 Steiner complexes also have an interesting interpretation, as the next result shows.

Proposition 10.2.5 *Let Σ be a Steiner complex containing the bitangent $[0, 7]$, let ω be the bitangent of Σ that is opposite to $[0, 7]$ in the sense of Definition 9.5.7, and let $\Sigma' = \Sigma \backslash \{[0, 7], \omega\}$.*

(i) *The elements of Σ' correspond to the set of ten lines (among the 27 lines on a cubic surface) that are incident to the line corresponding to the bitangent ω.*
(ii) *The elements of Σ' consist of five opposite pairs of bitangents within Σ, in the sense of Definition 9.5.7.*
(iii) *The map $\iota_{\mathcal{T}}$ sending one of the 27 lines to the set of ten lines incident to it is injective.*

Proof Lemma 10.1.6 shows that there are ten lines incident to the line corresponding to ω. Lemma 10.1.7 and Theorem 9.2.1 then show these ten lines correspond to those bitangents α for which $\{[0, 7], \omega, \alpha\}$ is an azygetic triple. Part (i) now follows by Theorem 9.5.8 (vi), and part (ii) is immediate from (i).

By Theorem 9.5.8 (vii) and the construction, there are 27 possibilities for the sets of ten lines mentioned in (i), and they form a single orbit under the action of $W(E_6)$. Part (iii) follows from this observation. □

Definition 10.2.6 We denote the collection of 27 10-tuples of lines defined in Proposition 10.2.5 by $\mathcal{T}$.

Proposition 10.2.7 *Let L_1 and L_2 be two distinct lines among the 27 lines on a cubic surface, and let $\iota_{\mathcal{T}}(L_1)$ and $\iota_{\mathcal{T}}(L_2)$ be the corresponding elements of $\mathcal{T}$, as in Proposition 10.2.5 (iii).*

(i) *The subgroup of $W(E_6)$ that stabilizes an element of $\mathcal{T}$ setwise is isomorphic to $W(D_5)$ acting in its natural permutation representation on the ten points*

$$\{1, 2, 3, 4, 5\} \cup \{\bar{1}, \bar{2}, \bar{3}, \bar{4}, \bar{5}\}.$$

Two of the ten points correspond to mutually incident lines if and only if they are of the form $\{i, \bar{i}\}$ for some $1 \leq i \leq 5$.

(ii) *Suppose that L_1 and L_2 are incident to each other. Then the intersection $\iota_{\mathcal{T}}(L_1) \cap \iota_{\mathcal{T}}(L_2)$ is a singleton, $\{\omega\}$, and $L_3 = \iota_{\mathcal{T}}(\omega)$ is the unique line that is incident to both of L_1 and L_2. Furthermore, the union $\iota_{\mathcal{T}}(L_1) \cup \iota_{\mathcal{T}}(L_2) \cup \iota_{\mathcal{T}}(L_3)$ is the entire set of 27 lines.*

(iii) *Suppose that L_1 and L_2 are skew to each other. If we consider $L_1 \cap L_2$ as a subset of L_1 and use the identifications of (i), then $\iota_{\mathcal{T}}(L_1) \cap \iota_{\mathcal{T}}(L_2)$ has size 5 and contains no pair of the form $\{i, \bar{i}\}$. Furthermore, each of the 16 lines L that is skew to L_1 gives rise to a different set $L_1 \cap L$.*

Proof Consider the parabolic subheap F of the full heap $\varepsilon : \mathrm{FH}(E_6^{(1)}(5)) \to E_6^{(1)}$ shown in Figure 6.12. The convex subheap $S_{5,1}$ of F, defined in Definition 8.2.7, consists of the eight elements lying between the two occurrences of 1, and is isomorphic (in **Heap**, after ignoring empty boxes and unused labels) to the parabolic subheap of the full heap $\varepsilon : \mathrm{FH}(D_5^{(1)}(1)) \to D_5^{(1)}$. The identifications of Lemma 10.1.7 and Exercise 3.3.14 then establish the first assertion of (i); the second assertion follows from Theorem 9.2.1.

By Proposition 10.2.5 (iii), the action of $W(E_6)$ on the 27 elements of $\mathcal{T}$ is equivalent to the action on the 27 lines, and the latter action is rank 3 by Theorem 8.2.22. It follows that, up to $W(E_6)$-symmetry, there are only two possible relative positions for the elements $\iota_{\mathcal{T}}(L_1)$ and $\iota_{\mathcal{T}}(L_2)$, according as L_1 and L_2 are incident or skew.

For $i \in \{1, 2\}$, let Σ_i denote the Steiner complex

$$\iota_{\mathcal{T}}(L_i) \cup \{\omega_i\} \cup \{[0, 7]\},$$

where ω_i is the bitangent corresponding to the line L_i. By Proposition 9.5.15, the set $\Sigma_1 \cap \Sigma_2$ has size 4 or 6, and the previous paragraph shows that these possibilities correspond to L_1 and L_2 being incident or skew (in some order).

Theorem 9.5.8 (v) shows that the triple $T_\omega = \{[0, 7], \omega_1, \omega_2\}$ is azygetic if and only if $\omega_2 \in \Sigma_1$.

Suppose first that T_ω is azygetic. By Corollary 9.4.11, there is a unique bitangent ω_3 such that $Q = \{[0, 7], \omega_1, \omega_2, \omega_3\}$ is an azygetic 4-tuple. It follows from Lemma 10.1.7 that L_1 and L_2 are incident, and that ω_3 corresponds to the unique line incident to both ω_1 and ω_2. By Lemma 9.5.17, we have $Q = \Sigma_\alpha \cap \Sigma_\beta$ for two Steiner complexes corresponding to the positive roots α and β respectively. As in Proposition 9.5.13, let γ be the unique positive root such that $\{\alpha, \beta, \gamma\}$ is a strongly orthogonal triple. Proposition 9.5.13 (iii) and Theorem 9.5.8 (v) now show that, after relabelling if necessary, Σ_α (respectively, Σ_β, Σ_γ) are the unique Steiner complexes in which ω_1

(respectively, ω_2, ω_3) and $[0, 7]$ form an opposite pair. It follows that

$$\iota_{\mathcal{T}}(L_1) \cap \iota_{\mathcal{T}}(L_2) = \{\omega_3\}.$$

It remains to prove the last assertion of (ii), and this follows from Proposition 9.5.13 (i).

The other possibility is that T_ω is syzygetic, L_1 and L_2 are skew, and $\Sigma_1 \cap \Sigma_2$ has size 6. In this case, we must have $[0, 7] \in \Sigma_1 \cap \Sigma_2$. This shows that $Q' = \iota_{\mathcal{T}}(L_1) \cap \iota_{\mathcal{T}}(L_2)$ has size 5. It follows from (i) and Proposition 9.5.15 (iv) that Q' contains no pair of the form $\{i, \bar{i}\}$.

If we regard Q' as a subset of the ten points giving the natural permutation representation of type $W(D_5)$, then the orbit of Q' under $W(D_5)$ contains 16 points, because the group acts by permutations and even numbers of sign changes. However, Lemma 10.1.6 shows that there are only 16 lines skew to L_1, and it follows from this that they must all have distinct intersections with L_1. This completes the proof of (iii). □

Exercise 10.2.8 Consider the ordered triples of lines (L_1, L_2, L_3) among the 27 lines such that L_1 is skew to both of L_2 and L_3, but L_2 and L_3 are incident with each other.

(i) Show, for example, by considering the ideals of the principal subheap of the full heap $\text{FH}(E_6^{(1)}(5))$, that the number of such triples is $27 \times 16 \times 5 = 2160$.
(ii) Show that the (pointwise) stabilizer in $W = W(E_6)$ of one such triple of lines is isomorphic to S_4 and conjugate to the parabolic subgroup $\langle s_2, s_3, s_6 \rangle$.
(iii) Show that (E_1, E_2, G_1) is an example of such a triple of lines, and describe its full point stabilizer explicitly.

Exercise 10.2.9 Let α be a root in type E_6. Show that the set of roots orthogonal to α has a unique highest element.

Proposition 10.2.10 *Let Σ' be the set of 27 lines, let Σ_α and Σ_β be distinct double sixes, and let $T = \Sigma' \backslash (\Sigma_\alpha \cup \Sigma_\beta)$.*

(i) *We have $|\Sigma_\alpha \cap \Sigma_\beta| \in \{4, 6\}$.*
(ii) *If $|\Sigma_\alpha \cap \Sigma_\beta| = 6$, then the symmetric difference*

$$\Sigma_\alpha \mathbin{\Delta} \Sigma_\beta = \Sigma_\gamma$$

is also a double six. Furthermore, $\{\pm\alpha, \pm\beta, \pm\gamma\}$ forms a root system of type A_2.
(iii) *Suppose $|\Sigma_\alpha \cap \Sigma_\beta| = 4$. Then there exists a unique line $b \in T$ such that*

$$T \cup \{\Sigma_\alpha \cap \Sigma_\beta\} = \{b\} \cup T',$$

where T' is the set of ten lines incident to b.

Proof Part (i) is immediate from Theorem 10.2.1 (iv) and Proposition 9.5.15 (i). Part (ii) is immediate from Theorem 10.2.1 (iv) and Proposition 9.5.15 (iii).

Suppose now that $|\Sigma_\alpha \cap \Sigma_\beta| = 4$. By Proposition 9.5.13 (ii),

$$T \cup (\Sigma_\alpha \cap \Sigma_\beta) \cup \{[0, 7]\}$$

is a Steiner complex, Σ_γ, where $\{\alpha, \beta, \gamma\}$ is a strongly orthogonal triple in type E_7. Now Σ_γ consists of six opposite pairs of bitangents, and two of these are contained in $\Sigma_\alpha \cap \Sigma_\beta$ by Lemma 9.5.12 (i). Let b be the element of Σ_γ opposite $[0, 7]$; by the preceding remarks, b cannot lie in $\Sigma_\alpha \cap \Sigma_\beta$. There are ten lines incident with b by Lemma 10.1.6, and it follows by Lemma 10.1.7 that these ten elements are the ones forming an azygetic triple with b and $[0, 7]$. Theorem 9.5.8 (v) now shows that these ten lines are precisely the elements of T. □

Example 10.2.11 An example of two double sixes that intersect in six elements is given by the pair

$$\begin{pmatrix} E_1 & E_2 & E_3 & E_4 & E_5 & E_6 \\ G_1 & G_2 & G_3 & G_4 & G_5 & G_6 \end{pmatrix} \text{ and } \begin{pmatrix} E_1 & E_2 & E_3 & F_{4,5} & F_{4,6} & F_{5,6} \\ F_{2,3} & F_{1,3} & F_{1,2} & G_6 & G_5 & G_4 \end{pmatrix}.$$

The symmetric difference of this pair is another double six, namely

$$\begin{pmatrix} E_4 & E_5 & E_6 & F_{2,3} & F_{1,3} & F_{1,2} \\ F_{5,6} & F_{4,6} & F_{4,5} & G_1 & G_2 & G_3 \end{pmatrix}.$$

Example 10.2.12 An example of two double sixes that intersect in four elements is given by the pair

$$\begin{pmatrix} E_1 & E_2 & E_3 & E_4 & E_5 & E_6 \\ G_1 & G_2 & G_3 & G_4 & G_5 & G_6 \end{pmatrix} \text{ and } \begin{pmatrix} E_1 & G_1 & F_{2,3} & F_{2,4} & F_{2,5} & F_{2,6} \\ E_2 & G_2 & F_{1,3} & F_{1,4} & F_{1,5} & F_{1,6} \end{pmatrix}.$$

In this case, the set $T \cup \{\Sigma_\alpha \cap \Sigma_\beta\}$ is the set of 11 lines

$$\{F_{1,2}, E_1, E_2, G_1, G_2, F_{3,4}, F_{3,5}, F_{3,6}, F_{4,5}, F_{4,6}, F_{5,6}\};$$

in other words, the line $F_{1,2}$ together with the ten lines incident to it.

Exercise 10.2.13 Exercise 8.4.11 shows that there are 36 positive roots in type E_6.

(i) Recall the description of the root system of type A_l from Exercise 5.3.20, and deduce that there are 15 positive roots in type A_5.
(ii) Recall from Exercise 4.4.7 that the group $W(E_6)$ acts transitively on pairs of orthogonal roots of type E_6. Use the previous facts to show that there are 270 unordered pairs of orthogonal roots in type E_6, and that $W(E_6)$ acts transitively on them.
(iii) Show that in the principal subheap F of the full heap $\varepsilon : \mathrm{FH}(E_6^{(1)}(5)) \to E_6^{(1)}$, there is a unique pair of ideals (I, I') with $\chi(I, I') = \alpha_3 + \alpha_5$, where $\{s_3, s_5\}$ is a pair of commuting generators.
(iv) Show that the pair of weights $\{\lambda_1, \lambda_2\}$ corresponding to the pair $\{I, I'\}$ in (iii) is a pair of weights at distance $\sqrt{2}D$ in the polytope 2_{21}, where D is the minimal nonzero distance between vertices.
(v) Use Exercise 8.6.5 and Theorem 9.2.1 to show that there is a two-to-one correspondence between (a) the set of 270 pairs of orthogonal reflections in type E_6 and (b) the set of 135 pairs of incident lines among the 27 lines.

Exercise 10.2.14 Let T be the set of ordered triples of weights $(\lambda_1, \lambda_2, \lambda_3)$ of $L(E_6, \omega_5)$ such that the λ_i are at mutual distance $\sqrt{2}D$, where D is the minimal nonzero distance between weights. Show that the triple

$$(\omega_4 - \omega_5 - \omega_6,\ \omega_6 - \omega_4,\ \omega_5)$$

is an element of T. Show that the diagonal action of $W(E_6)$ on T is transitive, and furthermore that the sum of the three weights in any triple of T is zero.

Exercise 10.2.15 Let F be the principal subheap of Exercise 10.2.13 (iii).

(i) Show (without using an exhaustive check) that for any pair of distinct commuting fundamental reflections s_i and s_j of $W(E_6)$, there is a unique pair of ideals (I, I') of F such that $\chi(I, I') = \alpha_i + \alpha_j$.

(ii) Deduce that there is a unique antichain $\{\alpha, \beta\}$ of F such that $\varepsilon(\alpha) = i$ and $\varepsilon(\beta) = j$.

Exercise 10.2.16 Consider the action of $W(E_6)$ on the 27 lines on a cubic surface.

(i) Show that product of the pair of orthogonal reflections $\{s_3, s_5\}$ exchanges the pair of incident lines $\{F_{3,5}, F_{4,6}\}$.

(ii) Show that the product of the pair $\{s_3, s_5\}$ does not exchange any other pair of incident lines.

(iii) Show that the pair of orthogonal reflections $\{s_\alpha, s_4\}$, where $\alpha = \alpha_3 + \alpha_4 + \alpha_5$, is the *only* other pair of orthogonal reflections whose product exchanges the pair of lines $\{F_{3,5}, F_{4,6}\}$.

10.3 2-graphs

Definition 10.3.1 Let Ω be a finite set and let $\Omega^{(n)}$ be the set of all n-element subsets of Ω. An *n-cochain* is a function

$$f : \Omega^{(n+1)} \to \mathbb{Z}_2.$$

The automorphism group of an n-cochain f is the subgroup of permutations π in $\Sigma(\Omega)$ for which $f(\Gamma^\pi) = f(\Gamma)$ for all $\Gamma \in \Omega^{(n+1)}$. The set C^n of all n-cochains forms an abelian group under pointwise addition. The *coboundary operator* is the homomorphism $\delta : C^n \to C^{n+1}$ given by

$$(\delta f)(\Gamma) = \sum_{\alpha \in \Gamma} f(\Gamma \backslash \{\alpha\}),$$

where $\Gamma \in \Omega^{(n+2)}$ and $f \in C^n$. An n-cochain of the form δf is called an *n-coboundary*, and an n-cochain f for which $\delta f = 0$ is called an *n-cocycle*.

A *2-graph* is a 2-cocycle $\Phi : \Omega^{(3)} \to \mathbb{Z}_2$. An element $\Gamma \in \Omega^{(3)}$ is called a *coherent triple* of a 2-graph Φ if $\Phi(\Gamma) = 1$, and an *incoherent triple* otherwise. A subset of $\Omega^{(k)}$ for $k > 3$ is called *coherent* (respectively, incoherent) if each of its 3-element subsets is coherent (respectively, incoherent). A 2-graph Φ is said to be *regular* if any 2-element subset of Ω is contained in the same number of coherent triples.

In this section, we will show how to use minuscule representations to gain insight into the following theorem of D. E. Taylor [83].

Theorem 10.3.2 (Taylor) *Up to isomorphism, there are precisely three regular 2-graphs with the property that every coherent triple is contained in a unique coherent 4-tuple.*

Proof This is part of [83, theorem 3.7]. □

Note that a graph with vertex set Ω can be regarded as a 1-cochain, where $f(v, w) = 1$ if v and w are adjacent, and $f(v, w) = 0$ otherwise. Although the concept of a 2-graph is not directly analogous to that of a graph, we may think of a graph on the set Ω as a collection of pairs of elements, and a 2-graph on the set Ω as a collection of distinguished triples of elements, namely the coherent triples.

Definition 10.3.3 Maintain the above notation. For each $\omega \in \Omega$, define the *contracting homotopy* $\Delta_\omega : C^{n+1} \to C^n$ by

$$(\Delta_\omega f)(\Gamma) = \begin{cases} f(\Gamma \cup \{\omega\}) & \text{if } \omega \notin \Gamma, \\ 0 & \text{otherwise,} \end{cases}$$

where $\Gamma \in \Omega^{(n+1)}$.

Lemma 10.3.4 *If $f \in C^n$ and $\omega \in \Omega$, then we have*

$$f = \Delta_\omega(\delta f) + \delta(\Delta_\omega f).$$

Proof The proof reduces to evaluating each side of the equation on an arbitrary element $\Gamma \in \Omega^{(n+1)}$.

Let us suppose first that $\omega \notin \Gamma$. We have

$$\begin{aligned}(\Delta_\omega(\delta f))(\Gamma) &= (\delta f)(\Gamma \cup \{\omega\}) \\ &= f(\Gamma) + \sum_{\alpha \in \Gamma} f(\Gamma \backslash \{\alpha\} \cup \omega)\end{aligned}$$

and

$$(\delta(\Delta_\omega f))(\Gamma) = \sum_{\alpha \in \Gamma} (\Delta_\omega f)(\Gamma \backslash \{\alpha\}),$$

because $\omega \notin \Gamma \backslash \{\alpha\}$. When these two expressions are added, all the terms cancel, except the extra term in the first expression, producing a total of $f(\Gamma)$, as required.

Finally, we consider the case where $\omega \in \Gamma$. It follows from the definitions that in this case we have

$$(\Delta_\omega(\delta f))(\Gamma) = 0.$$

On the other hand, we have

$$(\delta(\Delta_\omega f))(\Gamma) = \sum_{\alpha \in \Gamma} (\Delta_\omega f)(\Gamma \backslash \{\alpha\}).$$

Each term in the sum will be zero unless $\omega \notin \Gamma \backslash \{\alpha\}$, but because $\omega \in \Gamma$, this can only happen if $\alpha = \omega$. It follows again that

$$(\Delta_\omega(\delta f) + \delta(\Delta_\omega f))(\Gamma) = f(\Gamma),$$

as required. □

Lemma 10.3.5 *An n-cochain is an n-cocycle if and only if it is an n-coboundary.*

Proof Let $f \in C^n$ and $\Gamma \in \Omega^{(n+1)}$. It follows from the definitions that

$$(\delta^2 f)(\Gamma) = \sum_{\alpha,\beta\in\Gamma,\ \alpha\neq\beta} f(\Gamma\backslash\{\alpha,\beta\}) = 0,$$

which shows that every coboundary is a cocycle.

Conversely, if f is a cocycle, we have $\delta f = 0$, and Lemma 10.3.4 shows that $f = \delta(\Delta_\omega f)$ is a coboundary. □

Proposition 10.3.6 *Let Ω be a finite set and let $T \subseteq \Omega^{(3)}$.*

(i) *The set T is the set of coherent triples of a 2-graph if and only if every element of $\Omega^{(4)}$ contains an even number of elements of T among its four 3-element subsets.*
(ii) *If $T = \Omega^{(3)}$ or $T = \emptyset$, then T is the set of coherent triples of a 2-graph.*
(iii) *The set T is the set of coherent triples of a 2-graph Φ if and only if $\Omega^{(3)}\backslash T$ is the set of coherent triples of a 2-graph Φ^*.*

Proof A 2-cochain Φ is a 2-graph if and only if for any $\Gamma \in \Omega^{(4)}$, we have

$$0 = (\delta\Phi)(\Gamma) = \sum_{\alpha\in\Gamma} f(\Gamma\backslash\{\alpha\}).$$

This condition is equivalent to the condition in (i). Parts (ii) and (iii) are immediate from (i). □

Definition 10.3.7 The 2-graphs appearing in Proposition 10.3.6 (ii) are called *trivial*. The 2-graph Φ^* constructed in Proposition 10.3.6 (iii) is called the *complement* of Φ.

Lemma 10.3.8 *Any collection of four bitangents contains an even number of azygetic triples as subsets.*

Proof Let $B = \{b_1, b_2, b_3, b_4\}$ be a set of four bitangents. If there are no azygetic 3-element subsets of B, then we are done, so assume without loss of generality that $\{b_1, b_2, b_3\}$ is azygetic. By Proposition 9.4.5, it suffices to prove the result for one particular triple of azygetic bitangents, so by Proposition 9.4.9, we may assume that $b_1 = [0, 1]$, $b_2 = [2, 3]$ and $b_3 = [4, 5]$.

Up to permutations in S_8, there are three situations to consider regarding $b_4 = [i, j]$, depending on whether the number

$$k = |\{i, j\} \cap \{0, 1, 2, 3, 4, 5\}|$$

is equal to 0, 1 or 2.

If $k = 0$, we must have $b_4 = [6, 7]$. In this case, B is an azygetic 4-tuple and contains four azygetic triples.

If $k = 1$, then we may assume that $b_4 = [5, 6]$. In this case, there are precisely two azygetic 3-subsets of B, namely $\{b_1, b_2, b_3\}$ and $\{b_1, b_2, b_4\}$. This completes the proof in this case.

If $k = 2$, then we may assume that $b_4 = [2, 4]$. Again, there are precisely two azygetic 3-subsets of B in this case, namely $\{b_1, b_2, b_3\}$ and $\{b_2, b_3, b_4\}$. This completes the proof. □

Theorem 10.3.9 *Let Ω be the set of 28 bitangents. Suppose there exists a subgroup $H \leq W(E_7)$ and a subset $\Omega' \subseteq \Omega$ such that H acts doubly transitively on Ω'.*

(i) *There is a (unique) 2-graph structure Φ on Ω' in which the coherent triples are the azygetic triples in Ω'.*
(ii) *The complementary 2-graph Φ^* has as its coherent triples the syzygetic triples of bitangents.*
(iii) *Both Φ and Φ^* are regular 2-graphs.*

Proof The first assertion follows from the definitions, Proposition 10.3.6 (i) and Lemma 10.3.8. The second assertion is immediate from the first, and the last assertion follows by hypothesis. □

Corollary 10.3.10 *The 28 bitangents have the structure of a regular 2-graph, Ω_E, where the coherent triples are the azygetic triples. The automorphism group of Ω_E contains a subgroup isomorphic to $W(E_7)/Z(W(E_7))$.*

Proof The hypotheses of Theorem 10.3.9 are satisfied for $H = W(E_7)$ and $\Omega' = \Omega$ by Corollary 9.3.6, which proves the first assertion. By Lemma 9.3.4, the only elements of W that can fix all the bitangents are 1, and w_0 if w_0 acts as -1. These comprise the centre of W by Proposition 4.3.17 (ii), and this completes the proof. □

Proposition 10.3.11 *Let Ω be the set of 28 bitangents, and let $T \subset \Omega^{(3)}$ be the set of azygetic triples. The set*

$$\Omega_D = \{[0, 1]\} \cup \{[i, j] : 2 \leq i < j \leq 7\}$$

equipped with the coherent triples $T_D := T \cap \Omega_D^{(3)}$ is a regular 2-graph whose automorphism group contains a subgroup isomorphic to $W(D_6)/Z(W(D_6))$.

Proof It follows from Corollary 10.3.10 and Proposition 10.3.6 (i) that Ω_D forms a 2-graph, but it remains to show that Ω_D is regular.

Consider the type D_6 subgraph of the Dynkin diagram of type E_7 consisting of all vertices other than 1. Let H be the parabolic subgroup of $W(E_7)$ generated by these six generators.

When $L(E_7, \omega_6)$ is restricted to the Lie subalgebra corresponding to H, Example 8.2.11 shows that one of the direct summands appearing is $L(D_6, \omega_6)$. The weights of this submodule can be found by looking at the principal subheap F of $L(E_7, \omega_6)$: they consist of $\pm v_{0,1}$ together with all weights $\pm v_{i,j}$ for $2 \leq i < j \leq 7$. Recall from Lemma 9.3.2 (iii) that the spin modules in type D_6 have weights that are closed under negation, and the Weyl group acts as a rank 4 permutation group on the weights. It follows that Ω_D forms a single orbit under the action of H on Ω, and by Proposition 9.3.5, H acts doubly transitively on Ω_D. The regularity of Ω_D now follows from Theorem 10.3.9.

By construction, the group $W(D_6)$ acts as automorphisms of Ω_D. If an element $w \in W(D_6)$ fixes every element of Ω_D, it follows from Lemma 9.3.2 (iii) and Lemma 9.3.4 that $w \in \{1, w_0\}$. Proposition 4.3.17 (ii) then shows that $W(D_6)/Z(W(D_6))$ acts faithfully on Ω_D, completing the proof. □

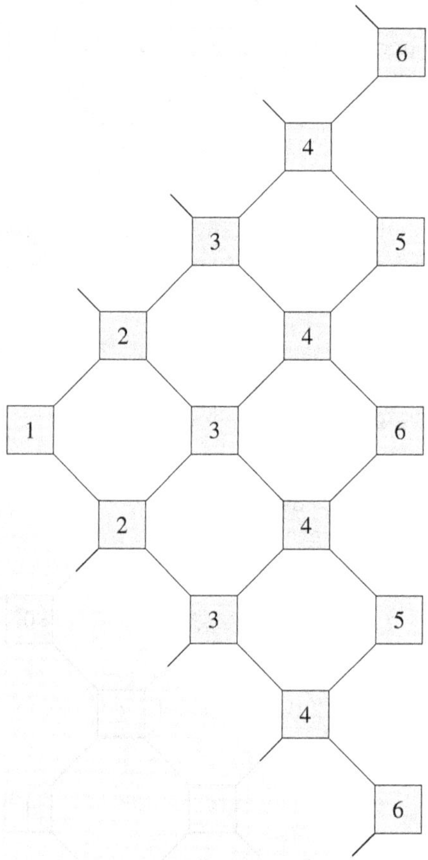

Figure 10.1 A principal subheap of $\varepsilon_6 : \mathrm{FH}(D_6^{(1)}(6)) \to D_6^{(1)}$

Exercise 10.3.12 Show that the stabilizer of the bitangent $[0, 1]$ in Ω_D is isomorphic to $S_6 \times \mathbb{Z}_2$. (Hint: think of S_6 as the symmetric group on the set $\{2, 3, 4, 5, 6, 7\}$ and $\mathbb{Z}_2$ as the subgroup of $W(D_6)$ generated by the longest element.)

Exercise 10.3.13 Consider the spin representation $L(D_6, \omega_6)$, whose principal subheap is shown in Figure 10.1. Let U be the set of ordered triples of weights (λ, μ, ν) at mutual distance D (the smallest nonzero distance between weights). Show that $W(D_6)$ acts transitively on U, and deduce that the syzygetic triples in the 2-graph Ω_D form a single orbit under $W(D_6)$.

Exercise 10.3.14 Consider the spin representation $L(D_6, \omega_6)$, and consider the set U of ordered triples of weights (λ, μ, ν) at mutual distance $\sqrt{2}D$. Show that $W(D_6)$

acts transitively on U, and deduce that the azygetic triples in the 2-graph Ω_D form a single orbit under $W(D_6)$.

Proposition 10.3.15 *Let Ω be the set of 28 bitangents, and let $T \subset \Omega^{(3)}$ be the set of azygetic triples. The set*

$$\Omega_A = \{[i, j] : 3 \leq i < j \leq 7\}$$

equipped with the coherent triples $T_A := T \cap \Omega_A^{(3)}$ is a regular 2-graph whose automorphism group contains a subgroup isomorphic to $W(A_5) \cong S_6 \cong W(A_5)/Z(W(A_5))$.

Proof The set Ω_A forms a 2-graph for the same reason as in the proof of Proposition 10.3.11; it remains to be shown that Ω_A is regular.

Consider the type A_5 subgraph of the Dynkin diagram of type D_6 consisting of all vertices other than 6. Let H be the parabolic subgroup of $W(D_6)$ generated by these five generators.

When $L(D_6, \omega_6)$ is restricted to the Lie subalgebra corresponding to H, Exercise 8.2.12 shows that one of the direct summands appearing is $L(A_5, \omega_3)$. The weights of this submodule can be found by looking at the principal subheap F of $L(A_5, \omega_3)$, which in turn corresponds to the subheap of $L(E_7, \omega_6)$ with maximal element $F(4, 4)$ and minimal element $F(4, 2)$. The corresponding set of weights is

$$\{\pm v_{i,j} : 3 \leq i < j \leq 7\}.$$

It follows that Ω_A forms a single orbit under the action of H on Ω, and by Proposition 9.3.5, H acts doubly transitively on Ω_A. The regularity of Ω_A now follows from Theorem 10.3.9.

By construction, the group $W(A_5)$ acts as automorphisms of Ω_A. If an element $w \in W(A_5)$ fixes every element of Ω_A, it follows from Lemma 9.3.2 (iii) and Lemma 9.3.4 that w is the identity, completing the proof. $\square$

Exercise 10.3.16 Show that the stabilizer in S_6 of a bitangent in Ω_A is isomorphic to $S_3 \wr \mathbb{Z}_2$, that is, the extension of the group $S_3 \times S_3$ by an involution exchanging the two factors. Show that such a stabilizer is conjugate to the subgroup of S_6 generated by the set

$$\{(1, 2), (2, 3), (4, 5), (5, 6), (14)(25)(36)\}.$$

Exercise 10.3.17 Consider the representation $L(A_5, \omega_3)$, and consider the set U of ordered triples of weights (λ, μ, ν) at mutual distance D (the smallest nonzero distance between weights).

(i) Show that $W(A_5)$ has *two* orbits in its action on U.
(ii) Show that the stabilizer group of Exercise 10.3.16 acts transitively on U.
(iii) Show that $W(A_5)$ acts transitively on the syzygetic triples in Ω_A.

Exercise 10.3.18 Consider the spin representation $L(A_5, \omega_3)$, and consider the set U of ordered triples of weights (λ, μ, ν) at mutual distance $\sqrt{2}D$. Show that $W(A_5)$ acts transitively on U, and deduce that the azygetic triples in the 2-graph Ω_A form a single orbit under $W(A_5)$.

Proposition 10.3.19 *The regular 2-graphs Ω_E (on 28 points), Ω_D (on 16 points) and Ω_A (on 10 points) each have the property that any of their coherent triples is contained in a unique coherent 4-tuple. They are the only regular 2-graphs with this property.*

Proof The second assertion follows from the first assertion together with Theorem 10.3.2.

Corollary 9.4.11 proves the first assertion in the case of Ω_E.

To prove the first assertion for Ω_D, it is enough by Proposition 9.4.9 to prove that if Ω_D contains three of a collection of four disjoint edges then it contains the fourth, and if it contains three edges of a square, then it contains the fourth. This property is easily checked from the explicit description of Ω_D in Proposition 10.3.11.

A similar check using Proposition 10.3.15 proves the first assertion for Ω_A. □

Proposition 10.3.20 *The full automorphism groups of Ω_E, Ω_D and Ω_A are $W(X)/Z(W(X))$, where $X = E_7$, D_6 and A_5 respectively.*

Proof Corollary 10.3.10 and Propositions 10.3.11 and 10.3.15 describe automorphism groups

$$W(X)/Z(W(X))$$

of Ω_E, Ω_D and Ω_A for $X = E_7$, D_6 and A_5 respectively, so it remains to show that these are the full automorphism groups.

Taylor proves [83, Theorem 3.7] that the automorphism groups of Ω_A and Ω_E are S_6 and Sp(6, 2) respectively, which proves the claim in these cases because $W(E_7) \cong$ Sp(6, 2) $\times \mathbb{Z}_2$. Taylor describes the automorphism group of Ω_D imprecisely as "the semidirect product of an elementary abelian group of order 16 by Sp(4, 2)". It is well-known to group theorists that Sp(4, 2) $\cong S_6$, giving an automorphism group of order $2^4.6!$. It follows from Exercise 5.3.30 that $W(D_6)$ has order $2^5.6!$, and it follows from the proof of Proposition 10.3.11 that $Z(W(D_6))$ has order 2, and this completes the proof. □

Exercise 10.3.21 Use Exercise 9.2.15 (with $n = 6$) and Proposition 10.3.19 to prove the following.

(i) There are two $W = W(D_6)$-orbits of incoherent 4-tuples in Ω_D: those that can be extended to incoherent 5-tuples, and those that cannot.
(ii) Every incoherent triple in Ω_D is contained in precisely three extendable incoherent 4-tuples, and in a unique nonextendable incoherent 4-tuple.

Exercise 10.3.22 Derive the following analogue of Exercise 10.3.21 using Proposition 9.2.8 and Exercise 9.2.15 (with $n = 5$).

(i) There are two $W = W(E_7)$-orbits of incoherent 6-tuples in Ω_D: those that can be extended to incoherent 7-tuples, and those that cannot.
(ii) Every incoherent 5-tuple in Ω_E is contained in precisely two extendable incoherent 6-tuples, and in a unique nonextendable incoherent 6-tuple.

10.4 Generalized quadrangles

Definition 10.4.1 An *incidence structure* is a triple of sets (P, B, I) with $I \subseteq P \times B$. The elements of P (respectively, B, I) are called the *points* (respectively, *lines*, *flags*) of the incidence structure. If $(p, L) \in I$, we say that the point p lies on the line L, or that the line L contains the point p. An isomorphism of two incidence structures consists of a bijection between the two sets of points and a bijection between the two sets of lines compatible with the two incidence relations.

Two points p_1 and p_2 are said to be *collinear* if there exists a line L such that $(p_i, L) \in I$ for $i \in \{1, 2\}$. A *k-coclique* (or *coclique* for short) is a collection of k points, no two of which are collinear.

A *generalized quadrangle* is an incidence structure (P, B, I) satisfying the following conditions for certain natural numbers s and t:

(i) it is not possible for each of two distinct points p_1 and p_2 to lie on each of two distinct lines L_1 and L_2;
(ii) every line contains $s + 1$ points;
(iii) every point lies on $t + 1$ lines;
(iv) if the point p is not on the line L, then there is a unique line M containing p, and a unique point q on L, such that q lies on M.

We call (s, t) the *parameters* of the generalized quadrangle.

In this section, we will use minuscule representations and 2-graphs to gain insight into the following well-known theorem in finite geometry.

Proposition 10.4.2 *Any generalized quadrangle with parameters* $(2, t)$ *must have* $t \in \{1, 2, 4\}$, *and each choice of* t *determines a unique isomorphism class of generalized quadrangle.*

Proof It follows from Payne and Thas [60, 1.2.2, 1.2.3] that we must have $t \in \{1, 2, 4\}$. The uniqueness in the case $t = 1$ (respectively, $t = 2$, $t = 4$) is proved in [60, section 1.1] (respectively, [60, 5.2.3], [60, 5.3.2 (ii)]). □

Theorem 10.4.3 *Let* Ω *be one of the 2-graphs* Ω_E, Ω_D *and* Ω_A. *We fix some element* $\omega \in \Omega$ *and define* $P = \Omega \backslash \{\omega\}$. *Let* B *be the set of azygetic 4-tuples* L *of* Ω *such that* $\omega \in L$, *and let* $I \subseteq P \times B$ *consist of those elements* (p, L) *for which* $p \in L$.

(i) *The incidence structure* (P, B, I) *is a generalized quadrangle.*
(ii) *The parameters of* (P, B, I) *are* $(2, 4)$ *in the case of* Ω_E, $(2, 2)$ *in the case of* Ω_D *and* $(2, 1)$ *in the case of* Ω_A. *Up to isomorphism, these are the only generalized quadrangles with* $s = 2$.

Proof It is immediate from the definitions that every line in B contains three points. Suppose that the points p_1 and p_2 lie on the line L. This means that $\{p_1, p_2, \omega, q\}$ is an azygetic quadruple for some point q. By Proposition 10.3.19, an azygetic quadruple is determined by any three of its elements. This shows that L is unique and proves condition (i) of Definition 10.4.1.

By Corollary 10.3.10, Proposition 10.3.11 and Proposition 10.3.15, Ω is a regular 2-graph. This shows that if ω' is an element of P, the number of lines containing the pair $\{\omega, \omega'\}$ is independent of the choice of ω'. This means that each point lies on the same number of lines.

In the case of Ω_E, we can argue as in the proof of Proposition 9.2.12 (iv) to show that any pair of bitangents is contained in precisely five azygetic 4-tuples. It follows that every point of Ω_E lies on five lines. In the case of Ω_D, we observe that the pair of bitangents $\{[0, 1], [2, 3]\}$ can be completed in three ways to an azygetic quadruple of four disjoint lines containing $[0, 1]$, corresponding to the three partitions of the set $\{4, 5, 6, 7\}$ into pairs. This shows that every point of Ω_D lies on three lines. Finally, in the case of Ω_A, we observe that the pair of bitangents $\{[3, 4], [4, 5]\}$ can be completed in two ways to an azygetic quadruple containing $[3, 4]$, by adding either $\{[5, 6], [3, 6]\}$ or $\{[5, 7], [3, 7]\}$. This shows that every point of Ω_A lies on two lines.

This completes the proof of conditions (ii) and (iii) of Definition 10.4.1, as well as the assertions in (ii) regarding the values of parameters. The uniqueness of the quadrangles follows from Proposition 10.4.2. It remains to check condition (iv) of Definition 10.4.1.

To prove condition (iv), let us first observe by Proposition 10.3.19 that points q_1 and q_2 are collinear if and only if $\{\omega, q_1, q_2\}$ is azygetic. Let L be a line and let p be a point not lying on L. We have $L = \{p_1, p_2, p_3, \omega\}$. By Proposition 10.3.19, the quadruples $L_2 = \{p_1, p_2, p_3, p\}$, $L_3 = \{\omega, p_1, p_2, p\}$ and $L_4 = \{\omega, p_1, p_3, p\}$ are not azygetic.

Since L_2 contains the azygetic triple $\{p_1, p_2, p_3\}$, it follows by Corollary 10.3.10 and Proposition 10.3.6 (i) that L_2 contains precisely two azygetic triples. By re-indexing if necessary, we may assume that these are $\{p_1, p_2, p_3\}$ and $\{p_1, p_2, p\}$; in particular, $\{p_1, p_3, p\}$ is syzygetic.

The same argument applied to L_3 shows that $\{\omega, p_1, p\}$ and $\{\omega, p_2, p\}$ are syzygetic, proving that p is not collinear to either of p_1 or p_2.

So far, we have seen that $\{\omega, p_1, p_3\}$ is azygetic but both $\{p_1, p_3, p\}$ and $\{\omega, p_1, p\}$ are syzygetic. It follows from Corollary 10.3.10 and Proposition 10.3.6 that $\{\omega, p_3, p\}$ is azygetic. Let L_5 be the unique azygetic quadruple of the form $\{\omega, p_3, p, p_4\}$; this exists by Proposition 10.3.19. We have now shown that p is collinear to p_3, but not to p_1 or p_2, and that L_5 is the unique line containing p and p_3. This proves condition (iv) and completes the proof. □

Definition 10.4.4 We denote the generalized quadrangles corresponding to Ω_E, Ω_D and Ω_A by GQ(2,4), GQ(2,2) and GQ(2,1), respectively.

Exercise 10.4.5 Recall the construction of the 2-graph Ω_A from Proposition 10.3.15.

(i) Show that the construction identifies the points of GQ(2,1) with the set of nine bitangents of the form

$$\{[i, j] : 3 \leq i < j \leq 7,\ [i, j] \neq [3, 4]\}.$$

(ii) Show that the construction identifies the lines of GQ(2,1) with the set of six quadruples of bitangents of the form

$$\{\{[3, 4], [4, i], [i, j], [3, j]\} : 5 \leq i \neq j \leq 7\}.$$

(iii) Show that GQ(2,1) is isomorphic (as an incidence structure) to the grid consisting of nine points $\{(i, j) : i, j \in \{0, 1, 2\}\}$ and six lines, where each line consists of the three points having a fixed x (or y) coordinate.
(iv) It follows from the construction and Exercise 10.3.16 that the group $S_3 \wr \mathbb{Z}_2$ acts as automorphisms of GQ(2,1). Describe the action of this group explicitly in terms of the grid. Show that the action is faithful and transitive and verify that $S_3 \wr \mathbb{Z}_2$ is the full automorphism group of GQ(2,1).

Exercise 10.4.6 Recall the construction of the 2-graph Ω_D from Proposition 10.3.11.

(i) Show that the construction identifies the points of GQ(2,2) with the set of 15 bitangents of the form

$$\{[i, j] : 2 \le i < j \le 7\},$$

in other words, the set of pairs of elements in a 6-element set (called "duads").
(ii) Show that the construction identifies the lines of GQ(2,2) with the set of 15 quadruples of bitangents of the form

$$\{\{[0, 1], [i, j], [k, l], [m, n]\} : \{i, j, k, l, m, n\} = \{2, 3, 4, 5, 6, 7\}\},$$

that is, the 15 ways to decompose a 6-element set into three disjoint pairs (called "synthemes").
(iii) Show that, in the above model, two points are collinear if and only they correspond to disjoint 2-element sets.
(iv) Recall from Exercise 10.3.12 that the group $S_6 \times \mathbb{Z}_2$ acts as automorphisms of GQ(2,2). Describe the action of this group explicitly in terms of the model above.
(v) Show that the kernel of the action in (iv) is precisely the subgroup $1 \times \mathbb{Z}_2$, and that the quotient group (isomorphic to S_6) acts transitively on the points and on the lines.

Exercise 10.4.7 Recall the construction of the 2-graph Ω_E in terms of the 28 bitangents. Denote by Q_1 (respectively, Q_2) the set of azygetic quadruples of bitangents corresponding as in Proposition 9.4.9 to collections of four disjoint edges (respectively, squares).

(i) Show that the construction identifies the points of GQ(2,4) with the set of 27 bitangents of the form

$$\{[i, j] : 0 \le i < j \le 7,\ [i, j] \ne [0, 7]\},$$

that is, the 27 lines E_i, F_{jk}, G_l on a cubic surface.
(ii) Show that the construction identifies the elements of Q_1 containing the bitangent $[0, 7]$ with the 30 triples of lines of the form

$$\{E_i, F_{ij}, G_j : 1 \le i \ne j \le 6\}.$$

(iii) Show that the construction identifies the elements of Q_2 containing the bitangent $[0, 7]$ with the 15 triples of lines of the form

$$\{\{F_{ij}, F_{kl}, F_{mn}\} : \{i, j, k, l, m, n\} = \{1, 2, 3, 4, 5, 6\}\}.$$

(iv) Deduce that GQ(2,4) has 45 lines. Find a direct proof of this fact using the parameters $s = 1$ and $t = 3$ and the fact that GQ(2,4) has 27 points.

Exercise 10.4.8 Let (p, L) be a flag in GQ(2, 4).

(i) Show that the elements of $L\backslash\{p\}$ correspond to a pair of incident lines among the 27 lines in the cubic surface.
(ii) Show that the correspondence of (i) is a bijection between the 135 flags and the 135 pairs of incident lines (see Exercise 10.2.13 (v)), and that the bijection is compatible with the actions of $W(E_6)$ on each set.

Exercise 10.4.9

(i) Show that condition (iv) of Definition 10.4.1 for GQ(4,2) is a consequence of Proposition 10.2.7 (ii), after identifying the points of GQ(4,2) with the 27 lines on a cubic surface.
(ii) Suppose that $\{p_1, p_2, p_3\}$ are the points of a line in GQ(2,2), and let P_i be the 6-tuple of points collinear with p_i, for $i \in \{1, 2, 3\}$. Show that the pairwise intersections of the P_i are singletons, and that $P_1 \cup P_2 \cup P_3$ is the set of all 15 points of GQ(2,2).
(iii) Suppose that $\{p_1, p_2, p_3\}$ are the points of a line in GQ(1,2), and let P_i be the 4-tuple of points collinear with p_i, for $i \in \{1, 2, 3\}$. Show that the pairwise intersections of the P_i are singletons, and that $P_1 \cup P_2 \cup P_3$ is the set of all nine points of GQ(1,2).

Exercise 10.4.10 Use the definition of 2-graph to show that a 3-coclique of GQ(2,4), regarded as a set of bitangents other than $[0, 7]$, corresponds to an incoherent triple of bitangents. (Note that this property is inherited by GQ(2,2) and GQ(2,1).)

Exercise 10.4.11 The following is an analogue of Exercise 10.3.21 for GQ(2,2).

(i) Show that there are two S_6-orbits of 3-cocliques GQ(2,2): those that can be extended to 4-cocliques, and those that cannot.
(ii) Show that in the duad–syntheme model of Exercise 10.4.6, the extendable 3-cocliques are those of the form

$$\{\{a, b\}, \{a, c\}, \{a, d\}\}$$

where a, b, c, d are distinct, and the nonextendable 3-cocliques are those of the form

$$\{\{a, b\}, \{a, c\}, \{a, b\}\},$$

where a, b, c are distinct.
(iii) Show that every pair of noncollinear points in GQ(2,2) is contained in precisely three extendable 3-cocliques, and in a unique nonextendable 3-coclique.

Exercise 10.4.12 A famous property of the symmetric group S_6 is that, unlike other finite symmetric groups, it has *outer automorphisms*; that is, automorphisms that do not arise as conjugation by a fixed element. Each outer automorphism sends the 15 transpositions (i, j) of S_6 bijectively to the 15 triple transpositions $(i, j)(k, l)(m, n)$ with $\{i, j, k, l, m, n\} = \{1, 2, 3, 4, 5, 6\}$. Such an automorphism can also be shown to induce a bijection between the points (duads) and the lines (synthemes) of GQ(2,2).

(i) Show that the stabilizer in S_6 of a duad is isomorphic to $S_4 \times S_2$.
(ii) Show that the stabilizer in S_6 of a syntheme is isomorphic to the wreath product $\mathbb{Z}_2 \wr S_3$.
(iii) Deduce from the properties of outer automorphisms of S_6 described above to show that the groups $S_4 \times S_2$ and $\mathbb{Z}_2 \wr S_3$ are isomorphic.
(iv) Show that the groups in (iii) are each isomorphic to the group $W(B_4)$.

Exercise 10.4.13 The following is an analogue of Exercise 10.3.22 for GQ(4,2).

(i) Show that there are two $W = W(E_6)$-orbits of 5-cocliques GQ(4,2): those that can be extended to 6-cocliques, and those that cannot.
(ii) Show that 4-coclique in GQ(2,2) is contained in precisely two extendable 5-cocliques, and in a unique nonextendable 5-coclique.

10.5 Higher invariant forms

In this section, we generalize the ideas of Section 5.6 to multilinear forms for minuscule representations. A complete treatment of this topic would take far more space than we have here, but we outline the main ideas.

Definition 10.5.1 If $\mathfrak{g}$ is a Lie algebra over a field k and V is a $\mathfrak{g}$-module, then an n-linear map

$$B : V^{\otimes n} \to k$$

is said to be a $\mathfrak{g}$-*invariant* n-*form* if for any $x \in \mathfrak{g}$ and any simple tensor $v_1 \otimes v_2 \otimes \cdots \otimes v_n \in V^{\otimes n}$, we have

$$\sum_{i=1}^{n} B(v_1, v_2, \ldots, v_{i-1}, x.v_i, v_{i+1}, v_{i+2}, \ldots, v_n) = 0.$$

If B is invariant under all permutations of the inputs, we call the n-form B *symmetric*.

Lemma 10.5.2 *There is a symplectic bilinear form, p, on the module $L(E_7, \omega_6)$.*

Proof The principal subheap F associated with $L(E_7, \omega_6)$ is self-dual and has an odd number of vertices, so the result follows from Theorem 5.6.4. □

Definition 10.5.1 will allow us to introduce and study certain 3-linear invariant forms on $L(E_6, \omega_k)$ (for $k \in \{1, 5\}$) and 4-linear invariant forms on $L(E_7, \omega_6)$. In order for this to seem natural, we will use an alternative construction of $L(E_7, \omega_6)$ in terms of the simple Lie algebra of type E_8 (even though the latter has no minuscule representation). We denote the simple Lie algebras of type E_7 and E_8 by $\mathfrak{g}(E_7)$ and $\mathfrak{g}(E_8)$ respectively. We also consider $\mathfrak{g}(E_7)$ as a subalgebra of $\mathfrak{g}(E_8)$ in a canonical way.

Proposition 10.5.3 *The highest root in type E_8 is*

$$\theta = 2\alpha_1 + 3\alpha_2 + 4\alpha_3 + 5\alpha_4 + 6\alpha_5 + 4\alpha_6 + 2\alpha_7 + 3\alpha_8,$$

and θ is the only root in which α_1 appears with coefficient 2.

Proof The value of θ follows from Proposition 4.3.18 (ii) combined with the data in Appendix B. It follows from Proposition 4.3.18 (iv) that θ is orthogonal to all fundamental roots except α_1, and a calculation shows that $s_1(\theta) = \theta - \alpha_1$. It now follows from Corollary 4.3.11 that every root $\alpha \neq \theta$ of type E_8 satisfies $\alpha \leq \theta - \alpha_1$, from which the result follows. □

Definition 10.5.4 If α is a root of E_8, we define the α_i*-height* of α to be the coefficient of α_i in α. For $i \in \{-2, -1, 1, 2\}$, we define $\mathfrak{g}_{(i)}$ to be the subspace of $\mathfrak{g}(E_8)$ spanned by the basis elements x_α of α_i-height i. We define $\mathfrak{g}_{(0)}$ to be the subspace spanned by the remaining basis elements, that is, the subalgebra $\mathfrak{g}(E_7)$ together with the remaining basis element h_8 of the Cartan subalgebra $\mathfrak{h}$ of $\mathfrak{g}(E_8)$.

We can make $\mathfrak{g}(E_8)$ into a left $\mathfrak{g}(E_7)$-submodule, by restricting the adjoint representation and defining $x.y := [x, y]$. Using the analogue of Theorem 7.2.15 for $\mathfrak{g}(E_8)$, we find that, in this situation, each of the $\mathfrak{g}_{(i)}$ is an $\mathfrak{g}(E_7)$-submodule of $\mathfrak{g}(E_8)$.

Proposition 10.5.5 *The subspaces $\mathfrak{g}_{(i)}$ for $i = -2, -1, 0, 1, 2$ have dimensions* 1, 56, 134, 56, 1 *respectively. Furthermore, the subspace $\mathfrak{g}_{(1)}$ above is isomorphic, as a $\mathfrak{g}(E_7)$-submodule, to $L(E_7, \omega_6)$.*

Proof It follows from Proposition 10.5.3 that the spaces $\mathfrak{g}_{(2)}$ and $\mathfrak{g}_{(-2)}$ are one-dimensional. Since $\mathfrak{g}(E_7)$ has 126 roots (Example 8.4.9), it has dimension $126 + 7 = 133$. It follows that $\mathfrak{g}_{(0)}$ has dimension 134. Recall from Exercise 8.5.19 that $\mathfrak{g}(E_8)$ has 240 roots; it follows that its dimension is 248. It follows from the properties of positive and negative roots that the submodules $\mathfrak{g}_{(-1)}$ and $\mathfrak{g}_{(1)}$ have the same dimension; this dimension must therefore be

$$\frac{248 - 2 - 134}{2} = 56.$$

The root $s_1(\theta)$ in the proof of Proposition 10.5.3 is the highest weight vector of $\mathfrak{g}_{(1)}$, and its highest weight is the minuscule weight of $\mathfrak{g}(E_7)$. The proof is completed by Theorem 5.1.2 (ii). □

This construction of $\mathfrak{g}(E_7)$ makes the definition of the 4-linear form more transparent, as follows.

Definition 10.5.6 Let $\beta_1, \beta_2, \beta_3, \beta_4$ be roots of α_1-height 1. We define the 4-linear function

$$q : \mathfrak{g}_{(1)}^{\otimes 4} \to \mathbb{C}$$

by

$$q(x_{\beta_1}, x_{\beta_2}, x_{\beta_3}, x_{\beta_4})x_\theta = \frac{1}{4!}\sum_{\pi \in S_4}(\operatorname{ad} x_{\beta_{\pi(1)}} \circ \operatorname{ad} x_{\beta_{\pi(2)}} \circ \operatorname{ad} x_{\beta_{\pi(3)}} \circ \operatorname{ad} x_{\beta_{\pi(4)}})(x_{-\theta}),$$

where S_4 is the symmetric group, and $\operatorname{ad} x$ is the linear map given by the left action of x.

Under the (canonical) identification of $\mathfrak{g}_{(1)}$ with $M = L(E_7, \omega_6)$ given above, we may regard q as a 4-linear function from $M^{\otimes 4}$ to $\mathbb{C}$. It is clear from the definition that q is symmetric in the sense of Definition 10.5.1.

Proposition 10.5.7 *Let $M = L(E_7, \omega_6)$. The map $q : M^{\otimes 4} \to \mathbb{C}$ is a symmetric $\mathfrak{g}(E_7)$-invariant 4-form; in other words, for all $g \in \mathfrak{g}(E_7)$ and all $m_1, m_2, m_3, m_4 \in M$, we have*

$$q(g.m_1, m_2, m_3, m_4) + q(m_1, g.m_2, m_3, m_4) + q(m_1, m_2, g.m_3, m_4) \\ + q(m_1, m_2, m_3, g.m_4) = 0.$$

Proof Let $\beta_1, \beta_2, \beta_3, \beta_4$ be roots of α_1-height 1. Consider the expression

$$(\operatorname{ad}[g, x_{\beta_1}] \circ \operatorname{ad} x_{\beta_2} \circ \operatorname{ad} x_{\beta_3} \circ \operatorname{ad} x_{\beta_4})(x_{-\theta}).$$

By Exercise 1.1.8, this is equal to

$$(\operatorname{ad} g \circ \operatorname{ad} x_{\beta_1} \circ \operatorname{ad} x_{\beta_2} \circ \operatorname{ad} x_{\beta_3} \circ \operatorname{ad} x_{\beta_4})(x_{-\theta}) \\ - (\operatorname{ad} x_{\beta_1} \circ \operatorname{ad} g \circ \operatorname{ad} x_{\beta_2} \circ \operatorname{ad} x_{\beta_3} \circ \operatorname{ad} x_{\beta_4})(x_{-\theta}).$$

There are three other similar expressions in which g acts in the second, or third, or fourth coordinate instead of the first. Adding these four expressions up yields

$$X = (\operatorname{ad} g \circ \operatorname{ad} x_{\beta_1} \circ \operatorname{ad} x_{\beta_2} \circ \operatorname{ad} x_{\beta_3} \circ \operatorname{ad} x_{\beta_4})(x_{-\theta}) \\ - (\operatorname{ad} x_{\beta_1} \circ \operatorname{ad} x_{\beta_2} \circ \operatorname{ad} x_{\beta_3} \circ \operatorname{ad} x_{\beta_4} \circ \operatorname{ad} g)(x_{-\theta}).$$

It is enough to prove that we have $X = 0$. By linearity, we may assume that g is a basis element. By Proposition 10.5.3, θ (respectively, $-\theta$) is the only root of α_1-height 2 (respectively, -2). This means that if g is one of the root basis vectors x_β, then none of $\pm\theta \pm \beta$ can be roots. In turn, this means that both halves of the expression defining X are zero, completing the proof in this case.

The other possibility is that $g = h_i$, where h_i is a basis element of the Cartan subalgebra $\mathfrak{h}$. This means that $i \neq 1$, which in turn means that $h_i.x_\theta = h_i.x_{-\theta} = 0$. This also implies that both halves of the expression defining X are zero, as required. □

Corollary 10.5.8 *If m_1, m_2, m_3, m_4 are basis vectors of $L(E_7, \omega_6)$ then $q(m_1, m_2, m_3, m_4)$ is zero unless the sum of the weights of the m_i is equal to zero.*

Proof Use the $\mathfrak{g}$-invariance of q established in Proposition 10.5.7 applied to each of the elements $h_i \in \mathfrak{g}(E_7)$. □

Exercise 10.5.9 Use Exercise 9.3.17 to show that if $m_1, \dots, m_4$ are weight vectors for $L(E_7, \omega_6)$ and $q(m_1, m_2, m_3, m_4) \neq 0$, then either (a) the m_i form a regular tetrahedron of side $\sqrt{2}D$, where D is the minimal distance between weights, or (b) the m_i consist of pairs of opposite weights.

Exercise 10.5.10 Let q' be any invariant 4-form on $L(E_7, \omega_6)$, and let v_λ (respectively, $v_{-\lambda}$) be a weight vector of $L(E_7, \omega_6)$ of weight λ (respectively, $-\lambda$). Suppose that

$\lambda - \alpha$ is a weight for some root α. Show that

$$q'(v_\lambda, v_\lambda, v_{-\lambda}, v_{-\lambda}) = 0.$$

(Hint: consider the action of Y_α on $v_\lambda \otimes v_\lambda \otimes v_{-\lambda} \otimes v_{-\lambda}$ and use the idea of Corollary 10.5.8.)

Exercise 10.5.11 Let q'' be any symmetric invariant 4-form on $L(E_7, \omega_6)$, and let v_λ (respectively, $v_{-\lambda}$, v_μ, $v_{-\mu}$) be a weight vector of $L(E_7, \omega_6)$ of weight λ (respectively, $-\lambda$, μ, $-\mu$). Assume that $\lambda \neq \pm\mu$.

(i) Use Lemma 9.3.3 to show that, after relabelling weights if necessary, we may assume that the quadruple $\{\lambda, -\lambda, \mu, -\mu\}$ is such that $\lambda - \mu = \alpha$ is a positive root.
(ii) Show that $\lambda - \alpha$ and $-\mu - \alpha$ are weights, but that $-\lambda - \alpha$ and $\mu - \alpha$ are not.
(iii) Show that $q''(v_\lambda, v_\mu, v_{-\mu}, v_{-\lambda}) = 0$. (Hint: consider the action of Y_α on $v_\lambda \otimes v_\lambda \otimes v_{-\mu} \otimes v_{-\lambda}$, apply the result of Exercise 10.5.10, and use the hypothesis that q'' is symmetric to show that two of the resulting summands are equal.)

The symplectic form p of Lemma 10.5.2 plays an important role in conjunction with the invariant 4-linear form in the context of the following theorem.

Theorem 10.5.12 (Lurie) *The Lie algebra of $\mathbb{C}$-linear endomorphisms of $L(E_7, \omega_6)$ that leaves invariant the symplectic form p and the 4-linear form q is precisely $\mathfrak{g}(E_7)$.*

Proof This is [47, Theorem 6.2.3]. □

Remark 10.5.13 Lurie in fact proves a more general result than 10.5.12 that is valid over arbitrary commutative rings. If 2 is not a unit in the ring, then the situation is considerably more complicated, but Lurie also shows how to modify Theorem 10.5.12 to work in this case too.

Lurie also shows [47, section 6.2] that, just as in Exercises 10.5.9, 10.5.10 and 10.5.11, the 4-linear form q takes nonzero values on quadruples of weights corresponding to the tetrahedra of Exercise 10.5.9. Lurie calls these quadruples the "general" ones, and refers to the corresponding part of the 4-linear form as the "interesting" part.

We shall now show that the 4-linear form on $L(E_7, \omega_6)$ gives rise to a 3-linear form on $L(E_6, \omega_5)$. To do this, we regard the simple Lie algebra $\mathfrak{g}(E_6)$ of type E_6 as a subalgebra of $\mathfrak{g}(E_7)$ in the obvious way, as in Exercise 8.2.18.

Proposition 10.5.14 *There is a $\mathfrak{g}(E_6)$-invariant 3-linear form on $L(E_6, \omega_5)$ given by*

$$q'(m_1, m_2, m_3) := q(v_{0,7}, m_1, m_2, m_3),$$

where $v_{0,7}$ is the lowest weight of $L(E_7, \omega_6)$, and where we identify $L(E_6, \omega_5)$ with a $\mathfrak{g}(E_6)$-submodule of $L(E_7, \omega_6)$ in the obvious way.

Proof Recall from Exercise 8.2.18 that if D is the minimal nonzero between weights of $L(E_7, \omega_6)$, then the vectors that are at distance D from the lowest weight $v_{0,7}$ become weights of $L(E_6, \omega_5)$ after omitting the Lie algebra generators corresponding to the Dynkin diagram vertex 6. The same exercise shows that the lowest and highest weight vectors of $L(E_7, \omega_6)$, namely $\pm v_{0,7}$, are annihilated by the action of the Lie subalgebra of type E_6. This, together with the invariance of q, establishes the invariance of q'. □

Remark 10.5.15 Although the form q' of Proposition 10.5.14 inherits the property of being symmetric from the form q, it is not immediately obvious that q' is nonzero. However, q takes nonzero values on the "general" quadruples of weights at mutual distance $\sqrt{2}D$ (see Remark 10.5.13), and there do exist such quadruples in which one of the elements is $v_{0,7}$. In contrast, the nongeneral quadruples do not play a role in the definition of q', because the weight opposite to $v_{0,7}$ (i.e., $-v_{0,7}$) has been excluded.

Exercise 10.5.16 State and prove a version of Proposition 10.5.14 involving the highest weight vector $-v_{0,7}$ and the representation $L(E_6, \omega_1)$.

If we identify the bitangent ω in Theorem 10.4.3 with the pair of highest and lowest weights of $L(E_7, \omega_6)$ and then apply Proposition 9.3.12 (ii), then the triples (m_1, m_2, m_3) for which $q'(m_1, m_2, m_3) \neq 0$ correspond to the lines in the generalized quadrangle GQ(2,4). It follows from Exercise 10.2.14 that if $q'(m_1, m_2, m_3) \neq 0$, then (considering the weights m_i as weights of $L(E_6, \omega_5)$, not of $L(E_7, \omega_6)$) we must have $m_1 + m_2 + m_3 = 0$.

10.6 Notes and references

1 Manin studied the 27 lines on a cubic surface from a modern point of view in [53], and wrote that

> "the configuration of the twenty-seven lines on a smooth cubic surface has been the subject of entire books... Their elegant symmetry is at once awe-inspiring and annoying."

An example of such a book is Henderson [35].

2 Our notation for the 27 lines given in Definition 10.1.1 follows Hartshorne [33].

3 Lemma 10.1.6 describes a partition of a 27 element set into three parts of sizes 16, 10 and 1. Another such partition appears in the song "27 Jennifers" by M. Doughty.

4 Manivel's paper [54] contains further information about the relationship between Lie algebras, the 27 lines on a cubic surface, and double sixes. Manivel also discusses related topics that we do not cover here, including Steiner triple systems and the octonions.

5 Our approach to regular 2-graphs follows that of Taylor [83]. Taylor attributes the notion of 2-graph to G. Higman, who used them to study the doubly transitive action of the sporadic Conway group Co_3 on 276 points.

6 A 2-graph is a special case of a *hypergraph*. A hypergraph is a pair $H = (X, E)$, where X is a set and E is a set of nonempty subsets of X called *hyperedges*. A hypergraph is called k*-uniform* if each of its hyperedges has size k. This means that a 2-uniform hypergraph is simply a graph, and that a 2-graph is a special kind of 3-uniform hypergraph, subject to the restrictions in Proposition 10.3.6 (i).

7 A good reference for generalized quadrangles is the book by Payne and Thas [60]. I thank S. E. Payne for explaining to me (a) the connection between outer automorphisms of S_6 and the points and lines of GQ(2,2) mentioned in Exercise 10.4.12, and (b) the construction of GQ(2, 4) of Exercise 10.4.7 (see [60, section 6.1]).

8 The duad–syntheme model of Exercise 10.4.6 was originally invented by J. J. Sylvester in 1844 [81].

9 Lurie's constructions in [47] of the invariant forms in types E_6 and E_7 are very explicit. This is especially true in type E_6, where the description is in terms of flex points and Eckard planes.

10 Lurie [47, section 3.2] uses a slightly different definition of "minuscule" by requiring that the Weyl group act transitively on the *nonzero* weights. By Lurie's definition, the zero weight is minuscule, and a minuscule weight has minimal length in its coset of the root lattice.

11 Helenius [34] generalizes the constructions of invariant 4-forms to types B, D, E and F, in the context of algebraic groups, except in characteristics 2 or 3. Definition 10.5.6 is [34, lemma 3.2]. Helenius works mainly with quartic forms rather than their linearizations; the quartic form $q(x)$ and the 4-linear invariant q are related by $q(x) = q(x, x, x, x)$. Helenius also investigates in detail an invariant quartic form associated with the simple Lie algebra of type D_4, and discusses how some of these phenomena can be understood in terms of the 27-dimensional *exceptional Jordan algebra* (or "Albert algebra"). Helenius [34, section 1.3] also gives a survey of other approaches to the 4-linear form in type E_7, including the work of Cooperstein [16].

12 It turns out that the invariant 3-linear form in type E_6 is unique up to scalar multiplication [47, section 5], but that there is a 4-dimensional vector space of 4-linear forms in type E_7 [47, section 6.2].

Over a field of characteristic different from 2, this 4-dimensional space has a basis consisting of q, Ψ_1, Ψ_2 and Ψ_3, where the Ψ_i are defined in terms of the symplectic form:

$$\Psi_1(w, x, y, z) = [w, x][y, z], \quad \Psi_2(w, x, y, z) = [w, y][x, z],$$
$$\Psi_3(w, x, y, z) = [w, z][x, y].$$

The Ψ_i themselves are not particularly interesting, because they annihilate the "general" quadruples of weights (see Remark 10.5.13). Furthermore, the Ψ_i become zero once they are symmetrized by the action of S_4. This means that the space of *symmetric* invariant 4-forms on $L(E_7, \omega_6)$ is one-dimensional.

Lurie also explains how to deal with the case of characteristic 2, which is considerably trickier.

11

Further topics

No discussion of the combinatorics of minuscule representations and their applications would be complete without mentioning the topics surveyed in this final chapter. Unlike the exceptional structures discussed in Chapters 9 and 10, the concepts discussed in Chapter 11 make sense for minuscule representations of arbitrary type.

Section 11.1 develops the theory of minuscule elements in Weyl groups. Theorem 11.1.18, which is due to Proctor, Stembridge, and Pfeiffer & Röhrle, describes how minuscule representations induce relationships between distinguished coset representatives, lattices, and the weak and strong Bruhat orders. Section 11.2 discusses Proctor's work on the combinatorics of the underlying posets of principal subheaps. This gives some insight into the nature of Gaussian posets and their connections with the theory of plane partitions, which is the subject of Section 11.3. We mention some of the points of contact between Gaussian posets and the cyclic sieving phenomenon of Reiner, Stanton and White [66]. Finally, Section 11.4 shows how the combinatorics of heaps can be used to calculate Schubert intersection numbers in algebraic geometry, following the work of Thomas and Yong [84].

11.1 Minuscule elements of Weyl groups

Let Γ be a Dynkin diagram. Recall from Lemma 3.1.10 that the Weyl group $W(\Gamma)$ is a quotient of the commutation monoid $\text{Co}(\Gamma)$. Matsumoto's Theorem (Theorem 4.3.1) shows that any two reduced expressions for an element $w \in W(\Gamma)$ are equivalent via a finite sequence of braid relations.

Proposition 11.1.1 *Let Γ be a doubly laced Dynkin diagram with commutation monoid* $\text{Co}(\Gamma)$ *and Weyl group $W = W(\Gamma)$, and let $w \in W$. Then the following are equivalent:*

(i) *no reduced expression for w has (a) a subword of the form $s_i s_j s_i$ with $a_{ij} a_{ji} = 1$, or (b) a subword of the form $s_i s_j s_i s_j$ with $a_{ij} a_{ji} = 2$;*
(ii) *one can pass between any two reduced expressions for w using a finite sequence of short braid relations;*
(iii) *any two reduced expressions for w are equal when considered as elements of* $\text{Co}(\Gamma)$.

Proof Suppose that (i) holds and that $\mathbf{u}_1$ and $\mathbf{u}_2$ are two reduced expressions for w. By Matsumoto's Theorem, there is a finite sequence of braid relations transforming $\mathbf{u}_1$ to $\mathbf{u}_2$. The hypotheses of (iii) mean that, during this sequence, there is never an opportunity to apply a long braid relation, and this proves (ii).

The implication (ii) $\Rightarrow$ (iii) follows from the definitions, so it remains to show that (iii) implies (i).

Suppose that (i) does not hold, and let $\mathbf{u}$ be a reduced expression for w containing one of the forbidden subwords involving s_i and s_j. Let $\mathbf{u}'$ be the reduced expression obtained by applying a braid relation to the forbidden subword. Let $[\mathbf{u}]$ (respectively, $[\mathbf{u}']$) be the element of $\text{Co}(\Gamma)$ corresponding to the word $\mathbf{u}$ (respectively, and $\mathbf{u}'$), and let $\varepsilon : E \to \Gamma$ (respectively, $\varepsilon' : E' \to \Gamma$) be the heaps associated with $[\mathbf{u}]$ (respectively, $[\mathbf{u}']$) by Theorem 2.1.20. Since (i) does not hold, the vertex chains $E \cap \varepsilon^{-1}\{i, j\}$ and $E' \cap \varepsilon'^{-1}\{i, j\}$ are not isomorphic in $\mathbf{Heap}(\Gamma)$, and this implies that $E \not\cong_{\mathbf{Heap}(\Gamma)} E'$. Using Theorem 2.1.20 again, we see that $[\mathbf{u}] \neq [\mathbf{u}']$. It now follows from Lemma 3.1.10 that (iii) does not hold, and this completes the proof. □

Definition 11.1.2 An element $w \in W$ satisfying the equivalent conditions of Proposition 11.1.1 is called *fully commutative*.

Condition (ii) of Proposition 11.1.1 makes the following definition meaningful.

Definition 11.1.3 Let $s_{i_1} s_{i_2} \cdots s_{i_r}$ be a reduced expression for a fully commutative element $w \in W$. We define the *content*, $c(w)$, of w to be the vector $\sum_i c_i \alpha_i$, where c_i is the number of occurrences of s_i in any reduced expression of w.

Definition 11.1.4 (Peterson) Let A be a generalized Cartan matrix with Weyl group W, and let λ be an integral weight. An element $w \in W$ is said to be λ-*minuscule* if there is a reduced expression $w = s_{i_1} \cdots s_{i_l}$ such that for all $1 \leq k \leq l$ we have

$$(s_{i_{k+1}} s_{i_{k+2}} \cdots s_{i_l} \lambda, \alpha_{i_k}^\vee) = 1$$

or, equivalently,

$$s_{i_{k+1}} s_{i_{k+2}} \cdots s_{i_l} \lambda = \lambda - \alpha_{i_k} - \alpha_{i_{k+1}} - \cdots - \alpha_{i_l}.$$

The above definition of minuscule element is somewhat inconvenient to use in practice. An easier test for the property is given by the following characterization.

Proposition 11.1.5 (Stembridge, [79, proposition 2.3]) *Let A be a generalized Cartan matrix with Weyl group W. If $s_{i_1} \cdots s_{i_l}$ is a reduced expression for $w \in W$, then w is minuscule if and only if between any consecutive pair of occurrences of a generator s_i, one of the following two situations occurs:*

(i) *there are two terms s_j and s_k (possibly with $j = k$) such that $a_{ij} = a_{ik} = -1$, or*
(ii) *there is one term s_j such that $a_{ij} = -2$.* □

Corollary 11.1.6 (Stembridge, [79, proposition 2.1]) *Any minuscule element is fully commutative.*

Proof A counterexample to condition (i) of Proposition 11.1.1 provides a counterexample to Proposition 11.1.5. □

There is also a simple characterization of minuscule elements in terms of heaps.

Proposition 11.1.7 (Stembridge) *Let Γ be a generalized Cartan matrix with Weyl group W, let $\varepsilon : E \to \Gamma$ be a finite heap, and let $[\mathbf{u}]$ be the trace associated with E by Theorem 2.1.20. Then the following are equivalent.*

(i) *Any representative word $\mathbf{u} \in S^*$ of the trace $[\mathbf{u}]$ is a reduced expression for a minuscule element $w \in W$.*
(ii) *The heap E satisfies Axiom (F3) of Definition 2.2.2.*

Proof By Lemma 3.1.10, all the words representing the trace $[\mathbf{u}]$ correspond to the same element of W. The result is now a restatement of [79, proposition 3.1 (b)]. (Stembridge's condition H1 is equivalent to the definition of a heap in the locally finite case, and Stembridge's condition H2 is our Axiom (F3).) □

Exercise 11.1.8 Let $\varepsilon : E \to \Gamma$ be a full heap. Show that the Weyl group $W(\Gamma)$ has infinitely many minuscule elements.

Theorem 11.1.9 *Let $\varepsilon : F \to \Gamma_0$ be a principal subheap of a full heap $\varepsilon : E \to \Gamma$ over an affine Dynkin diagram, and let V_F be the corresponding minuscule module of the simple Lie algebra $\mathfrak{g}$, of lowest weight $-\omega_p$. Let W be the finite Weyl group associated with $\mathfrak{g}$, let S be the set of generators for W, and let $S' = S\backslash\{s_p\}$.*

(i) *If I is an ideal of F, then $f_{<'}(I)$ is a reduced expression for a minuscule element of W, where $f_{<'}$ is as in Definition 2.1.18.*
(ii) *Let I be an ideal of F and $w = \pi_{\text{Co}(\Gamma)}([f_{<'}(I)])$, where $\pi_{\text{Co}(\Gamma)}$ is as in Lemma 3.1.10. Then, in the action of W on the ideals of F, we have $w^{-1}(\emptyset) = I$.*
(iii) *There is a bijection ψ from the set of ideals of F to the set of distinguished coset representatives of the right cosets $W_{S'}\backslash W$ given by*

$$\psi(I) = \pi_{\text{Co}(\Gamma)}([f_{<'}(I)]).$$

(iv) *The distinguished left and right coset representatives of $W_{S'}$ in W are minuscule elements.*

Proof It is immediate that an ideal I of F has property (H3), because I is a convex subset of a full heap. Proposition 11.1.7 now shows that $f_{<'}(I)$ is a reduced expression for a minuscule element, proving (i).

We prove (ii) by induction on $|I|$; the case $|I| = 0$ is trivial. Suppose that (ii) is true for ideals of size at most N, and suppose that $|I| = N + 1$. Let us write $f_{<'}(I) = s_{i_1} \cdots s_{i_{N+1}} \in S^*$. For $l < N + 1$, let $I_l \subset I \subseteq F$ be the ideal for which $f_{<'}(I_l) = s_{i_1} \cdots s_{i_l}$. Let $w' = \pi_{\text{Co}(\Gamma)}(f_{<'}(I_N))$. It follows from Definition 2.1.18 that we have $I_N \prec_{i_{N+1}} I_{N+1}$, and it follows from the inductive hypothesis that $w'^{-1}(\emptyset) = I_N$. Applying $s_{i_{N+1}}$ to both sides proves (ii).

Consider the action of W on the weights of V_F, which is known to be transitive (Corollary 6.6.4). Applying Theorem 4.3.10 (iv) to the weight ω_p, we see that the stabilizer of $-\omega_p$ is precisely $W_{S'}$. It follows that there is an isomorphism of (left) W-sets between $J(F)$ and the coset space $W_{S'}\backslash W$, and that this isomorphism identifies the empty ideal with the coset containing the identity.

By Lemma 5.5.1, F has a unique minimal element, and that element is labelled p. Theorem 2.1.20 then shows that every nonidentity expression of the form $f_{<'}(I)$ begins with s_p, independently of the choice of total order $<'$. Since no reduced expression for any of the elements $w = \pi_{\mathrm{Co}(\Gamma)}(f_{<'}(I))$ begins with a generator in S', it follows that $w \in {}^I W$. This shows that the image of ψ consists of distinguished coset representatives. The inverse of ψ is constructed in part (ii), so ψ is a bijection and (iii) follows.

The assertion of part (iv) for right cosets follows from (i) and (iii), and the assertion for left cosets follows from the left–right symmetry of Proposition 11.1.5. □

We also state the left coset version of Theorem 11.1.9 for convenience; its proof is left as an exercise.

Theorem 11.1.10 *Let $\varepsilon : F \to \Gamma_0$ be a principal subheap of a full heap $\varepsilon : E \to \Gamma$ over an affine Dynkin diagram, and let V_F be the corresponding minuscule module of the simple Lie algebra $\mathfrak{g}$, of highest weight ω_p. Let W be the finite Weyl group associated with $\mathfrak{g}$, let S be the set of generators for W, and let $S' = S \backslash \{s_p\}$.*

(i) *If I is a filter of F, then $f_{<'}(I)$ is a reduced expression for a minuscule element of W, where $f_{<'}$ is as in Definition 2.1.18.*
(ii) *Let I be a filter of F and $w = \pi_{\mathrm{Co}(\Gamma)}([f_{<'}(I)])$, where $\pi_{\mathrm{Co}(\Gamma)}$ is as in Lemma 3.1.10. Then, in the action of W on the ideals of F, we have $w(F) = F \backslash I$.*
(iii) *There is a bijection ψ from the set of filters of F to the set of distinguished coset representatives of the left cosets $W/W_{S'}$ given by*

$$\psi(I) = \pi_{\mathrm{Co}(\Gamma)}([f_{<'}(I)]).$$ □

Exercise 11.1.11 Prove Theorem 11.1.10.

Exercise 11.1.12 Let $\varepsilon : E \to \Gamma$ be a full heap over an affine Dynkin diagram, and let W be the associated affine Weyl group. Use Theorem 3.2.30 to show that if α is a (real or imaginary) positive root associated with Γ, then there is a minuscule element $w \in W$ with content $c(w) = \alpha$.

Exercise 11.1.13 Let $\varepsilon : E \to \Gamma$ be a full heap over an affine Dynkin diagram, and let W be the associated affine Weyl group. Let α and β be real roots associated with Γ. Use the techniques of Chapter 7 to prove the following.

(i) Show that if $s_\beta(\alpha) = \alpha + \beta$ then there exist minuscule elements u, v and $w = uv$ of W with contents α, β and $\alpha + \beta$, respectively.
(ii) Show that if $s_\beta(\alpha) = \alpha + 2\beta$ then there exist minuscule elements u, v, w and $x = uvw$ of W with contents β, α, β and $\alpha + 2\beta$, respectively.

Exercise 11.1.14 Let W be the Weyl group of type E_6 with the usual notational conventions. Let $w = s_1 s_2 s_3 s_6 s_4 s_5$.

(i) Show that w is minuscule.
(ii) Show that w is a distinguished right coset representative for the parabolic subgroup generated by $S \backslash \{s_1\}$.
(iii) Show that w is not a distinguished left coset representative for the parabolic subgroup generated by $S \backslash \{s_5\}$.

(iv) Show that the maximal possible length of a left or right coset representative for either of the parabolic subgroups generated by $S\backslash\{s_1\}$ or $S\backslash\{s_5\}$ is 16. Give an example of such a coset representative.

Exercise 11.1.15 Let W be the Weyl group of type $\Gamma = D_4$ with the usual notational conventions. Show that the element $w = s_1s_3s_4s_2$ is minuscule, but that we do not have $w = \pi_{\mathrm{Co}(\Gamma)}([f_{<'}(I)])$ for any convex subheap I of a principal subheap of a full heap F.

Lemma 11.1.16 *Let ω_p be a minuscule weight for a simple Lie algebra $\mathfrak{g}$ with Weyl group W generated by S. Let $S' = S\backslash\{s_p\}$.*

(i) *If x and w are distinguished left coset representatives for $W/W_{S'}$, then we have $x \leq w$ if and only if $x \leq_R w$.*
(ii) *If x and w are distinguished right coset representatives for $W_{S'}\backslash W$, then we have $x \leq w$ if and only if $x \leq_L w$.*

Proof We only prove (i), because the proof of (ii) is very similar.

It is immediate from the definitions that if $x \leq_R w$ then $x \leq w$, so suppose that $x \leq w$. Let V_F be a minuscule representation with lowest weight $-\omega_p$. In the action of W on the ideals of F, let I be such that $w^{-1}(\emptyset) = I$. Let $\alpha = \chi(I)$ and write $\alpha = \sum_{i\in S} c_i\alpha_i$.

By Theorem 11.1.9 (iii), c_i is equal to the number of occurrences of s_i in some reduced expression $s_{i_1}\cdots s_{i_k}$ for w. Some subexpression of this reduced expression is equal to a reduced expression for x. It follows that if we define I' to be the ideal for which $x^{-1}(\emptyset) = I'$, then we have $I' \subset I$. Since x is a distinguished coset representative, the length of x is equal to the size of $|I'|$. Applying Theorem 11.1.9 (iii) again shows that there is a reduced expression for w that contains a reduced expression for x as a suffix. This shows that $x \leq_R w$, as required. □

Remark 11.1.17 The result of Lemma 11.1.16 is in contrast to the typical relationships between the various Bruhat orders; see for example Exercises 4.3.6 and 4.3.7.

Theorem 11.1.18 (Proctor; Stembridge; Pfeiffer–Röhrle) *Let ω_p be a minuscule weight for a simple Lie algebra $\mathfrak{g}$ with Weyl group W generated by S. Let $S' = S\backslash\{s_p\}$.*

(i) *The distinguished left coset representatives of $W_{S'}$ in W form a distributive lattice under the Bruhat order, and under $\leq_R$.*
(ii) *The distinguished right coset representatives of $W_{S'}$ in W form a distributive lattice under the Bruhat order, and under $\leq_L$.*

Proof Let V_F be the corresponding minuscule representation with lowest weight $-\omega_p$. We know (Exercise A.1.9) that the ideals of F form a distributive lattice under the partial order induced by containment. This partial order induces, via the correspondence of Theorem 11.1.9, a partial order on the distinguished left coset representatives on $W_{S'}$. This latter order is the weak Bruhat order, $\leq_R$. The proof of (i) now follows from Lemma 11.1.16 (i).

The proof of (ii) is similar, but uses Theorem 11.1.10 and Lemma 11.1.16 (ii). □

Exercise 11.1.19 Let L and L' be a pair of skew lines among the 27 lines on a cubic surface. Let ω and ω' be the corresponding weights of $L(E_6, \omega_5)$, as in Section 10.1, let α be the positive root for which $\pm\alpha = \omega - \omega'$, and let k be the height of α.

(i) Show that k is the smallest integer for which there exists an element $w \in W(E_6)$ of length k such that $w(L) = L'$.
(ii) Prove that there is a unique w satisfying the conditions of (i).
(iii) Prove that w is minuscule.

11.2 Principal subheaps as abstract posets

The structure of principal subheaps of full heaps as abstract posets seems at first glance to be somewhat complicated, particularly in types E_6 and E_7. However, Proctor showed in [65] that the underlying posets of these principal subheaps have concise combinatorial descriptions and characterizations, as we saw in Section 8.3.

Definition 11.2.1 Let L be a finite poset, let $\rho : L \to \{0, 1, \ldots, r\}$ be a surjective rank function, and let $L_i := L \cap \rho^{-1}(i)$. We call the subsets L_i *ranks*. The *rank generating function* for L is the polynomial

$$\sum_{i=0}^{r} q^{|L_i|}.$$

If there exists an order-reversing involution $* : L \to L$ on L, then we say that L is *complemented*. If I is any subset of L, we write $I^* = \{x^* : x \in I\}$. An ideal $I \subseteq L$ is called *self-complementary* if we have $L \backslash I = I^*$. We say that L is *rank symmetric* if $|L_i| = |L_{r-i}|$ for all $0 \leq i \leq r$, and *rank unimodal* if there exists m such that

$$|L_0| \leq |L_1| \leq \cdots \leq |L_m| \geq |L_{m+1}| \geq \cdots \geq |L_r|.$$

We say that L is *strongly Sperner* if, for every $k \geq 1$, the largest union of k antichains is no larger than the largest union of k ranks. If this condition is satisfied for $k = 1$, then L is said to be *Sperner*. We say that L is *uniquely modular* if the posets $P = L$ and $P = L^*$ both satisfy the following condition: if $x, y, z \in P$ are such that y covers x and z covers x, then there is a unique $w \in P$ such that w covers each of y and z.

Exercise 11.2.2 Show that if P is a finite poset then the function $\rho : J(P) \to \{0, 1, \ldots, |P|\}$ sending I to $|I|$ is a surjective rank function.

Exercise 11.2.3 Show that, in the notation of Definition 11.2.1, L is rank symmetric if and only if the rank generating function $f(q)$ of L satisfies $f(q^{-1}) = q^{-r} f(q)$.

Exercise 11.2.4 Show that if L is both irreducible and complemented, then L is rank symmetric. Show also that the conclusion may fail if L is not assumed to be irreducible.

Exercise 11.2.5 Show that if P is a complemented poset then the distributive lattice $J(P)$ is also complemented, with respect to the involution $I \mapsto \{x \in P : x^* \notin I\}$.

Definition 11.2.6 A finite poset P is said to be *vertex labellable* if there exists a function $\pi : P \to \mathbb{Q}$ such that the equations

$$\sum_{x \in M_I} \pi(x) - |I| = \sum_{y \in m_I} \pi(y) - |P - I|$$

are satisfied for every ideal $I \subseteq P$, where M_I is the set of maximal elements in I, and m_I is the set of minimal elements in $P \backslash I$.

Definition 11.2.7 Let L be a uniquely modular ranked poset with surjective rank function $\rho : L \to \{0, 1, \ldots, r\}$. We say that L is *edge labellable* if each edge of the Hasse diagram for L can be labelled with a rational number in such a way that

(i) opposite edges of any square in the Hasse diagram have equal labels, and
(ii) if $b \in L_i$ then the sum of the labels on edges emanating below b minus the sum of the labels on edges emanating above b is equal to $2i - r$.

Theorem 11.2.8 *Let P be a finite poset, and let $[r]$ denote the set $\{1, 2, \ldots, r\}$ equipped with its usual total order.*

(i) *The poset P is vertex labellable if and only if the distributive lattice $J(P)$ is edge labellable.*
(ii) *Edge labellable lattices are rank symmetric, rank unimodal, and strongly Sperner.*
(iii) *The poset P is vertex labelled if and only if it is isomorphic to a direct sum of posets on the following list:*

$$\begin{aligned} &[s] \times [t] \quad (s, t \geq 1), \\ &J([2] \times [t]) \ (t \geq 1), \\ &J^k([2] \times [2]) \ (k \geq 0), \\ &J^2([2] \times [3]) \ \textit{and} \\ &J^3([2] \times [3]). \end{aligned}$$

(iv) *The posets in (iii) are respectively isomorphic to the underlying posets of the principal subheaps of full heaps corresponding to the minuscule representations of types A, D (spin), D (natural), E_6 and E_7. Every other principal subheap of a full heap is isomorphic, as an abstract poset, to one of these posets.*
(v) *A distributive lattice is edge labellable if and only if it is a product of lattices on the following list:*

$$\begin{aligned} &J([s] \times [t]) \ (s, t \geq 1), \\ &J^2([2] \times [t]) \ (t \geq 1), \\ &J^k([2] \times [2]) \ (k \geq 1), \\ &J^3([2] \times [3]) \ \textit{and} \\ &J^4([2] \times [3]). \end{aligned}$$

Proof Part (i) is Proctor [65, proposition 3]. It is stated without proof in [65], but the proof is not hard: an edge in the Hasse diagram of $J(P)$ corresponds to a covering relation $I \subset I'$, and the edge inherits the label of the unique $x \in P$ such that $I' \backslash I =$

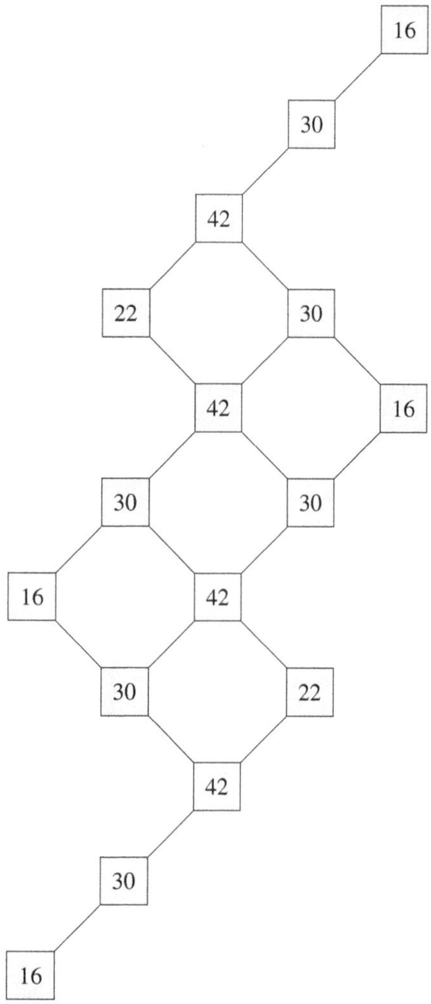

Figure 11.1 Vertex labelling of the principal subheap of $\varepsilon : E_6^{(1)}(5) \to E_6^{(1)}$

$\{x\}$. Under this correspondence, condition (i) of Definition 11.2.7 is automatically true for $J(P)$, and condition (ii) of Definition 11.2.7 is equivalent to Definition 11.2.6 applied to P.

Part (ii) is [65, proposition 2] and part (iii) follows from [65, theorem 2] (after translating the notation). Part (iv) is immediate from Theorem 8.3.10. Part (v) follows from (i), (iii) and the Fundamental Theorem for finite distributive lattices (Theorem A.1.11). □

Example 11.2.9 Figure 11.1 shows a vertex labelling for the underlying poset of the principal subheap of the full heap $\varepsilon : E_6^{(1)}(5) \to E_6^{(1)}$, which is isomorphic to $J^2([2] \times [3])$ by Theorem 8.3.10 (ii).

Exercise 11.2.10 Let P_a be the set

$$\{(i, j) : 1 \leq i \leq j \leq a\}$$

equipped with the partial order such that $(i, j) \leq (k, l)$ if $i \leq k$ and $j \leq l$. Let $G = S_2$ act on the poset $P = [a] \times [a]$ by exchanging the two coordinates.

(i) Show that P_a is isomorphic to $J([2] \times [a])$.
(ii) Show that G and P satisfy the hypotheses of Proposition A.2.4.
(iii) Show that the quotient poset resulting in (i) is is isomorphic to P_a, and thus to $J([2] \times [a])$.

Exercise 11.2.11 For each minuscule representation, find the corresponding poset in Theorem 11.2.8 (iii).

The next sequence of exercises shows that it is possible to construct the vertex labelling of a principal subheap directly from the heap structure with the aid of a simple auxiliary function.

Exercise 11.2.12 Let V_F be the minuscule representation $L(A_n, \omega_j)$, and let $\varepsilon : F \to A_n$ be the corresponding principal subheap. Define $f : [n] \to \mathbb{Z}$ by $f(a) = a(n + 1 - a)$.

(i) Show that $f \circ \varepsilon : F \to \mathbb{Z}$ is a vertex labelling of F.
(ii) Show how to modify the definition of f to describe a vertex labelling for the minuscule representation in type C_n.

Exercise 11.2.13 Define $f : [n] \to \mathbb{Z}$ by

$$f(a) = \begin{cases} a(2n - 1 - a) & \text{if } a \notin \{n - 1, n\}, \\ (n - 1)n/2 & \text{otherwise.} \end{cases}$$

(i) Let V_F be the minuscule representation $L(D_n, \omega_1)$, and let $\varepsilon : F \to D_n$ be the corresponding principal subheap. Show that $f \circ \varepsilon : F \to \mathbb{Z}$ is a vertex labelling of F.
(ii) Show that vertex labellings of the spin representations in type D may be constructed as in (i), using the *same* function f.
(iii) Show how to modify the definition of f to describe a vertex labelling for the spin representation in type B_n.

Exercise 11.2.14 Describe the vertex labelling in Figure 11.1 in terms of a function, as in Exercises 11.2.12 and 11.2.13.

Exercise 11.2.15 Let $f : [7] \to \mathbb{Z}$ be the function satisfying $f(1) = 34$, $f(2) = 66$, $f(3) = 96$, $f(4) = 75$, $f(5) = 52$, $f(6) = 27$ and $f(7) = 49$. Use this function and the techniques of Exercises 11.2.12 and 11.2.13 to construct a vertex labelling corresponding to the minuscule representation in type E_7.

Exercise 11.2.16 Let P be one of the posets appearing in Theorem 11.2.8 (iii), and let $\rho : P \to \mathbb{Z}$ be a rank function sending the minimal element of P to 0. Verify that at least half the elements of P are sent to even integers by ρ.

Exercise 11.2.17 Let P be one of the posets appearing in Theorem 11.2.8 (iii). Show that P is a distributive lattice. (This is immediate from Theorem A.1.11, except in the case of $[s] \times [t]$.)

11.3 Gaussian posets

Definition 11.3.1 Let P be a finite poset and let m be a nonnegative integer. Let $F_m(P, q)$ be the rank generating function for the poset $J(P \times [m])$. We say that the poset P is *Gaussian* if there exist strictly positive integers $a_1, \ldots, a_n$ such that for all $m \geq 0$, we have

$$F_m(P, q) = \prod_{i=1}^{n} \frac{1 - q^{m+a_i}}{1 - q^{a_i}}.$$

Exercise 11.3.2 Show that the ideals of the poset $P \times [m]$ are in natural bijection with multichains

$$I_1 \subseteq I_2 \subseteq \cdots \subseteq I_m$$

of ideals of P. Deduce that we have

$$F_m(P, q) = \sum_{I_1 \subseteq \cdots \subseteq I_m : I_j \in J(P)} q^{|I_1| + \cdots + |I_m|}.$$

Exercise 11.3.3 Let P be a Gaussian poset and let $a_1, \ldots, a_n$ be the associated sequence of integers. Show that at most half of the a_i can be even numbers. (Hint: take m to be odd and consider the product expression for $F_m(P, q)$ in Definition 11.3.1. Use the fact that this expression has a removable singularity at $q = -1$.)

The next result shows that minuscule representations give an interesting source of Gaussian posets. (In fact, these are the *only* known examples of connected Gaussian posets.)

Theorem 11.3.4 (Proctor) *Let $\mathfrak{g}$ be a simple Lie algebra with positive roots Φ^+, let λ be a minuscule weight for $\mathfrak{g}$, let P_λ be the underlying poset of the corresponding principal subheap, and let*

$$\rho = \frac{1}{2} \sum_{\alpha \in \Phi^+} \alpha.$$

Then we have

$$F_m(P_\lambda, q) = \prod_{\alpha \in \Phi^+} \frac{1 - q^{(m\lambda + \rho, \alpha^\vee)}}{1 - q^{(\rho, \alpha^\vee)}}.$$

In particular, P_λ is Gaussian.

Proof This is proved in [64]; another proof is given in Stembridge [77, theorem 3.4]. □

Proposition 11.3.5 (Proctor; Stanley) *Let P be a finite poset with n elements.*

(i) *The poset P is Gaussian if and only if every connected component of P is Gaussian.*
(ii) *If P is connected and Gaussian, then every maximal chain of P has the same length, $\ell(P)$, and P is ranked.*

Figure 11.2 A plane partition

(iii) *If the rank function in (ii) takes* 0 *as its minimal value, then the multisets* $\{a_1, a_2, \dots, a_n\}$ *and* $\{1 + \rho(x) : x \in P\}$ *agree.*
(iv) *If P is connected and Gaussian and we have $a_1 \leq a_2 \leq \cdots \leq a_n$, then $a_i + a_{n+1-i} = \ell(P) + 2$ for all $1 \leq i \leq n$.*
(v) *If P is connected and Gaussian, then every element of rank* 1 *covers exactly one minimal element of P.*

Proof This result and its proof appear in Stanley's book as [75, exercise 4.25 (a)–(e)]. A proof using the representation theory of Lie algebras may be found in Proctor's paper [64]. □

Definition 11.3.6 A *plane partition* with sum n is a two-dimensional array of non-negative integers $n_{i,j}$ (where $i, j \geq 0$) such that $n = \sum_{i,j} n_{i,j}$ is finite, and such that for all i and j, we have $n_{i,j} \geq n_{i,j+1}$ and $n_{i,j} \geq n_{i+1,j}$. It is convenient to identify the number $n_{i,j}$ with a stack of unit cubes placed on the plane at position $(i, -j)$.

Example 11.3.7 Figure 11.2 shows a picture of the plane partition

$$\begin{pmatrix} 6 & 3 & 2 & 2 & 1 \\ 4 & 2 & 2 & 1 & \\ 2 & 1 & & & \\ 1 & & & & \end{pmatrix}$$

of sum 27. The smallest cuboid containing this plane partition is of size $4 \times 5 \times 6$.

Figure 11.3 A symmetric plane partition

Example 11.3.8 Consider a plane partition whose box diagram fits inside a cube of size $a \times b \times c$, where c is the height. In this case, each horizontal slice of the plane partition corresponds to an ideal of the poset $a \times b$, and each successively higher slice corresponds to a smaller ideal with respect to containment. By Theorem 11.2.8 (iii) and Theorem 11.3.4, the poset $P = [a] \times [b]$ is Gaussian, and corresponds to a minuscule representation of type A. It follows from Exercise 11.3.2 that $F_c(P, q)$ is the generating function for plane partitions of sum n that fit in an $a \times b \times c$ box.

Definition 11.3.9 A plane partition is said to be *symmetric* if it is invariant under transposition; in other words, if its box diagram is symmetric about the plane $y = -x$.

Example 11.3.10 Figure 11.3 shows a picture of the symmetric plane partition

$$\begin{pmatrix} 5 & 4 & 2 & 1 \\ 4 & 3 & 2 & \\ 2 & 2 & & \\ 1 & & & \end{pmatrix}$$

of sum 26. The smallest cuboid containing this plane partition is of size $4 \times 4 \times 5$.

Example 11.3.11 Consider a symmetric plane partition whose box diagram fits inside a cube of size $a \times a \times c$, where c is the height. In this case, each horizontal slice of the plane partition corresponds to an ideal of the poset $a \times a$ that is symmetric under transposition; by Exercise 11.2.10, we may identify this with an ideal of the poset

$J([2] \times [a])$. By Theorem 11.2.8 (iii) and Theorem 11.3.4, the poset $P = J([2] \times [a])$ is Gaussian, and corresponds to a spin representation in type B or D. It follows from Exercise 11.3.2 that $F_c(P, q)$ is the generating function for symmetric plane partitions of sum n that fit in an $a \times a \times c$ box.

Theorem 11.3.12 (Stembridge) *Let P_λ be the underlying poset of a principal subheap corresponding to a minuscule representation of a simple Lie algebra, and let $F_m(P_\lambda, q)$ be as in Definition 11.3.1. Let $a_1, \ldots, a_n$ be the associated sequence of integers, where $n = |P_\lambda|$, and let $E = \{i : a_i \text{ is even}\}$.*

(i) *The integer $F_m(P_\lambda, -1)$ is the number of self-complementary ideals of $P_\lambda \times [m]$, or equivalently, the number of multichains*

$$I_1 \subseteq I_2 \subseteq \cdots \subseteq I_m$$

(where $I_j \in J(P_\lambda)$) such that $I_j^* = I_{m+1-j}$.

(ii) *We have*

$$F_m(P_\lambda, -1) = \begin{cases} \prod_{i \in E}(m + a_i)/a_i & \text{if } m \text{ is even,} \\ \prod_{i \notin E}(m + a_i)/\prod_{i \in E} a_i & \text{if } m \text{ is odd and } |E| = \frac{n}{2}, \\ 0 & \text{otherwise.} \end{cases}$$

Proof Part (i) is [77, theorem 4.1].

Part (ii) appears without proof as [77, (2.3)]. Since P_λ is Gaussian by Theorem 11.3.4, we use Definition 11.3.1 to show that

$$F_m(P_\lambda, -1) = \lim_{q \to -1} \prod_{i=1}^{n} \frac{1 - q^{m+a_i}}{1 - q^{a_i}}$$

and apply L'Hôpital's rule $|E|$ times. □

Remark 11.3.13 It is not immediately clear that the right hand sides of the formulae in Theorem 11.3.12 (ii) are integers.

Definition 11.3.14 A plane partition contained in an $a \times b \times c$ box is said to be *self-complementary* if it corresponds to a multichain

$$I_1 \subseteq I_2 \subseteq \cdots \subseteq I_m$$

such that $I_j^* = I_{m+1-j}$. Equivalently, the result of reflecting the box diagram of the plane partition in the three coordinate planes (in any order) is a translation of the set theoretic complement of the plane partition with respect to the $a \times b \times c$ box.

Example 11.3.15 An example of a self-complementary plane partition is

$$\begin{pmatrix} 3 & 3 & 2 & 2 & 2 \\ 3 & 2 & 2 & 1 & \\ 3 & 2 & 1 & 1 & \\ 1 & 1 & 1 & & \end{pmatrix},$$

of sum 26. It fits into a $5 \times 4 \times 3$ box, and is shown in Figure 11.4.

Figure 11.4 A self-complementary plane partition

Example 11.3.16 If $F_c(P, q)$ is the generating function of Example 11.3.8, then we may apply Theorem 11.3.12 to show that $F_c(P, -1)$ is the number of self-complementary plane partitions that fit inside an $a \times b \times c$ box.

Example 11.3.17 If $F_c(P, q)$ is the generating function of Example 11.3.11, then we may apply Theorem 11.3.12 to show that $F_c(P, -1)$ is the number of self-complementary symmetric plane partitions that fit inside an $a \times a \times c$ box.

The evaluations at $q = -1$ appearing in Theorem 11.3.12 are examples of *Stembridge's* $q = -1$ *phenomenon*, which is itself a special case of the mysterious "cyclic sieving phenomenon" of Reiner, Stanton and White [66], defined as follows.

Definition 11.3.18 Let X be a finite set on which a cyclic group $\langle c \rangle$ acts, and let $X(q) \in \mathbb{Z}[q]$ be a polynomial. We say that the triple $(X, X(q), \langle c \rangle)$ exhibits the *cyclic sieving phenomenon* if, for any integer d, the number of elements $x \in X$ that are fixed by c^d is obtained by evaluating $X(q)$ at $q = e^{2\pi i d/n}$.

Example 11.3.19 The order-reversing involution $*$ acts on $J(P_\lambda \times [m])$, so we may define $X = J(P_\lambda \times [m])$, $\langle c \rangle$ to be a cyclic group of order 2 generated by $*$, and $X(q) = F_m(P_\lambda, q)$. In this case, Exercise 11.3.2 shows that $X(1)$ is the number of ideals of $J(P_\lambda \times [m])$, and Theorem 11.3.12 shows that $X(-1)$ is the number of ideals of $J(P_\lambda \times [m])$ that are fixed by c. This is therefore an example of the cyclic sieving phenomenon with $n = 2$.

Gaussian posets also exhibit the cyclic sieving phenomenon with respect to the following action.

Definition 11.3.20 Let P be a finite poset and $J(P)$ the set of ideals of P. The *Fon-Der-Flaass action* Ψ maps an ideal $I \in J(P)$ to the unique ideal $\Psi(I)$ whose maximal elements are the minimal elements of $P \backslash I$.

Example 11.3.21 Let $P = [2] \times [3]$. The diagram below shows an orbit of ideals of length 5 under the Fon-Der-Flaass action, using the shorthand notation of Section 8.2. In each diagram, an asterisk represents an element of an ideal I, and a circled asterisk represents a maximal element of I. The minimal elements of $P \backslash I$ are denoted by empty circles.

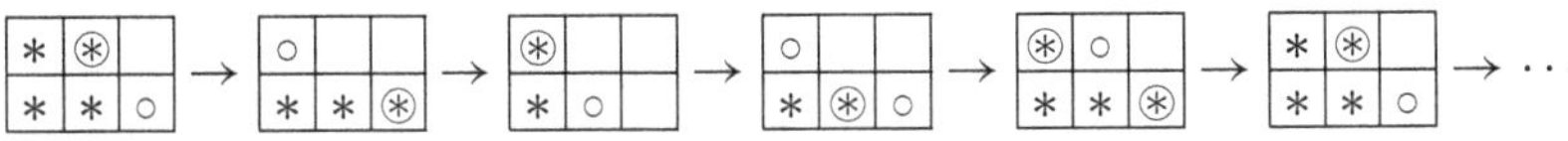

Theorem 11.3.22 (Rush, Shi) *Let $X = J(P_\lambda \times [m])$, let $\langle\Psi\rangle$ be the cyclic group generated by the Fon-Der-Flaass action, and let $X(q) = F_m(P_\lambda, q)$. If $m = 1$ or $m = 2$, then $(X, X(q), \langle\Psi\rangle)$ exhibits the cyclic sieving phenomenon.*

Proof See [67, theorems 1.1 and 1.2]. □

11.4 Jeu de taquin

Our final topic concerns the applications of the combinatorics of minuscule representations to computing intersection numbers of certain Schubert varieties in flag varieties. These results are due to Thomas and Yong [84], building on work of Proctor [63]. We will follow the approach of [84].

As in Chapters 9 and 10, it is not necessary for our purposes to understand the full geometric background of Schubert varieties, but we summarize it here for completeness.

Let G be a complex, connected, reductive Lie group with an opposite pair of Borel subgroups, B and B_-, and maximal torus $T = B \cap B_-$. The Weyl group W is $N(T)/T$, and this gives rise to a root system and a set of fundamental roots. Let P be a parabolic subgroup of G, and let W_P be the associated parabolic subgroup of W. The coset space G/P is called a *generalized flag variety*. The closures of the B_--orbits $\mathcal{X}_w := \overline{B_- wP/P}$ with $wW_P \in W/W_P$ are called *Schubert varieties*, and the generalized flag variety is a union of them. The cohomology ring $H^*(G/P, \mathbb{Q})$ has a basis, called the *Schubert basis*, which consists of the Poincaré duals $\{\sigma_w\}$ of the Schubert varieties.

When W_P is generated by all simple reflections other than a single s_p corresponding to a minuscule weight ω_p the Schubert varieties are much easier to understand than in the general case (see, for example, Perrin [61]). Such Schubert varieties are called *minuscule flag varieties*. In this case, the poset R_p (as in Theorem 8.3.10) turns out to be of key importance in the theory. Theorem 8.3.10 shows that the poset $J(R_p)$ is isomorphic as a poset to the set Λ_p of weights of the minuscule representation of highest weight ω_p. By Theorem A.1.11, this means that R_p is isomorphic as a poset to the principal subheap F corresponding to the minuscule representation. Thomas and Yong [84] denote the set $J(F) \cong J(R_p)$ by $\mathbb{Y}_{G/P}$. They refer to the elements of $J(F)$ as *shapes*, and to the elements of F itself as *boxes*. Shapes are naturally partially ordered by containment, and are usually denoted by lower case Greek letters, such as λ, μ, ν.

The correspondence between these spaces and minuscule representations is as follows.

Type	p	minuscule flag variety G/P
A_n	k	Grassmannian $Gr(k, \mathbb{C}^{n+1})$
B_n	n	orthogonal Grassmannian $OG(n+1, 2n+2)$
C_n	1	projective space $\mathbb{P}^{2n-1}$
D_n	1	even-dimensional quadric $\mathbb{Q}^{2n-2}$
D_n	$n-1, n$	orthogonal Grassmannian $OG(n, 2n)$
E_6	$1, 5$	Cayley plane $\mathbb{OP}^2$
E_7	6	Freudenthal variety $G_\omega(\mathbb{O}^3, \mathbb{O}^6)$

An important problem in the geometric theory is to compute the *Schubert intersection numbers*, $c^{\nu}_{\lambda\mu}(G/P)$, which count points of intersection of generically translated Schubert varieties. These numbers are positive integers that are invariant under permutation of the indices λ, μ, ν, and are given by the formula

$$\sigma_\lambda.\sigma_\mu = \sum_{\nu \in J(F)} c^{\nu}_{\lambda\mu}(G/P)\sigma_\nu.$$

The most important example of a minuscule flag variety is the Grassmannian, $Gr(k, \mathbb{C}^n)$. Each element of this space is a k-dimensional subspace of $\mathbb{C}^n$. In this case, the intersection numbers $c^{\nu}_{\lambda\mu}$ are given by the well-known *Littlewood–Richardson rule*. They can be calculated using a combinatorial method known as *jeu de taquin* ("teasing game"), which is the French name for the popular sliding puzzle known as the "15 puzzle". Proctor [63] generalized this algorithm to a class of posets known as *d-complete posets*, which include the underlying posets of principal subheaps of full heaps.

The basic technique in jeu de taquin is the "slide". In order to illustrate this, we will use the shorthand notation for labelled posets defined in Section 8.2.

Definition 11.4.1 If $\lambda \subseteq \nu$ are shapes in $J(F) \cong \mathbb{Y}_{G/P}$, we define the *skew shape* ν/λ to be their set-theoretic difference. A *standard filling* of a (skew) shape ν/λ is a bijective assignment $f : \nu/\lambda \to \{1, 2, \ldots, |\nu/\lambda|\}$ such that for all boxes x and y, we have $f(x) < f(y)$ whenever $x < y$. The result of this assignment is called a *standard tableau T of shape* $\text{shape}(T) := \nu/\lambda$. The set of all standard (skew) tableaux of shape ν/λ is denoted by $\text{SYT}_{G/P}(\nu/\lambda)$.

We may also refer to a shape (in the original definition) as a *straight shape*, and to tableaux of those shapes as *straight tableaux*, in order to contrast them with skew shapes and skew tableaux.

Example 11.4.2 An example of a skew tableau associated with the minuscule representation $L(E_6, \omega_5)$ (see Example 8.2.6) is

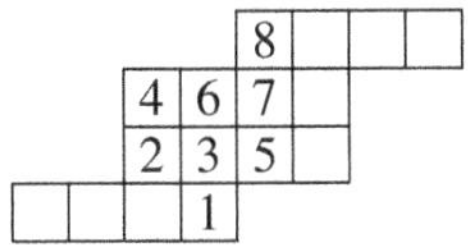

In this case, the shape λ is given by

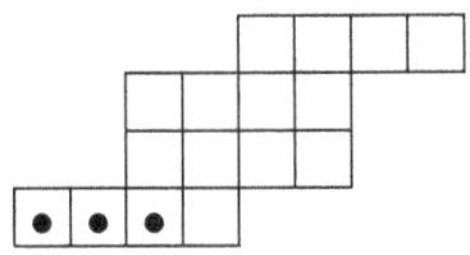

and the shape ν is given by

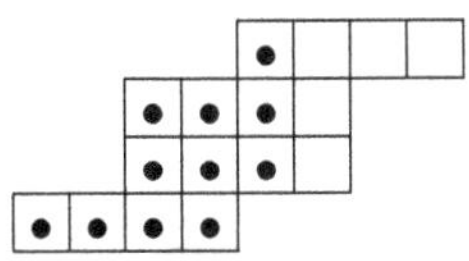

We will denote shapes by sequences of natural numbers representing the number of bullets in each column of the shape, reading from left to right, and omitting any trailing zeros. (The bullets necessarily appear as low as possible in the diagram.) This labelling is noncanonical and depends on how we chose to represent the underlying poset of the principal subheap as a planar diagram. In our notation, we have $\lambda = (1, 1, 1)$ and $\nu = (1, 1, 3, 3, 3)$.

Let $T \in \mathrm{SYT}_{G/P}(\nu/\lambda)$, and let x be a box that is not in ν/λ but maximal subject to the condition that it is below some element of ν/λ. We associate another standard tableau $\mathrm{jdt}_x(T)$, of a different skew shape, called the *jeu de taquin slide of T into x*, as follows. Among the boxes of ν/λ that cover x, let y be the one with the smallest label, $f(y)$. Move the label $f(y)$ to x, leaving y vacant. Look for boxes of ν/λ that cover y and repeat the process, moving into y the smallest label available among those boxes covering it. When no more such moves are possible, the tableau that results is by definition $\mathrm{jdt}_x(T)$. The order preserving nature of the slide means that $\mathrm{jdt}_x(T)$ is also a standard tableau. A *rectification* of T is the result of an iteration of jeu de taquin slides until we terminate at a standard tableau, rect(T) of a (straight) shape.

Example 11.4.3 Consider the skew tableau of Example 11.4.2, together with the box labelled x, as follows:

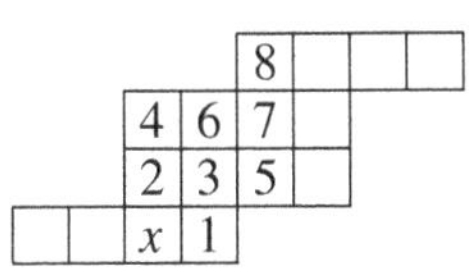

In this case, x is covered by two elements, labelled 1 and 2, so the jeu de taquin slide will exchange x with 1, giving the skew tableau

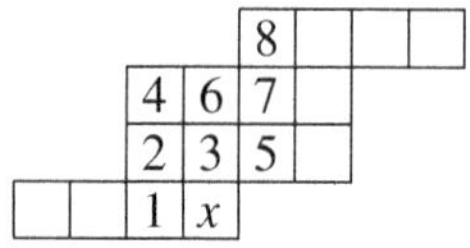

Repeating this process gives the following sequence of skew tableaux:

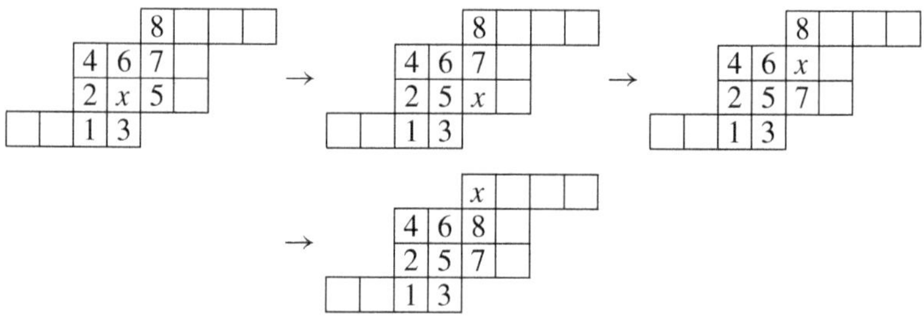

so that we have

$$\mathrm{jdt}_x(T) =$$

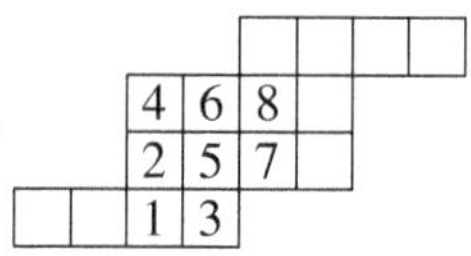

We then perform another jeu de taquin slide involving the box y given by

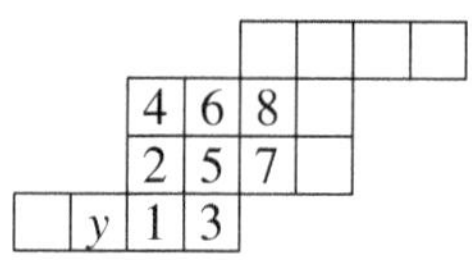

which gives

$$\mathrm{jdt}_y(\mathrm{jdt}_x(T)) =$$

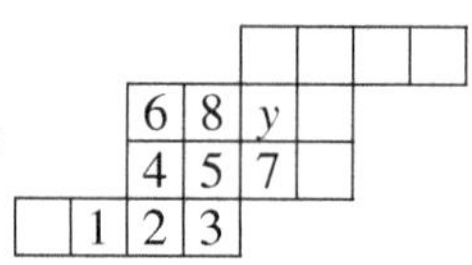

Finally, producing a jeu de taquin slide involving the box z given by

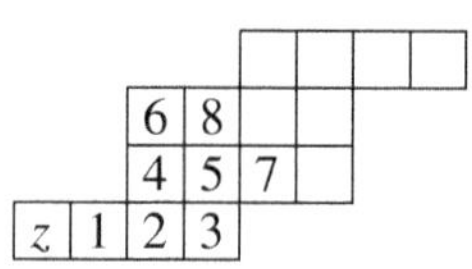

gives a rectification of T,

$$\mathrm{rect}(T) =$$

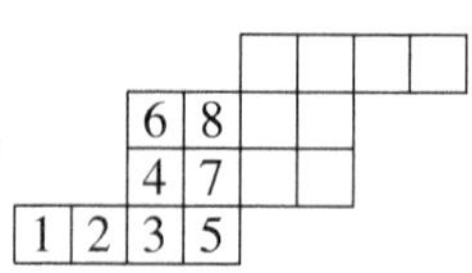

In Example 11.4.3, the sequence of boxes x, y, z is forced. In other examples, this may not be the case. Remarkably, the choice of sequence does not matter, thanks to the following result.

Theorem 11.4.4 (Proctor) *Let ν/λ be a skew shape associated with a minuscule flag variety, and let $T \in \mathrm{SYT}_{G/P}(\nu/\lambda)$. Then the standard tableau* $\mathrm{rect}(T)$ *is independent of the choices of jeu de taquin slides.*

Proof As mentioned in Thomas and Yong [84, theorem 4.1], this result is proved in Proctor [63] in the more general context of d-complete posets. □

Thanks to Theorem 11.4.4, we may speak of *the* rectification of a skew tableau.

An important property of jeu de taquin is that it is reversible, as we now explain. Let G/P be a minuscule flag variety, and let $T \in \mathrm{SYT}_{G/P}(\nu/\lambda)$. Let x be a box that is not in ν/λ, but minimal subject to being above some element of ν/λ. The reverse jeu de taquin slide $\mathrm{revjdt}_x(T)$ of T into x is defined similarly to a jeu de taquin slide, except that we move into x the largest of the labels among boxes in ν/λ covered by x. A reverse rectification $\mathrm{revrect}(T)$ is the end result of the iterated application of these slides, ending with a standard filling of a rotated shape. The key properties of reverse jeu de taquin and reverse rectification are the following.

Proposition 11.4.5 (Thomas, Yong) *Let ν/λ be a skew shape associated with a minuscule flag variety, and let $T \in \mathrm{SYT}_{G/P}(\nu/\lambda)$.*

(i) *The standard tableau* $\mathrm{revrect}(T)$ *is independent of the choices of reverse jeu de taquin slides.*
(ii) *If the box y is vacated by* $\mathrm{jdt}_x(T)$, *then* $\mathrm{revjdt}_y(\mathrm{jdt}_x(T)) = T$.
(iii) *If the box y is vacated by* $\mathrm{revjdt}_x(T)$, *then* $\mathrm{jdt}_y(\mathrm{revjdt}_x(T)) = T$.

Proof This is Thomas and Yong [84, proposition 4.2]. □

We are now ready to define a remarkable involution called "infusion". Consider a minuscule flag variety G/P, a straight tableau $T \in \mathrm{SYT}_{G/P}(\lambda)$ and a skew tableau $U \in \mathrm{SYT}_{G/P}(\nu/\lambda)$. The *infusion*, $\mathrm{infus}(T, U)$, of T and U is a certain pair (X, Y), where $X \in \mathrm{SYT}_{G/P}(\gamma)$ for some straight shape γ with $|\gamma| = |\nu/\lambda|$, and $Y \in \mathrm{SYT}_{G/P}(\nu/\gamma)$.

More precisely, X and Y are obtained as follows. Place T and U inside a straight tableau of shape ν, according to their shapes. Now remove the largest label that appears in T; say this is label m in box x of λ. Since x necessarily lies next to ν/λ, we may apply the slide $\mathrm{jdt}_x(U)$, leaving a hole at the other side of ν/λ. Place a label m in that hole and repeat moving the labels that were originally in T, in decreasing order, until they are exhausted. In particular, we declare that the labels placed in the created holes at each step never move for the duration of the procedure. The resulting straight shape tableau of shape γ and skew tableau of shape ν/γ are X and Y respectively.

Example 11.4.6 Let $G/P = Gr(3, \mathbb{C}^7)$, which corresponds to the minuscule representation $L(A_6, \omega_3)$. Let $\lambda = (2, 1)$ and $\nu = (3, 3, 2)$. We set

$$T = \begin{array}{|c|c} \hline 3 & \multicolumn{1}{|c}{} \\ \hline 1 & \multicolumn{1}{c|}{2} \\ \hline \end{array} \quad \text{and} \quad U = \begin{array}{|c|c|c|} \hline 4 & 5 & \\ \hline & 2 & 3 \\ \hline & & 1 \\ \hline \end{array}$$

denoting the pair (T, U) by the shorthand notation

$$\begin{array}{ccc} 4 & 5 & \\ \mathbf{3} & 2 & 3 \\ \mathbf{1} & \mathbf{2} & 1 \end{array}$$

where the bold labels denote the original labels of T. The sequence of jeu de taquin slides required to compute infus(T, U) proceeds as follows:

$$\begin{array}{cccc} 4 & 5 & \; & \; \\ \mathbf{3} & 2 & 3 & \\ \mathbf{1} & \mathbf{2} & 1 & \end{array} \rightarrow \begin{array}{cccc} 4 & 5 & \; & \; \\ 2 & 3 & \mathbf{3} & \\ \mathbf{1} & \mathbf{2} & 1 & \end{array} \rightarrow \begin{array}{cccc} 4 & 5 & \; & \; \\ 2 & 3 & \mathbf{3} & \\ \mathbf{1} & 1 & \mathbf{2} & \end{array} \rightarrow \begin{array}{cccc} 4 & \mathbf{1} & \; & \; \\ 2 & 5 & \mathbf{3} & \\ 1 & 3 & \mathbf{2} & \end{array}$$

Unpacking the result, we obtain

$$X = \begin{array}{cc} 4 & \\ 2 & 5 \\ 1 & 3 \end{array} \quad \text{and} \quad Y = \begin{array}{ccc} \; & 1 & \\ \; & \; & 3 \\ \; & \; & 2 \end{array}$$

meaning that $\gamma = (3, 2)$.

Remark 11.4.7 Without performing the algorithm, it is not obvious which shape γ will eventually appear, based on the input parameters, λ, ν, T and U.

Exercise 11.4.8 Let T, U, X, Y be as in Example 11.4.6. Verify that infus$(X, Y) = (T, U)$.

In the context of the following result, it is convenient to define $\mathcal{I}^{\nu}_{\lambda,\mu}(G/P)$ to be the set of all pairs (T, U) where T has shape λ, U has shape ν/λ, and infus$(T, U) = (X, Y)$ is such that X has shape μ.

Proposition 11.4.9 (Thomas, Yong) *Let G/P be a minuscule flag variety corresponding to the principal subheap F. The procedure* infus *defines an involution on the set*

$$L(G/P) := \bigcup_{\nu,\gamma \in J(F)} \mathrm{SYT}_{G/P}(\gamma) \times \mathrm{SYT}_{G/P}(\nu/\gamma).$$

In particular, infusion induces a bijection between the sets $\mathcal{I}^{\nu}_{\lambda,\mu}(G/P)$ and $\mathcal{I}^{\nu}_{\mu,\lambda}(G/P)$.

Proof This is Thomas and Yong [84, theorem 4.4]. □

Corollary 11.4.10 (Thomas, Yong) *Let G/P be a minuscule flag variety corresponding to the principal subheap F. If $\mu \in J(F)$ and $U \in \mathrm{SYT}_{G/P}(\mu)$, then the number of skew tableaux $W \in \mathrm{SYT}(\nu/\lambda)$ whose rectification is U is independent of the choice of $U \in \mathrm{SYT}_{G/P}(\mu)$.*

Proof Consider the set $I_{\mu,\lambda}(G/P)$, and fix $V \in \mathrm{SYT}_{G/P}(\nu/\mu)$. These choices (in conjunction with the notion of reverse jeu de taquin) give a natural bijection between the set $\mathrm{SYT}_{G/P}(\mu)$ and the sequences of possible jeu de taquin slides that rectify $V \in \mathrm{SYT}_{G/P}(\nu/\mu)$. By Theorem 11.4.4, the resulting rectification of V is independent of the choice of slides. It follows that the number of times that a particular tableau $U \in \mathrm{SYT}_{G/P}(\mu)$ appears as the first component of a pair in $I_{\mu,\lambda}(G/P)$ is independent of U; call this number N_1. More specifically, N_1 is the number of skew tableaux in $\mathrm{SYT}_{G/P}(\nu/\mu)$ whose rectification has shape λ.

On the other hand, the number of elements of $I_{\lambda,\mu}(G/P)$ that are carried by infusion to a pair of tableaux beginning with U is equal to $|\mathrm{SYT}_{G/P}(\lambda)|N_2$, where N_2 is the number of skew tableaux of shape ν/λ whose rectification is U. Proposition 11.4.9 shows that

$$N_1 = |\mathrm{SYT}_{G/P}(\lambda)|N_2,$$

from which it follows that N_2 is independent of the choice of $U \in \mathrm{SYT}_{G/P}(\mu)$, as required. □

We can now state the main result of this section. Note that Corollary 11.4.10 shows that the number $c^{\nu}_{\lambda,\mu}$ below is well-defined.

Theorem 11.4.11 (Thomas, Yong) *Let G/P be a minuscule flag variety associated with the principal subheap F, let λ, μ, $\nu \in J(F)$, and fix a tableau T_μ of shape μ. The Schubert intersection number $c^{\nu}_{\lambda,\mu}$ is equal to the number of standard skew tableaux of shape ν/λ whose rectification is T_μ.*

Proof This is Thomas and Yong [84, main theorem]. □

Example 11.4.12 Consider the minuscule flag variety G/P associated with the representation $L(D_5, \omega_1)$. In this case, the principal subheap is of shape

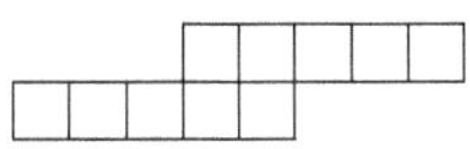

which we denote by $\nu = (1, 1, 1, 2, 2, 1, 1, 1)$. Let λ be the shape $1^5 = (1, 1, 1, 1, 1)$. There is only one standard tableau of shape λ, and only one standard skew tableau of shape ν/λ. Applying infusion to the pair

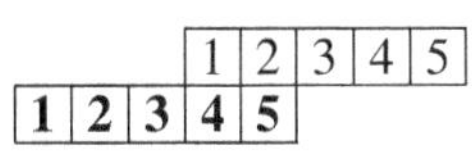

yields

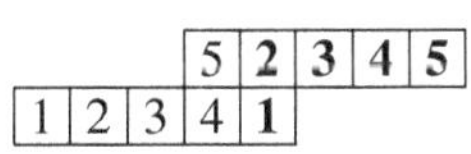

It follows that $c^{\nu}_{\lambda,\mu}$ is equal to 1 if $\mu = (1, 1, 1, 2)$, and equal to zero otherwise; in particular, $c^{\nu}_{\lambda,\lambda} = 0$.

Exercise 11.4.13 Consider the minuscule flag variety G/P associated with the representation $L(D_6, \omega_1)$, with principal subheap of shape $\nu = (1, 1, 1, 1, 2, 2, 1, 1, 1, 1)$. Let $\lambda = 1^6$. Show that in this case, $c^{\nu}_{\lambda,\mu}$ is equal to 1 if $\mu = \lambda$, and is equal to 0 otherwise.

Exercise 11.4.14 Show that the minuscule flag variety G/P associated with $L(D_n, \omega_1)$ behaves like Example 11.4.12 if n is odd, and behaves like Exercise 11.4.13 if n is even.

Exercise 11.4.15 Consider the case where G/P is the Grassmannian $Gr(4, \mathbb{C}^7)$, corresponding to the representation $L(A_6, \omega_4)$. In this case, the principal subheap is of shape $(4, 4, 4)$. Let $\nu = (4, 2, 1)$ and $\lambda = (3, 1)$. Show that there are six skew tableaux of shape ν/λ, two of which rectify to a tableau of shape λ. Deduce that $c^{\nu}_{\lambda,\mu} = 2$, where $\mu = (2, 1)$.

11.5 Notes and references

1 The notion of full commutativity (Definition 11.1.2) is due to Stembridge [78].

2 The definition of λ-minuscule (Definition 11.1.4) is attributed to D. Peterson (unpublished).

3 Littelmann [46] defines a generalization of minuscule weights called braidless weights. We call a fundamental weight ω *braidless* if for any $w \in W$ and any fundamental roots α_i and α_j such that $(w(\omega), \alpha_i^\vee) > 0$ and $(w(\omega), \alpha_j^\vee) > 0$, we also have $(\alpha_j, \alpha_i^\vee) = 0$. Littelmann proves in [46, lemma 3.1] that ω is braidless if and only if it satisfies one of the following conditions:
(i) ω is minuscule;
(ii) $\omega = \omega_1$ and W is of type B_n;
(iii) $\omega = \omega_n$ and W is of type C_n; or
(iv) ω is arbitrary and W has rank at most 2.
If the generator s_p corresponds to a braidless fundamental weight, Littelmann [46, lemma 3.2] proves an analogue of Theorem 11.1.9 (iv) for braidless weights, with "minuscule" replaced by "fully commutative".

4 Theorem 11.1.18 was first proved for Weyl groups by Proctor [64, propositions 3.1, 3.2] using a case analysis. The following more general result was proved by Stembridge [78, theorem 7.1]. Given a finite irreducible Weyl group W, Stembridge calls a set W^J of distinguished coset representatives "minuscule" if there is a simple Lie algebra $\mathfrak{g}$ with Weyl group W and a minuscule weight ω_p such that $S = J \dot{\cup} \{p\}$. Stembridge goes on to prove that the following are equivalent:
(i) W^J is a lattice under the Bruhat order;
(ii) W^J is a distributive lattice under the Bruhat order;
(iii) W^J is a distributive lattice under $\leq_L$;
(iv) the lattices $(W^J, \leq)$ and $(W^J, \leq_L)$ are equal; and
(v) W^J is minuscule.
For the more general case of an irreducible Coxeter group, there are two more examples satisfying conditions (i)–(iv), namely a parabolic subgroup of $W(H_3)$ of type $I_2(5)$, or a parabolic subgroup of $I_2(m)$ of rank 1 (possibly with $m = \infty$). Pfeiffer and Röhrle [62] give a conceptual proof of part (i) of this result in the Coxeter group case using descent algebras and Mackey induction.

5 The uniquely modular posets of Definition 11.2.1 are closely related to "wonderful posets". A poset $(P, \leq)$ with a maximal and minimal element is called *wonderful* if whenever $x, y, z, u \in P$ are such that z covers x, z covers y, $x \leq u$ and $y \leq u$, there exists an element $w \in P$ such that x covers w, y covers w, and $w \leq u$.

6 Gaussian posets appear in an exercise in Stanley's book [75, exercise 4.25].

7 Proctor's paper [65] contains explicit vertex labellings for all finite connected vertex labellable posets. Although the labels are only required to be rational numbers, they always turn out to be positive integers. There is a deep geometric reason for this given in [65]: the numbers can be interpreted as the coefficients of the Hodge adjoint of cup product multiplication with a hyperplane section in the cohomology ring of a minuscule flag manifold.

8 Theorem 11.3.4 is formally similar to Weyl's character formula in the theory of semisimple Lie algebras, and indeed, this is how Proctor proves the result.

9 The enumeration of plane partitions is also discussed by Macdonald: see [51, example I.5.13] for the ordinary (type A) case, and [51, example I.5.19] for the symmetric (type B) case. These enumeration results were first proved by P. A. MacMahon [52].
10 Examples 11.3.8, 11.3.11, 11.3.16 and 11.3.17 are taken from Stembridge's paper [77]. Example 11.3.21 comes from Rush and Shi's paper [67].
11 The analogue of Theorem 11.3.22 for $m = 3$ is false; a counterexample is given in this case by the poset $[3] \times [3]$ (see [67]).
12 Stanton [76] defines a *fake Gaussian sequence* to be a sequence $(a_1, a_2, \ldots, a_k)$ of positive integers for which $a_i = a_{k+1-i}$ for all i and the rational function $F_m(P, q)$ of Definition 11.3.1 is a polynomial for all nonnegative integers m. The paper [76] offers some conjectures on fake Gaussian sequences.
13 Minuscule weights also play an important role in the theory of affine Grassmannians; see Kamnitzer [38] for more details.
14 Table 1 of Thomas and Yong [84] contains a mistake: the Grassmannians associated with the Weyl group A_n involve subspaces of $\mathbb{C}^{n+1}$, not $\mathbb{C}^n$.
15 There are other types of symmetry for plane partitions. A plane partition is said to be *cyclically symmetric* if its box diagram is invariant under rotation of the axes, and *totally symmetric* if it is symmetric and cyclically symmetric, meaning that the full symmetric group S_3 leaves the box diagram invariant. Quite remarkably, the number of totally symmetric self-complementary plane partitions (TSSCPPs) in a box of size $2n \times 2n \times 2n$ is equal to the number of alternating sign matrices of size n. The latter are $n \times n$ matrices with entries in the set $\{0, 1, -1\}$, subject to the conditions that (a) the row and column sums are all $+1$, and (b) the nonzero entries alternate in sign in each row. Such matrices were enumerated by Zeilberger [91].

Appendix A

Posets, graphs and categories

In Appendix A, we recall some of the basic concepts associated with partially ordered sets, graphs and categories.

A.1 Posets and graphs

A *partial order* on a set P is a binary relation $\leq$ satisfying the following three properties:

(i) reflexivity: for all $x \in P$, we have $x \leq x$;
(ii) antisymmetry: for all $x, y \in P$, if we have both $x \leq y$ and $y \leq x$, then $x = y$;
(iii) transitivity: for all $x, y, z \in P$, if we have both $x \leq y$ and $y \leq z$, then $x \leq z$.

A set P equipped with a partial order $\leq$ is known as a *partially ordered set* or *poset*. If Q is a subset of a poset P, then Q inherits a poset structure from P by restricting the relation $\leq$ to Q.

Strictly speaking, a partial order is the subset of $P \times P$ given by

$$\{(x, y) \in P \times P : x \leq y\}.$$

Every partial order, $\leq$, on P has an *opposite order* on P, denoted by $\geq$. This is the subset of $P \times P$ with the property that $(y, x) \in \geq$ if and only if $(x, y) \in \leq$. It turns out (Exercise A.1.4) that $\geq$ is also a partial order. We write P^* to refer to the set P equipped with the opposite partial order.

If $\leq$ and $\leq'$ are two partial orders on a set P, we say that $\leq$ is *smaller than* $\leq'$ if it corresponds to a smaller subset of $P \times P$. Another way to say the same thing is that $\leq'$ is a *refinement* (or *extension*) of $\leq$.

Elements x and y are said to be *comparable* in the partial order P if we have either $x \leq y$ or $y \leq x$ (or both). A partial order in which every pair of elements is comparable is called a *total* (or *linear*) order. Szpilrajn's Theorem [82] shows that every partial order is the intersection of its set of linear extensions; in particular, any partial order can be refined to a total order. When $m \in \mathbb{N}$, we write $[m]$ for the totally ordered set $\{1, 2, \ldots, m\}$ equipped with its natural total order, $\leq$.

A *chain* of P is a subset C of P such that any two elements of C are comparable in P. A *multichain* of P is a sequence

$$x_1 \leq x_2 \leq \cdots \leq x_m$$

of elements of P. An *antichain* of P is a subset A of P such that no two elements of A are comparable in P.

A subset Q of P is said to be *convex* if whenever we have $x \leq z \leq y$ and $x, y \in Q$, then we also have $z \in Q$.

If $x, y \in P$, we write $x < y$ whenever $x \leq y$ and $x \neq y$.

If $x, y \in P$, we define the *closed interval* $[x, y]$ to be the subset of P given by

$$\{z \in P : x \leq z \leq y\}.$$

The corresponding *open interval* (x, y) is the subset of P given by

$$\{z \in P : x < z < y\}.$$

If $x, y \in P$, we say that y *covers* x if $x < y$ and whenever $x \leq z \leq y$, we have either $z = x$ or $z = y$. If there is a function $\rho : P \to \mathbb{Z}$ such that $\rho(y) = \rho(x) + 1$ whenever y covers x, then P is said to be *ranked*, and ρ is called a *rank function* for P.

We say P is *locally finite* if every interval of P is finite. If P is locally finite, we construct the *Hasse diagram* of P by representing each element of P as a vertex in the plane, and drawing a line segment or curve upwards from x to y whenever y covers x.

A locally finite poset is called *connected* or *irreducible* if its Hasse diagram is connected.

Exercise A.1.1 Prove that intervals are convex.

Exercise A.1.2 Prove that the partial order in a locally finite poset is determined by the covering relations, that is, that the partial order is the reflexive, transitive closure of the covering relations. Show that $\mathbb{Q}$ and $\mathbb{R}$, equipped with their usual partial orders, are nonisomorphic as partially ordered sets even though their covering relations are the same.

Exercise A.1.3 Explain how a locally finite partial order on a set may be reconstructed from its Hasse diagram.

Exercise A.1.4 Prove that the opposite order of a partial order is itself a partial order.

Exercise A.1.5 Suppose that $(P_1, \leq_1)$ and $(P_2, \leq_2)$ are partially ordered sets. Define a relation $\leq$ on the disjoint union $P = P_1 \dot{\cup} P_2$ by setting $\leq = \leq_1 \dot{\cup} \leq_2$. Prove that $\leq$ is a partial order on P.

Exercise A.1.6 Suppose that $(P_1, \leq_1)$ and $(P_2, \leq_2)$ are partially ordered sets. Define a relation $\leq$ on the Cartesian product $P = P_1 \times P_2$ by stipulating that $(x_1, y_1) \leq (x_2, y_2)$ if and only if we have both $x_1 \leq x_2$ and $y_1 \leq y_2$. Prove that $\leq$ is a partial order on P.

If x and y are elements of the poset P, then an *upper bound* of x and y is an element $z \in P$ satisfying $z \geq x$ and $z \geq y$. A *least upper bound* of x and y is an upper bound z such that for every other upper bound w of x and y we have $w \geq z$. A least upper bound is unique if it exists, and we denote the least upper bound (or *join*) of x and y by $x \vee y$ (pronounced "x join y"). The terms *lower bound* and *greatest lower bound* are defined analogously, and the greatest lower bound (or *meet*) of x and y, if it exists, is denoted by $x \wedge y$, and pronounced "x meet y". All these terms may be generalized in an obvious way to apply to arbitrary nonempty subsets of P, rather than pairs of elements.

If the poset P has the property that any pair of elements has a least upper bound and a greatest lower bound, then P is called a *lattice*. If, in addition, the operations $\wedge$ and $\vee$ distribute over each other, that is, we have

$$x \vee (y \wedge z) = (x \vee y) \wedge (x \vee z)$$
$$x \wedge (y \vee z) = (x \wedge y) \vee (x \wedge z)$$

for all $x, y, z \in P$, then P is called a *distributive lattice*.

An element x of a lattice P is called *join irreducible* (respectively, *meet irreducible*) if it is not the bottom (respectively, top) element, and if whenever $x = a \vee b$ (respectively, $x = a \wedge b$), then either $x = a$ or $x = b$.

An *order ideal* (or "ideal" for short) of a poset P is a subset I of P such that whenever $x \in I$ and $y \leq x$, we have $y \in I$. Dually, an *order filter* (or "filter" for short) of a poset P is a subset I of P such that whenever $x \in I$ and $y \geq x$, we have $y \in I$.

Exercise A.1.7 Prove that if I_1 and I_2 are ideals of P and $I_1 \subset I_2$, then $I_2 \backslash I_1$ is a convex subset of P.

The maximal elements of an ideal I form an antichain, and each antichain A in P determines an ideal

$$I = \{x \in P : x \leq y \text{ for some } y \in A\}.$$

Lemma A.1.8 *Let P be a locally finite poset with the property that every antichain of P is finite. The correspondence above induces a bijection between (a) the antichains of P and (b) the ideals I of P with the property that I contains no infinite ascending chains $x_1 < x_2 < x_3 < \cdots$.*

Proof Let A be an antichain of P, and let I be the corresponding ideal as defined above. Suppose for a contradiction that I contains an infinite ascending chain $x_1 < x_2 < x_3 < \cdots$. By definition of I, each x_i has the property that $x_i < y$ for some $y \in A$. Because A is finite by assumption and the chain is infinite, we may assume without loss of generality (by considering subsequences) that there exists a single y such that $x_i < y$ for all i. This shows that the interval (x_i, y) contains infinitely many elements, which contradicts the assumption that P be locally finite.

Conversely, suppose that I is an ideal of P with no infinite ascending chains. This implies that for each $x \in I$, there exists a maximal element $y \in I$ with $x \leq y$. Letting A be the set of maximal elements of I, we see that

$$I = \{x \in P : x \leq y \text{ for some } y \in A\}.$$

This shows that the ideals described in the statement are in bijection with their antichains of maximal elements, as required. □

Suppose that P satisfies the hypotheses of Lemma A.1.8. If the maximal elements of an ideal I are $\{x_1, x_2, \ldots, x_k\}$, then we write $I = \langle x_1, x_2, \ldots, x_k\rangle$. If $k = 1$, then we call I a *principal* ideal.

The ideals of the poset P form a distributive lattice, $J(P)$, where we define $I \leq I'$ if and only if $I \subseteq I'$. It follows that the operations $\vee$ and $\wedge$ are given by set-theoretic union and intersection, respectively.

Exercise A.1.9 Prove that $J(P)$ as defined above is a distributive lattice.

Lemma A.1.10 *Let P be a poset and let $J(P)$ be the associated distributive lattice. Then I' covers I in $J(P)$ if and only if $I' = I \cup \{x\}$ for some $x \in P \backslash I$.*

Proof The fact that the order relation in $J(P)$ is set-theoretic inclusion implies that if $I \cup \{x\} \in J(P)$ is such that $x \in P\backslash I$, then $I \cup \{x\}$ covers I. Conversely, suppose that I' covers I (and thus that $I \subset I'$), but that $I' \backslash I$ contains more than one element. Let x be any element of $I'\backslash I$ other than the (unique, if it exists) element y such that $y \geq z$ for all $z \in I'\backslash I$, and let x' be any element of $I'\backslash I$ such that $x \not\geq x'$. Let $I'' = I \cup \langle x \rangle$ be the smallest ideal containing I and x. Then we have $I \subsetneq I'' \subsetneq I'$, because $x \in I''\backslash I$ and $x' \in I'\backslash I''$. This contradicts the hypothesis that I' covers I, as required. □

The next result is known as the *Fundamental Theorem for finite distributive lattices*, and is proved in Stanley [75, theorem 3.4.1].

Theorem A.1.11 (Birkhoff) *Any finite distributive lattice L is isomorphic to a lattice of the form $J(P)$, where P is a finite poset. In this case, P is isomorphic to the subset of join irreducible elements of L under the restriction of the partial order on L. This gives a bijection between isomorphism types of finite posets and isomorphism types of finite distributive lattices.* □

A *graph* G is an ordered pair (V, E) where V is a set and E is a set of unordered pairs of elements of V. The elements of V and E are called the *vertices* and *edges* of G, respectively. We optionally allow edges of the form $\{x, x\}$; such edges are called *loops*.

A *path* in a graph G is a finite sequence $v_1, v_2, \ldots, v_n$ of vertices of G such that for each $1 \leq i < n$, $\{v_i, v_{i+1}\}$ is an edge of G. There is an equivalence relation $\sim$ on G defined by stipulating that $x \sim y$ if there is a (possibly trivial) path between x and y; the equivalence classes of $\sim$ are called *connected components*. If G has a single connected component, we say that G is *connected*; otherwise, G is called *disconnected*.

A.2 Categories

A *category* **C** consists of a class of objects and a set of morphisms for each pair of objects. If A and B are objects of **C**, we denote the set of morphisms from A to B by $\mathrm{Hom}_{\mathbf{C}}(A, B)$, or $\mathrm{Hom}(A, B)$ for short if the category **C** is understood. If A, B and C are objects of **C**, then there is a function

$$\mathrm{Hom}_{\mathbf{C}}(A, B) \times \mathrm{Hom}_{\mathbf{C}}(B, C) \to \mathrm{Hom}_{\mathbf{C}}(A, C)$$

called *composition of morphisms*. If $f \in \mathrm{Hom}_{\mathbf{C}}(A, B)$ and $g \in \mathrm{Hom}_{\mathbf{C}}(B, C)$, we write $g \circ f$ or gf for the corresponding morphism in $\mathrm{Hom}_{\mathbf{C}}(A, C)$.

We require the following conditions to be satisfied whenever A, B, C and D are objects of $\mathbf{C}$.

(i) If $A \neq C$ or $B \neq D$ then $\mathrm{Hom}_{\mathbf{C}}(A, B)$ and $\mathrm{Hom}_{\mathbf{C}}(C, D)$ are disjoint sets.
(ii) Whenever we have $f \in \mathrm{Hom}_{\mathbf{C}}(A, B)$, $g \in \mathrm{Hom}_{\mathbf{C}}(B, C)$ and $h \in \mathrm{Hom}_{\mathbf{C}}(C, D)$, the morphisms $h(gf)$ and $(hg)f$ are the same element of $\mathrm{Hom}_{\mathbf{C}}(A, D)$.
(iii) There is a (necessarily unique) morphism $1_A \in \mathrm{Hom}_{\mathbf{C}}(A, A)$ such that $f 1_A = f$ for every $f \in \mathrm{Hom}_{\mathbf{C}}(A, B)$ and $1_A g = g$ for every $g \in \mathrm{Hom}_{\mathbf{C}}(B, A)$.

It is not necessary for the objects of a category to form a set. A category whose objects do form a set is known as a *small* category. It is also not necessary for the morphisms between two objects to take the form of functions between sets, although, if this is the case, the category is known as a *concrete* category. The categories in the following two examples are concrete, but not small.

Example A.2.1 The category of partially ordered sets, **Poset**, has as objects all partially ordered sets. A morphism between two partially ordered sets P_1 and P_2 is a function $f : P_1 \to P_2$ such that whenever $a, b \in P_1$ satisfy $a \leq b$, we have $f(a) \leq f(b)$.

Example A.2.2 The category of graphs, **Graph**, has as objects all graphs in the sense of Section A.1. A morphism between two graphs (V_1, E_1) and (V_2, E_2) consists of a function $f : V_1 \to V_2$ such that for each edge $\{a, b\} \in E_1$, either we have $\{f(a), f(b)\} \in E_2$ or we have $f(a) = f(b)$.

Remark A.2.3 A more usual definition for the category **Graph** would be to disallow the possibility $f(a) = f(b)$ above. However, the definition of Example A.2.2 is more useful for applications involving heaps.

A morphism $f \in \mathrm{Hom}_{\mathbf{C}}(A, B)$ is called an *isomorphism* if there exists a morphism $g \in \mathrm{Hom}_{\mathbf{C}}(B, A)$ such that $gf = 1_A$ and $fg = 1_B$. In this case, we call g the *inverse* of f. If, in addition, we have $A = B$, then we call f an *automorphism* of A. The set of automorphisms of A, denoted by $\mathrm{Aut}_{\mathbf{c}}(A)$, forms a group under composition of morphisms.

A *subcategory* $\mathbf{S}$ of $\mathbf{C}$ is a category whose objects are objects of $\mathbf{C}$, with the property that whenever A and B are objects of $\mathbf{S}$, $\mathrm{Hom}_{\mathbf{S}}(A, B)$ is a subset of $\mathrm{Hom}_{\mathbf{C}}(A, B)$. If we always have $\mathrm{Hom}_{\mathbf{S}}(A, B) = \mathrm{Hom}_{\mathbf{C}}(A, B)$ in this situation, then $\mathbf{S}$ is said to be a *full* subcategory of $\mathbf{C}$.

Suppose that G is a group, P is a partially ordered set, and $\phi : G \to \mathrm{Aut}(P)$ is a homomorphism of groups. Then G acts on P via $g.p := \phi(g)(p)$; we abbreviate this by saying that G acts on P by poset automorphisms, and we write $g(p)$ for $\phi(g)(p)$ if ϕ is understood.

The next result is stated without proof by Hagiwara [32, proposition 3.3]; a proof appears in McGregor-Dorsey [57, proposition 1.2.8].

Proposition A.2.4 *Let $(P, \leq)$ be a poset on which a group G acts by poset automorphisms. Assume that whenever p and $g(p)$ are comparable for some $p \in P$ and $g \in G$, then we have $p = g(p)$. Let $G \backslash P$ be the set of all orbits of P under the action* *of G, and denote the orbit containing p by $\overline{p}$. Define a relation $\preceq$ on $G\backslash P$ by $\overline{p} \preceq \overline{q}$ if and only if there exists $g \in G$ such that $p \leq g(q)$. Then $(G\backslash P, \preceq)$ is a poset.*

Proof We first show that $\preceq$ is well-defined. Suppose that $p \leq g(q)$, $a \in \overline{p}$ and $b \in \overline{q}$; we need to show that $a \leq g'(b)$ for some $g' \in G$. By hypothesis, there exist $h_1, h_2 \in G$ such that $a = h_1(p)$ and $b = h_2(q)$. Since $p \leq g(q)$, we have $a = h_1(p) \leq h_1 g(q)$, and the result follows by taking $g' = h_1 g h_2^{-1}$.

Reflexivity of $\preceq$ is immediate by taking $g = 1$, and transitivity follows from the fact that if $p \leq g(q)$ and $q \leq g'(r)$, then $p \leq gg'(r)$. It remains to prove antisymmetry, so suppose that both $\overline{p} \preceq \overline{q}$ and $\overline{q} \preceq \overline{p}$. It follows that there exist $g, h \in G$ for which $p \leq g(q)$ and $q \leq h(p)$, from which it follows that $p \leq gh(p)$. Since p and $gh(p)$ are comparable, the hypotheses show that $p = gh(p)$, which means that $p = g(q)$ and $\overline{p} = \overline{q}$, as required. □

Definition A.2.5 If $(P, \leq)$, G, $G\backslash P$ and $\preceq$ satisfy the hypotheses of Proposition A.2.4, we call $(G\backslash P, \preceq)$ the *quotient poset* of P by G.

Appendix B

Lie theoretic data

The untwisted affine Dynkin diagrams are those of type $A_l^{(1)}(l \geq 2)$, $B_l^{(1)}(l \geq 3)$, $C_l^{(1)}(l \geq 2)$, $D_l^{(1)}(l \geq 4)$, $E_6^{(1)}$, $E_7^{(1)}$, $E_8^{(1)}$, $F_4^{(1)}$ and $G_2^{(1)}$. The diagrams and their vertex numberings are shown below.

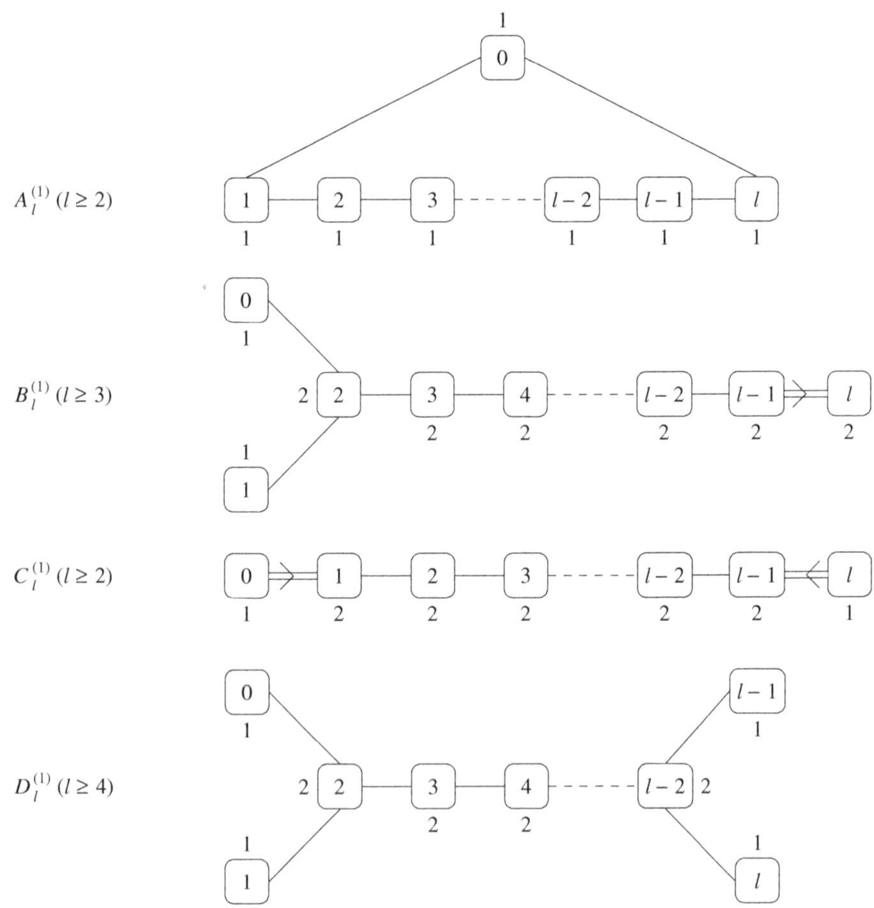

$E_6^{(1)}$

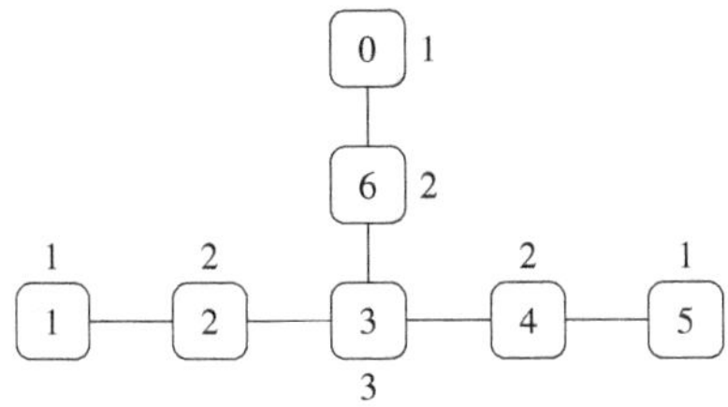

$E_7^{(1)}$

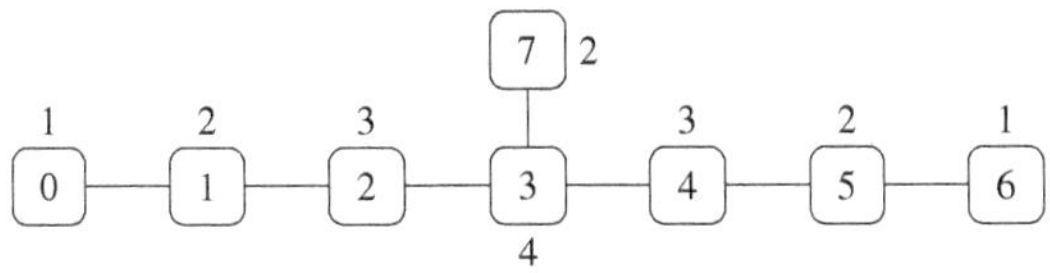

$E_8^{(1)}$

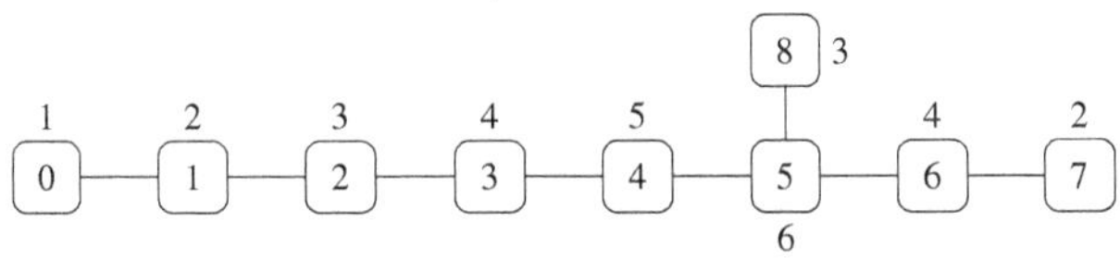

$F_4^{(1)}$

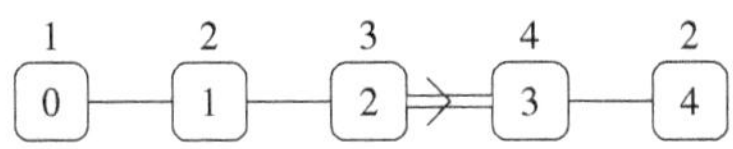

$G_2^{(1)}$

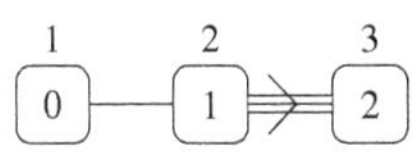

The twisted affine Dynkin diagrams are those of types $A_{2l}^{(2)}(l \geq 1)$, $A_{2l-1}^{(2)}(l \geq 3)$, $D_{l+1}^{(2)}(l \geq 2)$, $E_6^{(2)}$ and $D_4^{(3)}$. The diagrams, together with their vertex labellings, are shown below.

$A_2^{(2)}$

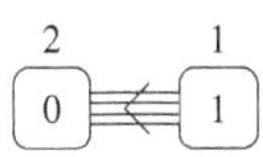

$A_{2l}^{(2)}\ (l \geq 2)$

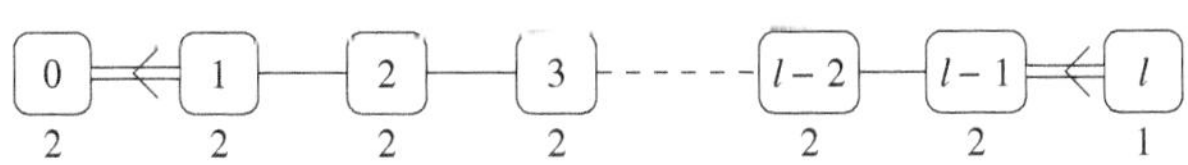

$A_{2l-1}^{(2)}\ (l \geq 3)$

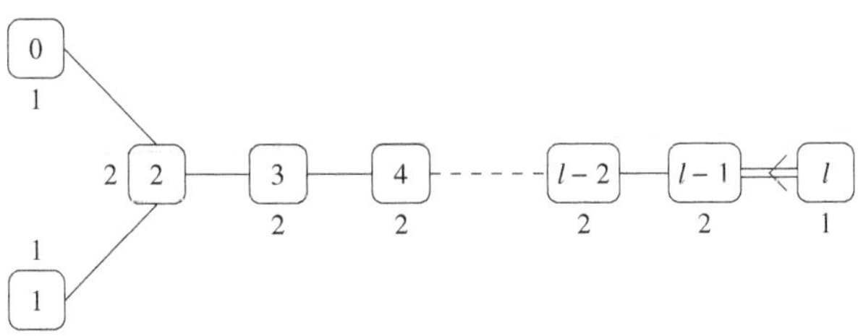

$D_{l+1}^{(2)}\ (l \geq 2)$

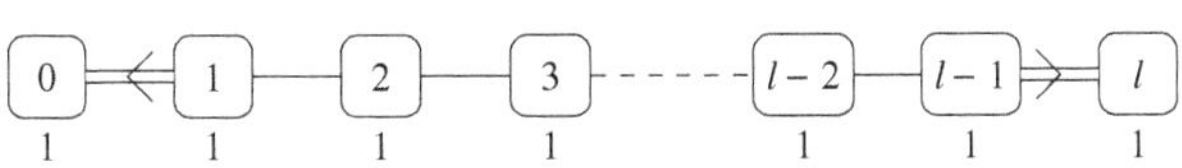

$E_6^{(2)}$

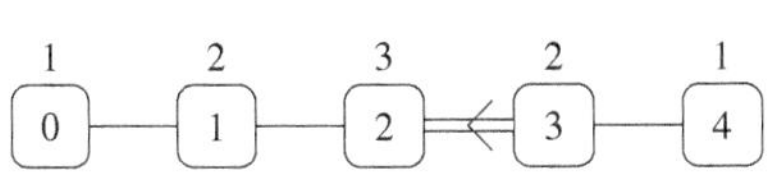

$D_4^{(3)}$

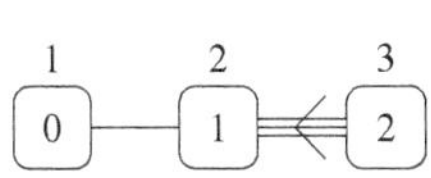

The table below shows the Coxeter numbers, Weyl group orders, numbers of positive roots and dimensions of the simple Lie algebras for each generalized Cartan matrix of finite type.

Type	Coxeter number h	Weyl group order	Number of positive roots	Dimension of Lie algebra
A_l	$l+1$	$l!$	$l(l+1)/2$	$l(l+2)$
B_l	$2l$	$2^l \times l!$	l^2	$l(2l+1)$
C_l	$2l$	$2^l \times l!$	l^2	$l(2l+1)$
D_l	$2l-2$	$2^{l-1} \times l!$	$l(l-1)$	$l(2l-1)$
E_6	12	$72 \times 6! = 51480$	36	78
E_7	18	$8 \times 9! = 2903040$	63	133
E_8	30	$192 \times 10! = 696729600$	120	248
F_4	12	1152	24	52
G_2	6	12	6	14

"... de tout autre façon... sans doute, car si les choses se répètent, c'est avec de grandes variations ..." Proust, *La Prisonnière*

References

[1] R. Bédard, Cells for two Coxeter groups, *Comm. Alg.* **14** (1986), 1253–1286.

[2] P. Biane, Minuscule weights and random walks on lattices. In *Quantum Probability and Related Topics*, vol. 7, pp. 51–65, River Edge, NJ, World Scientific, 1992.

[3] A. Björner and F. Brenti, Affine permutations of type A, *Electron. J. Combin.* **3** (1996), research paper 18.

[4] A. Björner and F. Brenti, *Combinatorics of Coxeter Groups*, New York, Springer, 2005.

[5] R. J. Blok, Highest weight modules and polarized embeddings of shadow spaces, *J. Algebraic Combin.* **34** (2011), 67–113.

[6] A. Borel, Compact Clifford–Klein forms of symmetric spaces, *Topology* **2** (1963), 111–122.

[7] A. Borel and J. Tits, Groupes réductifs, *Publ. Math. I.H.E.S.* **27** (1965), 55–151.

[8] N. Bourbaki, *Groupes et Algèbres de Lie*, Paris, Hermann, 1975.

[9] P. J. Cameron, Finite permutation groups and finite simple groups, *Bull. London Math. Soc.* **13** (1981), 1–22.

[10] R. W. Carter, Simple groups and simple Lie algebras, *J. London Math. Soc.* **40** (1965), 193–240.

[11] R. W. Carter, *Lie Algebras of Finite and Affine Type*, Cambridge, UK, Cambridge University Press, 2005.

[12] W. Casselman, Geometric rationality of Satake compactifications. In *Algebraic Groups and Lie Groups*, Austral. Math. Soc. Lect. Ser., vol. 9, pp. 81–103, Cambridge, UK, Cambridge University Press, 1997.

[13] V. Chepoi, Basis graphs of even Delta-matroids, *J. Combin. Theory Ser. B* **97** (2007), 175–192.

[14] C. Chevalley, Sur certains groupes simples, *Tôhoku Math. J.* **2** (1955), 14–66.

[15] M. J. Collins, Bounds for primitive complex linear groups, *J. Algebra* **319** (2008), 759–776.

[16] B. N. Cooperstein, The fifty-six-dimensional module for E_7, I. A four form for E_7, *J. Algebra* **173** (1995), 361–389.

[17] H. S. M. Coxeter, *Regular Polytopes*, New York, Pitman, 1947.

[18] C. de Concini, Normality and non normality of certain semigroups and orbit closures. In *Algebraic Transformation Groups and Algebraic Varieties*: proceedings of the conference "Interesting algebraic varieties arising in algebraic transformation group theory", pp. 15–35, Berlin–New York, Springer, 2004.

[19] V. Diekert and G. Rozenberg (eds.), *The Book of Traces*, Singapore, World Scientific, 1995.

[20] I. Dolgachev, *Topics in Classical Algebraic Geometry, part I*, in preparation, 2010.

[21] P. du Val, On the directrices of a set of points in a plane, *Proc. London Math. Soc.* **2:35** (1931), 23–74.

[22] T. Eguchi, N. P. Warner and S.-K. Yang, ADE singularities and coset models, *Nucl. Phys. B* **607** (2001), 3–37.

[23] K. Erdmann and M. J. Wildon, *Introduction to Lie algebras*, Springer Undergraduate Mathematics Series, London, Springer, 2006.

[24] H. Eriksson, *Computational and Combinatorial Aspects of Coxeter Groups*, Ph.D. thesis, KTH, Stockholm, 1994.

[25] M. Formichella, R. M. Green and E. Stade, Coxeter group actions on ${}_4F_3(1)$ hypergeometric series, *Ramanujan J.* **24** (2011), 93–128.

[26] M. Geck and G. Pfeiffer, *Characters of Finite Coxeter Groups and Iwahori–Hecke Algebras*, London Math. Soc. Monographs, New Series, vol. 21, Oxford, Oxford University Press, 2000.

[27] V. Ginzburg, Variations on themes of Kostant, *Transform. Groups* **13** (2008), 557–573.

[28] R. M. Green, Full heaps and representations of affine Kac–Moody algebras, *Int. Electron. J. Algebra* **2** (2007), 138–188.

[29] R. M. Green, Full heaps and representations of affine Weyl groups, *Int. Electron. J. Algebra* **3** (2008), 1–42.

[30] R. M. Green, Representations of Lie algebras arising from polytopes, *Int. Electron. J. Algebra* **4** (2008), 27–52.

[31] B. H. Gross, On minuscule representations and the principal SL_2, *Represent. Theory* **4** (2000), 225–244.

[32] M. Hagiwara, Minuscule heaps over Dynkin diagrams of type $\widetilde{A}$, *Electron. J. Combin.* **11** (2004), research paper 31.

[33] R. Hartshorne, *Algebraic Geometry*, New York, Springer-Verlag, 1977.

[34] F. W. Helenius, *Freudenthal Triple Systems via Root System Methods*, Ph.D. thesis, Emory University, 2009, pid.emory.edu/ark:/25593/1dtbq.

[35] A. Henderson, *The Twenty-seven Lines upon the Cubic Surface*, Cambridge, UK, Cambridge University Press, 1911.

[36] J. E. Humphreys, *Reflection Groups and Coxeter groups*, Cambridge, UK, Cambridge University Press, 1990.

[37] V. G. Kac, *Infinite Dimensional Lie Algebras*, third edn., Cambridge, UK, Cambridge University Press, 1990.

[38] J. Kamnitzer, The Beilinson–Drinfeld Grassmannian and symplectic knot homology, *Grassmannians, Moduli Spaces and Vector Bundles*, Clay Math. Proc., vol. 14, Providence, RI, American Mathematical Society, 2011, pp. 81–94.

[39] M. Kashiwara, On crystal bases of the q-analogue of universal enveloping algebras, *Duke Math. J.* **63** (1991), 465–516.

[40] D. Kazhdan and G. Lusztig, Representations of Coxeter groups and Hecke algebras, *Inventiones Math.* **53** (1979), 165–184.

[41] S. Kumar, A refinement of the PRV conjecture, *Invent. Math.* **97** (1989), 305–311.

[42] V. Lakshmibai and C. S. Seshadri, Standard monomial theory, *Proceedings of the Hyderabad Conference on Algebraic Groups* (Madras), Manoj Prakashan, 1991, pp. 279–322.

[43] M. Lassalle, An (inverse) Pieri formula for Macdonald polynomials of type C, *Transform. Groups* **15** (2010), 154–183.

[44] C. Lecouvey, E. Lesigne and M. Peigné, Random walks in Weyl chambers and crystals, Proc. London Math. Soc. **104**(2) (2012), 323–358.

[45] P. Littelmann, Paths and root operators in representation theory, *Ann. Math.* **142** (1995), 499–525.

[46] P. Littelmann, Cones, crystals and patterns, *Transform. Groups* **3** (1998), 145–179.

[47] J. Lurie, On simply laced Lie algebras and their minuscule representations, *Comment. Math. Helv.* **76** (2001), 515–575.

[48] G. Lusztig, Some examples of square integrable representations of semisimple p-adic groups, *Trans. Amer. Math. Soc.* **277** (1983), 623–653.

[49] G. Lusztig, Cells in affine Weyl groups I, *Adv. Studies in Pure Math.* **6** (1985), 225–287.

[50] G. Lusztig, Canonical bases arising from quantized enveloping algebras, *J. Amer. Math. Soc.* **3** (1990), 447–498.

[51] I. G. Macdonald, *Symmetric Functions and Hall Polynomials*, second edn., Oxford, Clarendon Press, 1995.

[52] P. A. MacMahon, *Combinatory Analysis*, vol. 1–2, Cambridge, UK, Cambridge University Press, 1915–16.

[53] Yu. I. Manin, *Cubic Forms: Algebra, Geometry, Arithmetic*, Moscow, Nauka, 1972, English translation (2nd edn.), Amsterdam, North-Holland, 1986.

[54] L. Manivel, Configurations of lines and models of Lie algebras, *J. Algebra* **304** (2006), 457–486.

[55] R. J. Marsh, Canonical bases for fundamental modules of quantized enveloping algebras of type A, *Comm. Alg.* **25** (1997), 1365–1386.

[56] R. J. Marsh, Canonical bases for the miniscule modules of the quantized enveloping algebras of types B and D, *Comm. Alg.* **25** (1997), 1387–1411.

[57] Z. S. McGregor-Dorsey, *Full Heaps over Dynkin Diagrams of Type* $\widetilde{A}$, Master's thesis, University of Colorado Boulder, 2008.

[58] G. A. Miller, H. F. Blichfeldt and L. E. Dickson, *Theory and Applications of Finite Groups*, New York, Wiley, 1916.

[59] C. Parker and G. Röhrle, The restriction of minuscule representations to parabolic subgroups, *Math. Proc. Camb. Phil. Soc.* **135** (2003), 59–79.

[60] S. E. Payne and J. A. Thas, *Finite Generalized Quadrangles*, Research Notes in Mathematics, vol. 110, Boston/London/Melbourne, Pitman Advanced Publishing Program, 1984.

[61] N. Perrin, The Gorenstein locus of minuscule Schubert varieties, *Adv. in Math.* **220** (2009), 505–522.

[62] G. Pfeiffer and G. Röhrle, Distributive coset graphs of finite Coxeter groups, *J. Group Theory* **6** (2003), 311–320.

[63] R. A. Proctor, *D-complete Posets Generalize Young Diagrams for the Jeu de Taquin Property*, preprint, arXiv:0905.3716.

[64] R. A. Proctor, Bruhat lattices, plane partition generating functions, and minuscule representations, *Europ. J. Combin.* **5** (1984), 331–350.

[65] R. A. Proctor, A Dynkin diagram classification theorem arising from a combinatorial problem, *Adv. Math.* **62** (1986), 103–117.

[66] V. Reiner, D. Stanton and D. White, The cyclic sieving phenomenon, *J. Combin. Theory Ser. A* **108** (2004), 17–50.

[67] D. B. Rush and X. Shi, *On Orbits of Order Ideals of Minuscule Posets*, preprint, arXiv:1108.5245 to appear in the *Journal of Algebraic Combinatorics*.

[68] I. Satake, On representations and compactifications of symmetric Riemannian symmetric spaces, *Ann. Math.* **71** (1960), 77–110.

[69] V. V. Serganova and A. N. Skorobogatov, Del Pezzo surfaces and representation theory, *J. Algebra and Number Theory* **1** (2007), 393–419.

[70] C. S. Seshadri, Theory of standard monomials for minuscule representations. In *C. P. Ramanujam—A Tribute*, pp. 207–239, Berlin–New York, Springer, 1978.

[71] R. Shankar, F. J. Burnell and S. L. Sondhi, Flux Hamiltonians, Lie algebras, and root lattices with minuscule decorations, *Ann. Physics* **324** (2009), 267–295.

[72] J. Y. Shi, *The Kazhdan–Lusztig Cells in Certain Affine Weyl Groups*, Lecture Notes in Mathematics, vol. 1179, Berlin, Springer, 1986.

[73] J. Y. Shi, The verification of a conjecture on left cells of certain Coxeter groups, *Hiroshima Math. J.* **23** (1994), 627–646.

[74] J. Y. Shi, Fully commutative elements and Kazhdan–Lusztig cells in the finite and affine Coxeter groups, *Proc. Amer. Math. Soc.* **131** (2003), 3371–3378.

[75] R. P. Stanley, *Enumerative Combinatorics*, Volume 1, Cambridge, UK, Cambridge University Press, 2002.

[76] D. Stanton, Open positivity conjectures for integer partitions, *Trends Math.* **2** (1999), 19–25.

[77] J. R. Stembridge, On minuscule representations, plane partitions and involutions in complex Lie groups, *Duke Math. J.* **73** (1994), 469–490.

[78] J. R. Stembridge, On the fully commutative elements of Coxeter groups, *J. Algebraic Combin.* **5** (1996), 353–385.

[79] J. R. Stembridge, Minuscule elements of Weyl groups, *J. Algebra* **235** (2001), 722–743.

[80] I. D. Suprunenko and A. E. Zalesski, Fixed vectors for elements in modules for algebraic groups, *Int. J. Algebr. Comput.* **17** (2007), 1249–1261.

[81] J. J. Sylvester, *The Collected Mathematical Papers of James Joseph Sylvester*, vol. 1, Cambridge, UK, Cambridge University Press, 1904.

[82] E. Szpilrajn, Sur l'extension de l'ordre partiel, *Fund. Math.* **16** (1930), 386–389.

[83] D. E. Taylor, Regular 2-graphs, *Proc. London Math. Soc.* **3:35** (1977), 257–274.

[84] H. Thomas and A. Yong, A combinatorial rule for (co)minuscule Schubert calculus, *Adv. Math.* **222** (2009), 596–620.

[85] J. F. van Diejen, L. Lapointe and J. Morse, Determinantal construction of orthogonal polynomials associated with root systems, *Compositio Math.* **140** (2004), 255–273.

[86] N. A. Vavilov, Subgroups of Chevalley groups containing a maximal torus, *Amer. Math. Soc. Transl.* **155** (1993), 59–100.

[87] N. A. Vavilov, Can one see the signs of structure constants?, *St Petersburg Math. J.* **19** (2008), 519–543.

[88] G. X. Viennot, Heaps of pieces, I: basic definitions and combinatorial lemmas. In *Combinatoire Énumérative*, eds. G. Labelle and P. Leroux, pp. 321–350, Berlin, Springer-Verlag, 1986.

[89] H. Wenzl, On tensor categories of Lie type E_N, $N \neq 9$, *Adv. Math.* **177** (2003), 66–104.

[90] N. J. Wildberger, A combinatorial construction for simply-laced Lie algebras, *Adv. Appl. Alg.* **30** (2003), 385–396.

[91] D. Zeilberger, Proof of the alternating sign matrix conjecture, *Electron. J. Combin.* **3** (1996), research paper 13.

[92] G. M. Ziegler, *Lectures on Polytopes*, New York, Springer-Verlag, 1995.

Index

For EU product safety concerns, contact us at Calle de José Abascal, 56–1°,
28003 Madrid, Spain or eugpsr@cambridge.org.

www.ingramcontent.com/pod-product-compliance
Lightning Source LLC
LaVergne TN
LVHW011527200926
848484LV00001B/174